Innovative Approaches to Plant Disease Control

WILEY SERIES IN ECOLOGICAL AND APPLIED MICROBIOLOGY

Edited by *Ralph Mitchell*

MICROBIAL LECTINS AND AGGLUTININS: Properties and Biological Activity
David Mirelman Editor

THERMOPHILES: General, Molecular, and Applied Microbiology
Thomas D. Brock Editor

INNOVATIVE APPROACHES TO PLANT DISEASE CONTROL
Ilan Chet Editor

Innovative Approaches to Plant Disease Control

edited by

ILAN CHET

The Hebrew University of Jerusalem
Faculty of Agriculture
Rehovot, Israel

A Wiley-Interscience Publication

JOHN WILEY & SONS

New York • Chichester • Brisbane • Toronto • Singapore

Library of Congress Cataloging in Publication Data:

Innovative approaches to plant disease control.

(Wiley series in ecological and applied microbiology)
"A Wiley-Interscience publication."
Includes index.
1. Pests—Biological control. 2. Plant diseases.
I. Chet, Ilan, 1939- . II. Series.

SB975.I26 1987 632′.3 86-15829
ISBN 0-471-80962-4

Printed in the United States of America

10 9 8 7 6 5 4 3 2 1

CONTRIBUTORS

RALPH BAKER, Department of Plant Pathology and Weed Science, Colorado State University, Fort Collins, Colorado

ILAN CHET, Department of Plant Pathology and Microbiology, The Hebrew University of Jerusalem, Faculty of Agriculture, Rehovot, Israel

JOHN R. COLEY-SMITH, Department of Plant Biology and Genetics, University of Hull, Hull, United Kingdom

R. JAMES COOK, USDA-ARS, Root Disease and Biological Control Research Unit, Pullman, Washington

BERNADETTE DUBOS, Station de Pathologie Végétale, Institut National de la Recherche Agronomique, INRA—Centre de Bordeaux, France

J. DUNEZ, Station de Pathologie Végétale, Institut National de la Recherche Agronomique, Centre de Bordeaux, France

ANNEGRET HAARS, Forstbotanisches Institut, Universität Göttingen, Gottingen-Weende, West Germany

KLAUS HAHLBROCK, Max Planck Institut für Züchtungsforschung, Cologne, West Germany

TERUYOSHI HASHIBA, National Institute of Agro-Environmental Sciences, Division of Microbiology, Yatabe, Tsukuba, Japan

ALOYS HÜTTERMANN, Forstbotanisches Institut, Universität Göttingen, West Germany

JAACOV KATAN, Department of Plant Pathology and Microbiology, The Hebrew University of Jerusalem, Faculty of Agriculture, Rehovot, Israel

JOSEPH KUĆ, Department of Plant Pathology, University of Kentucky, Lexington, Kentucky

B. Lugtenberg, Department of Molecular Botany, University of Leiden, Leiden, The Netherlands

G. C. Papavizas, Soilborne Diseases Laboratory, Plant Protection Institute, USDA-ARS, Beltsville, Maryland

Dierk Scheel, Max Planck Institut für Züchtungsforschung, Cologne, West Germany

N. C. Schenck, Plant Pathology Department, University of Florida, Gainesville, Florida

Fran M. Scher, Allelix Inc., Mississauga, Ontario, Canada

B. Schippers, Willie Commelin Scholten Phytopathological Laboratory, Department of Plant Pathology, University of Utrecht, University of Amsterdam, Baarn, The Netherlands

Ilan Sela, Virus Laboratory, The Hebrew University of Jerusalem, Faculty of Agriculture, Rehovot, Israel

Jennifer A. Thomson, CSIR Laboratory for Molecular and Cell Biology, University of the Witwatersrand, Johannesburg, South Africa

P. J. Weisbeek, Department of Molecular Cell Biology, University of Utrecht, Utrecht, The Netherlands

David M. Weller, USDA-ARS, Root Disease and Biological Control Research Unit, Pullman, Washington

Lothar Willmitzer, Max Planck Institut für Züchtungsforschung, Cologne, West Germany

For the tree of the field is man's life . . . .

Deuteronomy 20:19

SERIES PREFACE

The Ecological and Applied Microbiology series of monographs and edited volumes is being produced to facilitate the exchange of information relating to the microbiology of specific habitats, biochemical processes of importance in microbial ecology, and evolutionary microbiology. The series will also publish texts in applied microbiology, including biotechnology, medicine, and engineering, and will include such diverse subjects as the biology of anaerobes and thermophiles, paleomicrobiology, and the importance of biofilms in process engineering.

During the past decade we have seen dramatic advances in the study of microbial ecology. It is gratifying that today's microbial ecologists not only cooperate with colleagues in other disciplines but also study the comparative biology of different habitats. Modern microbial ecologists, investigating ecosystems, gain insights into previously unknown biochemical processes, comparative ecology, and evolutionary theory. They also isolate new microorganisms with application to medicine, industry and agriculture.

Applied microbiology has also undergone a revolution in the past decade. The field of industrial microbiology has been transformed by new techniques in molecular genetics. Because of these advances, we now have the potential to utilize microorganisms for industrial processes in ways microbiologists could not have imagined 20 years ago. At the same time, we face the challenge of determining the consequences of releasing genetically engineered microorganisms into the natural environment.

New concepts and methods to study this extraordinary range of exciting problems in microbiology are now available. Young microbiologists are increasingly being trained in ecological theory, mathematics, biochemistry, and genetics. The barriers between the disciplines essential to the study of modern microbiology are disappearing. It is my hope that this series in Ecological and Applied Microbiology will facilitate the reintegration of microbiology and stimulate research in the tradition of Louis Pasteur.

Modern control measures for plant diseases depend on the application of new concepts in microbial ecology. It is appropriate that this volume should appear in a series devoted to ecological and applied microbiology. The methods described in this book involve inoculation with antagonistic microorganisms, use of genetic techniques, and amplification of natural soil processes. Complex ecological processes are important factors in most of these control measures.

Ilan Chet's research has resulted in the development of radical new solutions to the eradication of microbial pathogens of agricultural crops. His work has been built on a strong foundation of concepts in microbial ecology. In this volume he has brought together some of the leading authorities on ecological approaches to control of plant diseases. Chapters emphasize biochemical, immunological, and genetic research. In all cases the complex and often poorly understood ecological component of the system is considered. This book should provide a new perspective on plant disease control which will integrate research from a wide range of disciplines.

RALPH MITCHELL

Cambridge, Massachusetts
July 1986

PREFACE

Plant diseases are as old as agriculture itself. Since God evicted Adam and Eve from the Garden of Eden saying ". . . cursed is the ground for thy sake; in sorrow shalt thou eat of it all the days of thy life" (Gen. 3:17), man has been harassed by plant diseases—"The Lord shall smite thee with . . . blasting, and with mildew; and they shall pursue thee until thou perish" (Deut. 28:22).

In their age-old quest for high-yield crops, farmers have created severe epidemics and, in certain cases, lost some of the natural resistance of the wild types. Together with developments in farming, pesticides have become the most common tool for reducing losses due to plant diseases. In the United States alone these losses are estimated at 18 billion dollars annually, of which only 25% are controlled by such conventional methods as chemicals and breeding for resistant varieties.

In recent years concern has grown regarding the increasing use of potentially hazardous chemicals. Pesticides that accumulate in the soil and ground water may interfere with the biotic activity and food chain. Moreover, intensive use of pesticides has, through natural selection, led to the increase in pathogens' resistant to these chemicals. These problems, and others, have instigated the search for new approaches to plant disease control. Fascinating progress has been achieved in agricultural biotechnology, of which preparations of biocontrol agents, cell fusions and genetic manipulations, and genetic engineering of microorganisms and plants are but a small portion. Some excellent and comprehensive books are devoted to biological control, one of the most promising of these new approaches.

The idea behind this book was to elaborate on the approaches and methods that have the potential of becoming alternative means for controlling diseases caused by fungi, bacteria, and viruses. Three major concepts of disease control are discussed—different biological control systems and their possible mechanisms, potential applications, and genetic improvement; biochemical and physiological manipulations in plants to induce resistance and reduce disease damage; and the use of molecular biology and the potential of genetic engineering to induce plant resistance by the introduction of foreign genes. There is no doubt in my mind that some of these approaches can be combined into an integrated pest management.

The authors, wishing to share with the reader some of the deliberations and excitement involved in discovering novel ideas and approaches and reassessing old ones, disclose updated findings alongside their views and speculations on plant disease control. I sincerely hope that the readers, be they researchers, students, biotechnologists, or extension specialists, find the following pages as stimulating and exciting as I did.

I would like to thank Professor David Mirelman and Dr. Yitzhak Hadar for their long hours of discussion, comments and advice and last, but by no means least, Professor Ralph Mitchell for his most valuable suggestions and above all, encouragement.

ILAN CHET

Rehovot, Israel
September 1986

CONTENTS

1

ENHANCING THE ACTIVITY OF BIOLOGICAL CONTROL AGENTS

RALPH BAKER

Department of Plant Pathology and Weed Science, Colorado State University, Fort Collins, Colorado

FRAN M. SCHER

Allelix Inc., Mississauga, Ontario, Canada

It is a well-recognized observation that a substantial amount of biological control operates in nature. In addition, increased or even spectacular suppression of plant disease may occur in some agricultural areas (Baker, 1982), and it is possible to detect and isolate the living agents responsible for such control (e.g., Baker, 1983). In most instances the agents are microorganisms commonly found in nature and are well-known antagonists. It follows that the suppressiveness induced by such agents is related to factors that enhance their activity in a particular agricultural habitat. Therefore, enhancing the activity biocontrol agents could be the most important factor in their success in controlling plant disease and their ultimate acceptance in commercial disease management.

The potential for enhancement has not been exploited systematically for most biological control systems. In this chapter we attempt to bring together, from the widely scattered literature, various strategies that have been employed and the logical extension of others not yet tested. These adumbrations are selective, not exhaustive, but, perchance, will form a basis for the future development of enhanced efficiency of biocontrol agents.

1. MANIPULATION OF ENVIRONMENT

1.1. Management of Potentially Limiting Essential Elements

Soilborne plant pathogens conventionally exist as dormant propagules until substrates are introduced that trigger germination. When the host is present, root exudates provide the stimulus. Thus certain elements are usually required for germination and subsequent penetration and infection of the host. The nutrients required for successful completion of this process for a number of pathogens are carbon and nitrogen (Griffin, 1981); more recently, it was demonstrated that iron (Fe) is necessary for normal germ-tube development from chlamydospores of *Fusarium* spp. (Scher and Baker, 1982).

Competition for nitrogen in soil is best illustrated by the disease system bean root rot, which is induced by *F. solani* f. sp. *phaseoli* (Baker, 1968). Nitrogen limits germination and successful infection by soilborne chlamydospores. Thus crop residues (such as barley straw) or cellulose added to soil and acted upon by the microbiological community immobilize nitrogen and decrease efficiency of the pathogen to induce infection. Competition must occur in the infection court for it to be effectively in control. There is evidence that adding cellulose to soil infested with *Rhizoctonia solani* shrinks the usual spermosphere influence on inoculum around a seed to a rhizoplane effect (Rouse and Baker, 1978). Thus only those propagules touching the seed where nutrients have not been immobilized have a probability of successfully completing the infection process. This lowers inoculum potential (Baker, 1978) by reducing the amount of inoculum capable of initiating infection.

The potential limiting nutritional factors in such a system may only occur in fixed infection courts. Guy and Baker (1977) attempted, without success, to reduce the incidence of pea wilt by adding cellulose to soil. Such treatment reduces the efficiency of inoculum but does not kill the pathogen (Baker and Nash, 1966). Thus substantial reduction of disease is more likely to occur when inoculum is within fixed infection courts, such as a seed (Rouse and Baker, 1978) or hypocotyl (Baker and Nash, 1966). However, *F. oxysporum* f. sp. *pisi* penetrates through root-tip infection courts that move through soil infested with inoculum. Therefore, even though inoculum efficiency may be reduced by limiting the nutrients available to propagules of the pathogen, moving infection courts (root tips) encounter much more inoculum than fixed infection courts (e.g., seed).

A reduction in efficiency of infection, and subsequent substantial decrease in disease for a pathogen penetrating through root tips, may be realized by the application of a rhizosphere-competent antagonist. Fluorescent pseudomonads induced suppressiveness in soil to Fusarium wilt pathogens (Kloepper et al., 1980a; Scher and Baker, 1980, 1982). Evidence suggests that these common inhabitants of the rhizosphere compete for available iron, which is essential for successful germination, penetration, and infection by the pathogen (Scher and Baker, 1982). Fluorescent pseudomonads produce iron-acquiring compounds, called siderophores, of the catechol-hydroxamate type (Kloepper et al., 1980a).

Siderophores have a higher iron-binding (stability) constant than the hydroxamate siderophores produced by many fungal pathogens (Garraway and Evans, 1984). Therefore, in alkaline soils with low availability of iron, rhizosphere-competent fluorescent pseudomonads may compete with and deprive Fusarium wilt pathogens of this essential nutrient.

To enhance iron competition, FeEDDHA (which has a higher stability constant than the hydroxamate siderophores of Fusarium wilt pathogen) was introduced with a fluorescent pseudomonad. Using this combination, control was additive (Scher and Baker, 1982).

By managing iron availability in such a system, control could be synergistic. Given an alkaline soil in which iron availability is low, pseudomonads could be favored in growth and activity by adding a chelate that binds iron so that it is capable of being procured by the siderophores of the antagonist but not by those of many other microorganisms, including fungal pathogens. Thus Scher et al. (1984a) demonstrated that the addition of FeEDDHA stimulated fluorescent pseudomonad growth in the rhizosphere of cucumber grown in monoculture. This is a particularly attractive strategy because the plant can utilize the iron in the FeEDDHA; consequently, the ligand, deprived of this element, specifically binds more iron from the Fe^{3+} pool (Scher and Baker, 1982). In such a system, therefore, the plant itself participates in depriving the pathogen of iron.

1.2. The Limitations Imposed by Temperatures Required for Optimum Activity of Biocontrol Agents

The activity of biocontrol agents is influenced profoundly by extrinsic factors of the environment. Pesticides used for chemical control are much less affected by such factors and this explains, in part, why biologicals are more erratic in disease control than chemicals in some systems. Although such physical factors as soil moisture and pH influence the activity of biocontrol agents, unfavorable temperatures may be an even more important limiting factor.

If the activity of a pathogen is limited by the same environment that is unfavorable for the biocontrol agent, there is no disease and, consequently, no disadvantage to the agent. Such a favorable situation may not be common, however. For example, when Harman et al. (1981) coated seeds with *Trichoderma* conidia, at temperatures unfavorable for the growth of the biocontrol agent, the incidence of Pythium or Rhizoctonia damping-off was not different from nontreated controls. In contrast, almost complete control occurred at temperatures near optimum for growth of the antagonists.

Two strategies may be employed to solve this problem. First, biocontrol agents may find initial commercial application in environmentally controlled covered structures. In such operations, plants are usually grown at near optimum conditions to insure the most efficient production. Temperatures in these structures often coincide with those insuring optimum activity of biocontrol agents. For crops grown in the field, biocontrol agents like *Trichoderma* spp., with relatively high optimum temperature requirements, may function in warm climates but may be ineffective in cool soils (personal communication from researchers who have used our agents in various sections of the country). Assuming that other factors are favorable, it follows that late plantings, when temperatures are optimum for the activity of biologicals, can insure more efficient control.

The other strategy involves acquisition of biological control agents that function at temperatures found in field soils when plants are susceptible to pathogens. The search for such agents has only just begun; however, Dubos recently reported at the first international workshop on *Trichoderma* and *Gliocladium* that cold-tolerant *Trichoderma* spp. controlled Botrytis bunch rot of grape at low temperatures.

1.3. Soil Reaction

The pH value of soil has a profound effect on biological control systems. For example, soil suppressiveness to *R. solani* occurred in alkaline soils with radish and cucumber monoculture but not sugar beet, alfalfa, and wheat (Lui and Baker, 1980). In another study (Chet and Baker, 1980), the induction of suppressiveness in alkaline soils was not accomplished in sugar beet, alfalfa, or radish after six weekly replantings; however, when soil pH was lowered to 6.5 or 5.7, suppressiveness was observed in 4 wk. An acid soil suppressive to *R. solani* was found in Columbia, South America. In all cases the agents responsible for suppression were *Trichoderma* spp. The germination of propagules of fungi belonging to this genus is inhibited at neutral pH, and they do not germinate in alkaline conditions (Chet and Baker, 1980).

The examples described above are systems in which biocontrol operates most efficiently in acid soils. In contrast, when the virginal soil pH of 8.1 of a *Fusarium*-suppressive soil was lowered to 7.0, much of the suppressiveness was lost (Scher and Baker, 1980). At pH 6.0, the soil was conducive. If fluorescent pseudomonads are involved in suppressiveness, as discussed above, this phenomenon is explained by theory. Siderophores are produced only under iron-deficient conditions (Neilands, 1973); acid soils have abundant available iron (Lindsay, 1974). It follows that there is no competition for iron in acid soils and pathogens have adequate supplies for successful infection (Scher and Baker, 1982).

Recently, a hitherto undescribed antagonist was found and named *Pythium nunn* (Lifshitz et al., 1984b). By adding bean leaf meal to alkaline soil, this antagonist increased in population density and the soil became suppressive to *Pythium* spp. pathogenic to higher plants (Lifshitz et al., 1984a). Using the same treatment, suppressiveness was not generated in acid soils.

In intensive agricultural operations, such as floriculture, growers can routinely manipulate such environmental parameters as soil pH. Thus adjustment to favor biocontrol agents is commercially feasible. An example illustrating this principle may have been discovered for lily root rot. Whereas the cause of this disorder is thought to be complex, *Pythium* spp. appear to be the predominant causal agents. In lily culture, growers add calcium (to prevent scorch), which has a tendency to raise the soil pH. In such situations one third or more of the plants may display chlorotic symptoms on the lower leaves and are unsalable. Application of sulphuric acid to lower the pH to 6–6.5 provided complete control of root rot, and this treatment is currently used routinely in the industry where indigenous *Trichoderma* spp. are present in the soil. We hypothesize that these antagonists become active biocontrol agents in acid soil but perform poorly under alkaline conditions.

1.4. Nutrition and Formulation

Antagonists may function by decreasing the efficiency of infection of pathogens (Baker and Drury, 1981), that is, inoculum potential (Baker, 1978), in infection courts, or may hasten the death of propagules of pathogens (Baker, 1981). The quantity and quality of nutrients certainly influence these processes.

Scientists have been intrigued by the potential use of crop residues and/or organic matter to favor saprophytic increase of biocontrol agents. Inherent in this approach is the application to soil of those substrates capable of being digested and utilized by the agent. For example, *Trichoderma* spp. commonly attack substrates high in cellulose (Garraway and Evans, 1984). Extremely high population densities of these agents were induced after corn stalks were added to soil at temperatures favorable for saprophytic activity of *Trichoderma* spp. (McFadden and Sutton, 1975).

Saprophytic activity of *Pythium* spp. apparently is favored by the addition of bean straw to soil (Lifshitz and Hancock, 1983). However, amendments of this substrate increased the population density of a mycoparasite, *P. nunn*, to the detriment of a *Pythium* sp. pathogenic to higher plants (Lifshitz et al., 1984a).

Enhancement of activity and survival of biocontrol agents may be realized when the product is formulated. For example, plant growth-promoting rhizobacteria (PGPR) survived best in dried formulations containing mixtures of xanthan gum and talc (Kloepper and Schroth, 1981a). The formulation could be applied to seed pieces of potatoes and such treatment resulted in significant increases in early plant development.

It would be desirable, however, not only to insure survival, but also to promote activity of agents by incorporating appropriate substrates in formulations. Baker et al. (1984) compared the efficiency of conidia in growth promotion of radish with thallus units of *T. harzianum* produced in a peat-bran formulation (Sivan et al., 1984). Although growth promotion was realized in treatments involving either of these propagules, at equal population densities growth was significantly greater in the peat-bran application. Theory supports these results; when propagules of fungi are added to soil, a substantial decrease in propagule density usually occurs (Baker, 1981). In contrast, even if a proportion of the thallus units embedded in a substrate are rendered nonviable, the particle still contains viable units. Again, propagules embedded in a particle of peat-bran have increased inoculum potential in contrast to a single conidium without substrate. Similar increases in the efficiency of biocontrol agents were obtained when they were incorporated with a food base into soil (Chet et al., 1979; Backman and Rodriguez-Kabana, 1975; Wells, et al., 1972).

Since substrates may be incorporated into seed pellets, this offers a particularly attractive method for enhancing biocontrol activity. For example, in laboratory trials, chitin or peat in pellets enhanced control of damping-off, whereas cellulose increased seed rot (Harman et al., 1981).

2. MANIPULATION OF THE PATHOGEN

Acquired physiological immunity or resistance in plants by protection against a fungal pathogen by the same pathogen has been reported (Cruickshank and Mandryk, 1960; Kuc et al., 1975; Randall and Helton, 1976; Yarwood, 1954). For example, inoculation of the cotyledons or first true leaf with an anthracnose pathogen of watermelon or muskmelon protected plants from disease induced by subsequent inoculation with the same pathogen (Caruso and Kuc, 1977). In such phenomena, host cultivars resistant to a race of the pathogen and susceptible to others may be protected from these by prior inoculation with the first race. Protection may be increased as the concentration of spores of the race used in cross protection is increased, and booster inoculations given after the first inoculation also may extend the time of protection (Kuc and Richmond, 1977). Similar protection was induced in beans by prior inoculation with cultivar-nonpathogenic races of *Colletotrichum lindemuthianum* (Elliston et al., 1975) and protection of pear or apple against fireblight was reported with avirulent strains of *Erwinia amylovora*, *E. herbicola*, and *Pseudomonas tabaci* (McIntyre et al., 1973). These examples illustrate the potential for enhancement of biocontrol by management of inoculation procedures with various avirulent strains of the pathogen.

Agrobacterium radiobacter strain 84 was effective against various biotypes of the related bacterium *A. tumefaciens*, the causal agent of crown gall in various crops (Kerr, 1980). A factor in control is believed to be the production of an antibiotic, agrocin 84, by the biocontrol agents, as reviewed by Vidaver (1981). Agrocin 84-insensitive *A. tumefacien* strains exist, however, and the development of new agents for control of resistant crown gall pathogens, such as a strain of *A. radiobacter* with a deficient plasmid transfer or mobilization system (Kerr, 1980), may be achieved in the future.

3. MANIPULATION OF THE HOST

Not only may cultivar nonpathogenic races and avirulent or attenuated pathogens provide cross protection, but nonpathogens (Baker et al., 1978), heat-killed (Bell and Presley, 1969), or even extracts from biocontrol agents (McIntyre et al., 1975) may also induce control. Such phenomena are believed to be brought about by host resistance reactions induced by recognition and signals originating from the cross-protection agent (Kuc, 1982). It is curious that this type of biocontrol strategy has not been exploited extensively in commercial applications.

Wymore and Baker (1982) introduced various formae specialis of *F. oxysporum* into the vascular system of tomato as cross-protection agents against *F. oxysporum* f. sp. *lycopersici*. Cross protection was achieved but the effect was temporary and soon the agents were not detected in the vascular system of the host. This suggests that only efficient biocontrol agents should be capable of occupying substrates and retaining activity in the host for an extended period. Also, since the living tissues of the host produce the resistance responses, the agent should occupy or be exposed to viable cells and not those that function in dead tissue such as certain xylem cells.

In an entirely different strategy of enhancement, soil suppressiveness to *Phymatotrichum omnivorum* in alfalfa and cotton, *Gaeumannomyces graminis* in wheat and barley (Cook, 1981), *Streptomyces scabes* in potatoes (Shipton, 1977), *R. solani* damping-off in radish (Baker, 1983), and *Cephalosporium gramineum* in wheat (Wiese and Ravenscroft, 1978) was generated through monoculture of the hosts. Because there is a consistent observation in the literature suggesting that interruption of monoculture with plants that are not hosts of the pathogen involved upsets the development of suppressiveness (e.g., Baker, 1982; Cook, 1982; Zogg and Amiet, 1980), enhancement in such a system involves crop management to insure that the appropriate crops are involved in monoculture and/or rotations.

4. MANIPULATION OF THE BIOCONTROL AGENT

Manipulation of some biocontrol agents is achieved by management of the environment; however, this has not resulted in consistent enhancement. Direct manipulation of a biocontrol agent, and resulting enhancement in disease control, is rarely reported. It is likely, however, that such manipulation will become commonplace in the near future. Advances in molecular biology and its intense study by plant pathologists may soon result in a number of unique biocontrol agents. The following characteristics of biocontrol agents are thought to account for their beneficial activity. By listing these desirable attributes of biocontrol agents, it is our goal to stimulate creative ideas and suggest avenues for their enhancement.

4.1. Desirable Attributes of a Biocontrol Agent

4.1.1. Active in an Environment Conducive for the Pathogen

Natural environments for pathogens extremes in pH, moisture, or temperature or low availability of nutrients or oxygen—may be unnatural for biocontrol agents. An organism adapted to the optimum environment of the pathogen will have the best chance for success in controlling the disease. An organism similar to the pathogen, ideally a nonvirulent variant, could be sought (see discussion above). A newly described mycoparasitic species of *Pythium* (Lifshitz et al., 1984b) induced early lysis of germinating sporangia of a plant pathogenic *Pythium* sp. *in vitro* and suppressed its capacity to induce damping-off of cucumber (Lifshitz et al., 1984a). Biocontrol of brown blotch of mushroom, caused by *Pseudomonas tolaasii*, was achieved by the application of a related *Pseudomonas* sp. (Nair and Fahy, 1972).

4.1.2. Compatible with Other Agents

Where plant pathogens are active under different environmental conditions, or where numerous pathogen types are a problem, more than one type of biocontrol agent or integration with fungicides may be necessary. Integration of fungicides and biocontrol agents is discussed below.

Ideally, a synergistic effect would occur when two biocontrol agents are applied together. This is not necessarily the case however. When *Trichoderma hamatum* was added to soil containing a high *Pseudomonas* sp. population density and low iron availability, *T. hamatum* was sensitive to iron deprivation induced by the *Pseudomonas* sp. and was ineffective as a biocontrol agent (Hubbard et al., 1983). Thus *Trichoderma* and *Pseudomonas* would not likely be compatible agents.

Kloepper (1983) reported that mixtures of two or more rhizobacteria resulted in significant reductions in daughter tuber infestation by *Erwinia carotovora*; however, control by mixed strains was not greater than that obtained by any single strain. It is not known whether the biocontrol agents were truly compatible, or if biocontrol was a result of one dominant strain.

4.1.3. Survive Agricultural Introduction

Conventional agricultural methods for pesticide introduction, such as seed application, overhead watering, and furrow application, can allow for the introduction of biocontrol agents as well. But living agents must be protected from conditions that render them ineffective; these may include desiccation, temperature extremes, and ultraviolet light.

Fungi applied in a dormant state are more naturally resistant to harsh environmental factors than are most bacteria. Reported bacterial biocontrol agents are commonly gram-negative and, therefore, do not produce resistant spores, as do gram-positive bacteria. Kloepper et al. (1980b) and Suslow and Schroth (1982) found that adding $MgSO_4$ ($0.1\,M$) to aqueous bacterial preparations enhanced their recovery and survival. They also described dry gum–talc formulations that permitted the successful introduction of pseudomonads onto seeds.

Dupler and Baker (1983) reported that *P. putida* did not survive addition to dry soil (-100 bars matric potential), but did survive in moist soil (-0.3 bars) when it was slowly air-dried to -100 bars. They hypothesized that the bacteria survived in microsites between clay particles. In this case, irrigation upon application would result in better survival. The selection and use of desiccation-resistant bacterial agents would enhance their survival upon introduction.

4.1.4. Colonize the Appropriate Substrate or Plant Part

A biocontrol agent applied to control epiphytic pathogens must colonize leaf areas to be successful. Likewise, root pathogen controlling agents must competitively colonize the rhizosphere.

An initiating mechanism for bacterial colonization may be recognition and subsequent binding between the bacterial and plant surfaces. *Rhizobium trifolii* bound to root hairs of clover and was agglutinated by plant lectins (Dazzo et al., 1976). Jasalavich and Anderson (1981) reported agglutination of saprophytic pseudomonads to legume extracts, suggesting that pseudomonads may bind to roots. Concentration-dependent binding of soil bacteria to radish roots was observed by James et al. (1984).

Bacterial chemotaxis toward plant exudates is suggested as a mechanism for initiation of colonization by both pathogens and beneficial bacteria. The angular leaf spot pathogen *P. lachrymans* exhibited chemotaxis toward water collected from leaf surfaces; chemotaxis was related to the concentration of amino acids and carbohydrates in the water (Chet et al., 1973). Zoospores of the fungal pathogen *Phytophthora cinnamoni* exhibited chemotaxis to roots of avocado seedlings (Zentmyer, 1961). Scher et al. (1984a, 1985) demonstrated that saprophytic root- and seed-colonizing pseudomonads were attracted to soybean exudates and amino acids. The bacteria swam 2 cm through moist nonsterile soil to colonize germinating soybean seeds. Bacteria may exhibit negative chemotaxis (migrate away from) to extremes in pH, toxins, hydrocarbons, and some amino acids (Chet and Mitchell, 1976). Positive and negative chemotactic responses would be advantageous for biocontrol agents in that they could actively migrate to areas rich in nutrients, migrate away from regions likely to be destructive, and compete more effectively with chemotactic pathogens.

Extensive root colonization by seed-applied gram-negative bacteria has been reported (Kloepper et al. 1980; Suslow and Schroth, 1982; Weller, 1983; Scher et al., 1984c). It is believed that root colonization by pseudomonads is a prerequisite for their success as biocontrol agents. Indeed, a fluorescent pseudomonad biocontrol agent of take-all colonized the entire length of wheat roots, with a population density gradient highest at the base of the roots and declining toward their tips (Weller, 1983). The biocontrol agent for Fusarium wilts, *P. putida*, colonized cucumber and radish roots (Dupler and Baker, 1983); colonization of radish roots by this agent was enhanced by the addition of FeEDDHA (Scher et al., 1984a).

Other measures to enhance rhizosphere colonization may simply involve adequate irrigation during seedling development. Howie and Cook (1983) reported that maximum colonization by a fluorescent pseudomonad suppressive to the take-all pathogen was reached at −0.1 to −0.3 bars. The population density of a fluorescent pseudomonad decreased on potato seedpieces as they were air-dried, but subsequent planting in moist soil led to increases in population densities immediately after planting (Geels and Schippers, 1983). They concluded that "environmental conditions, rather than initial density of *Pseudomonas* cells, determine the maximal densities that can be reached on the tuber surface."

4.1.5. Efficiently Suppress the Pathogen

Once in place, the successful biocontrol agent efficiently suppresses the pathogen and reduces disease incidence. Biocontrol agents suppress pathogens by such forms of antagonism as competition, antibiosis, and/or exploitation. Activity is not restricted to any one of these, and, indeed, the best biocontrol agents may suppress pathogens by a combination of methods.

Competition between the biocontrol agent and pathogen can result in displacement of the pathogen. By keeping soil wet (−0.4 bars) during tuber formation, fewer potatoes become infected with *Streptomyces scabies*, owing to early colonization of the lenticels (infection court) by soil bacteria, which out-competed the slower pathogen for nutrients and oxygen (Lapwood and Hering, 1970).

Competition for iron between *Fusarium* and siderophores of the biocontrol agent *Pseudomonas* was believed to be the mechanism of *Fusarium* suppression by *Pseudomonas* (Kloepper et al., 1980a; Scher and Baker, 1982). Kloepper and Schroth (1981b) reported that nonsiderophore producing mutants of PGPR were not inhibitory to soil microbes *in vitro* and did not promote growth, whereas siderophore producers did. They concluded that the ability to produce siderophores *in vitro* was directly related to a strain's ability to increase plant growth, and Suslow et al. (1979) likewise suggested that siderophore activity against a wide spectrum of microorganisms is a consistent property of PGPR.

A strain of *P. fluorescens* producing pyrrolnitrin, or the antibiotic itself, increased the emergence of cotton in soil infested with *R. solani* (Howell and Stipanovic, 1979), whereas *P. fluorescens* (strain Pf-5), which produced the antibiotic pyoluteorin, controlled *Pythium* but not *R. solani* (Howell and Stipanovic, 1980).

There has been much controversy as to the merits of prescreening potential biocontrol agents *in vitro* for antibiotic production against pathogens. Not all successful biocontrol agents show antibiosis *in vitro*. Wong and Baker (1984) reported that there was no correlation between take-all suppression by a fluorescent pseudomonad and *in vitro* antibiosis on agar. The agent *P. putida* A12 (Scher and Baker, 1982) controlled Fusarium wilt but did not show antibiosis against *Fusarium in vitro*. Conventional tests for zones of inhibition on agar may cause microbes that produce nondiffusable antibiotics, or antibiotics that are produced only under *in situ* conditions, to be overlooked. Only the development of techniques to monitor the production of antibiotics and their effects on pathogens *in situ* will confirm the role of antibiosis in disease suppression.

An example whereby such techniques were utilized to establish direct evidence for antibiosis in soil is the recent work by Rothrock and Gottlieb (1984).

They reported that *Streptomyces hygroscopicus* var. *geldanus* (shg) produces the antibiotic geldanomycin in soil and that the antibiotic inhibits the growth of *R. solani* in soil. Geldanomycin (20–80 μg/g soil) alone controlled Rhizoctonia root rot of pea in greenhouse pot trials. Antibiosis was suggested to be the mechanism for biocontrol of root rot by shg.

Parasitism by a biocontrol agent as a mechanism of disease control is an attractive concept because it could reduce inoculum density (Baker, 1968). Mycoparasitism has been shown *in vitro*, and is suspected to be, at least in part, the mode of action for biocontrol of a number of plant disease (Ayers and Adams, 1981).

Trichoderma, a biocontrol of several soilborne plant pathogens, was shown to actively invade *R. solani* and *Pythium* sp. mycelium (Chet et al., 1981) and to produce the lytic extracellular enzymes chitinase and β-1,3-glucanase *in vitro* (Elad et al., 1982).

Elad et al. (1983) presented evidence that *Trichoderma* hyphae bind to *R. solani* hyphae through a lectin interaction, suggesting that binding may be a precursor to parasitism. The bacterial biocontrol agent *Enterobacter cloacae* reduced *Pythium*-induced rot of pea, beet, and cucumber (Hadar et al., 1983). The bacteria were believed to attach to the hyphae through a lectin interaction (Nelson et al., 1984) and formed sheaths around hyphae of *P. ultimum*, resulting in lysis of the enclosed hyphae (Hadar et al., 1983).

Sneh (1981) utilized chitinase-producing bacteria for biological control of *oxysporum* on carnation, which were capable of lysing *Fusarium* hyphae *in vitro*. Recently, the chitinase gene for a *Serratia* sp. was transferred to *Pseudomonas* sp. rhizobacteria, which became chitinolytic (Jones et al., 1984). Such unique strains could be utilized to assess whether lytic enzyme production alone is sufficient to render a bacterial strain suppressive and/or whether such activity could enhance their biocontrol potential.

5. BIOLOGICAL CONTROL AND INTEGRATED PEST MANAGEMENT

Fungicides that are active against a relatively narrow spectrum of plant parasites, but not against biocontrol agents, may afford an opportunity for integration of biological and chemical control. This potential was realized for control of Rhizoctonia damping-off of radishes (Lifshitz et al., 1985). Whereas the ED_{50} value for control of damping-off was 1.3 g a.i./g of soil for benadonil, the ED_{50} for inhibition of radial growth of *T. harzianum* was 38 g a.i./g of medium. Control by the use of a seed treatment with the biocontrol agent and a soil mix of benadonil was additive.

Even if the biocontrol agent is sensitive to a fungicide, it is possible to produce mutants that are tolerant (Papavizas et al., 1982). Thus it was possible to produce a benomyl-tolerant mutant of a *Trichoderma* sp. isolate from Colombia (labeled T 95). Conidia of this isolate applied in rooting hormone to carnation cuttings reduced the incidence of Rhizoctonia rot 50% or more during propagation (Fig. 1). A similar amount of control was obtained by adding benomyl to the hormone solution. When both were added, 100% control was achieved.

When soils are steamed or fumigated, reinfestation by pathogens surviving the treatment often occurs. This may be prevented by reinfesting the treated soil with an appropriate antagonist (Elad et al., 1982).

In spite of the recent interest in integrated pest management (in which biocontrol agents could be an important component), there are few examples where biological control is deliberately exploited in such systems. Perhaps the best example of the potential for such strategy is illustrated by the integrated approach initiated by Cook and associates for the control of take-all of wheat (Cook, 1982). Suppresiveness of soil to the take-all pathogen is achieved with long-term wheat monoculture; however, this suppressiveness may be lost if non-susceptible host crops like alfalfa, oats, or soybeans are grown in rotation with wheat (Cook, 1981). Another aspect of cultural disease control involves fragmentation of the inoculum of the pathogen (*G. graminis*). No-till or direct-drilled planting of wheat results in large fragments of inoculum embedded in the wheat straw and crowns from the previous crop. These have greater inoculum potential than small fragments (Wilkinson and Cook, 1981), and are ideally po-

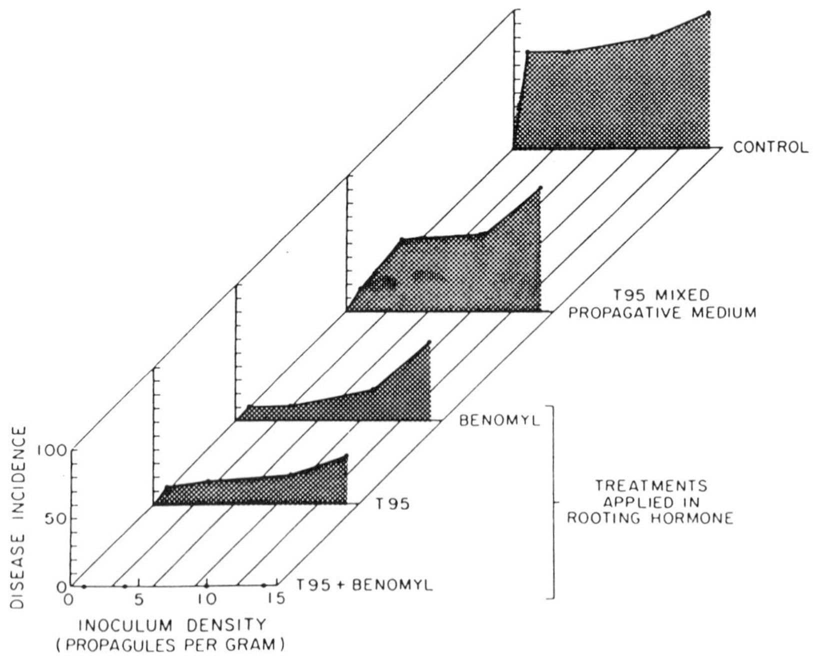

Fig. 1. Integrated chemical and biological control of Rhizoctonia stem rot during propagation of carnation cuttings. The T-95 is an isolate of *Trichoderma harzianum* mutated so that it was tolerant to benomyl. Benomyl (50% a.i.) was applied at 5% suspension in the rooting hormone.

sitioned for infection of the new wheat crop unless reduced in size by conventionally plowed seedbeds (Moore and Cook, 1984). Among the many additional strategies for integrated control is the possibility of seed treatment with fluorescent pseudomonads to induce suppressiveness to the take-all pathogen (Zhang et al., 1984).

6. CONCLUSIONS

Clearly, once researchers have identified agents capable of suppressing plant disease pathogens, they may exploit methods of enhancing their efficiency. The opportunities are limited only by the imagination of the initiator. We believe that the ultimate application of biological control in agricultural management systems rests on the success of efforts toward enhancement.

ACKNOWLEDGMENTS

Portions of the research done at Colorado State University and treated in this chapter were supported by funds from the National Science Foundation, the Science and Education Administration, Western Regional Project W-147, The United States–Israel Binational Agricultural Research and Development Fund, Allied Chemical Company, and Mallinckrodt, Inc. We especially thank Drs. P. Hanchey-Bauer, G. A. McIntyre, and E. Ruppel for their review of the manuscript.

REFERENCES

Ayers, W. A. and P. B. Adams (1981). Mycoparasitism and its application to biological control of plant diseases. In *Biological Control in Crop Protection* (G. C. Papavizas, ed.). Allanheld, Osmun, Granada, pp. 91–103.

Backman, P. A. and R. Rodriguez-Kabana (1975). A system for the growth and delivery of biological control agents to the soil. *Phytopathology*, **65**:819–821.

Baker, R. (1968). Mechanisms of biological control of soil-borne pathogens. *Ann. Rev. Phytopathol.*, **6**:263–294.

Baker, R. (1978). Inoculum potential. In *Plant Disease: An Advanced Treatise*, Vol. 2 (J. G. Horsfall and E. B. Cowling, eds.). Academic, New York, pp. 137–157.

Baker, R. (1981). Biological control: Eradication of plant pathogens by adding organic amendments to soil. In *Handbook of Pest Management in Agriculture*, Vol. 2 (D. Pimental, ed.). Chemical Rubber Company Press, Boca Raton, FL, pp. 137–157.

Baker, R. (1982). Induction of suppressiveness. In *Suppressive Soils and Plant Disease* (R. W. Schneider, ed.). American Phytopathology Society, St. Paul, MN, pp. 35–50.

Baker, R. (1983). State of the art: Plant diseases. In *Proceedings of the National Interdisciplinary Biological Control Conference* (S. L. Battenfield, ed.). Cooperative State Reservation Service, U.S. Department of Agriculture, Washington, D.C., pp. 14–22.

Baker, R. and R. Drury (1981). Inoculum potential and soil-borne pathogens: the essence of every model is within the frame. *Phytopathology*, **71:**363–372.

Baker, R. and S. Nash (1966). Ecology of plant pathogens in soil. VI. Inoculum density of *Fusarium solani* f. sp. *phaseoli* in bean rhizosphere as affected by cellulose and supplemental nitrogen. *Phytopathology*, **55:**1381–1382.

Baker, R., P. Hanchey, and S. D. Dottarar (1978). Protection of carnation against Fusarium stem rot by fungi. *Phytopathology*, **68:**1495–1501.

Baker, R., Y. Elad, and I. Chet (1984). The controlled experiment in the scientific method with special emphasis on biological control. *Phytopathology*, **74:**1019–1021.

Bell, A. A. and J. T. Presley (1969). Heat inhibited or heat-killed conidia of *Verticillium albo-atrum* induce disease resistance and phytoalexin synthesis in cotton. *Phytopathology*, **59:**1147–1151.

Caruso, F. L. and J. Kuc (1977). Protection of watermelon and muskmelon against *Colletotrichum lagenarium* by *Colletotrichum lagenarium*. *Phytopathology*, **67:**1285–1289.

Chet, I. and R. Baker (1980). Indication of suppressiveness to *Rhizoctonia solani* in soil. *Phytopathology*, **70:**994–998.

Chet, I. and R. Mitchell (1976). Ecological aspects of microbial chemotactic behavior. *Ann. Rev. Microbiol.*, **30:**221–239.

Chet, I., Y. Zilberstein, and Y. Henis (1973). Chemotaxis of *Pseudomonas lachrymans* to plant extracts and to water droplets collected from the leaf surfaces of resistant and susceptible plants. *Phys. Plant Pathol.*, **3:**474–479.

Chet. I., Y. Hadar, J. Katan, and Y. Henis (1979). Biological control of soil-borne plant pathogens by *Trichoderma harzianum*. In *Soil-Borne Plant Pathogens* (B. Schippers and W. Gams, eds.). Academic, New York, pp. 585–591.

Chet, I., G. E. Harman, and R. Baker (1981). *Trichoderma hamatum*: its hyphal interactions with *Rhizoctonia solani* and *Phythium* spp. *Microb. Ecol.*, **7:**29–38.

Cook, R. J. (1981). The influence of rotation crops on take-all decline phenomenon. *Phytopathology*, **71:**189–192.

Cook, R. J. (1982). Use of pathogen-suppressive soils for disease control. In *Suppressive Soils and Plant Disease* (R. W. Schneider, ed.). American Phytopathology Society, St. Paul, MN.

Cruickshank, I. A. M. and M. Mandryk (1960). The effect of stem infestation of tobacco with *Peronospora tabacina. Adam. J. Aust. Inst. Agric. Sci.*, **26:**369–372.

Dazzo, F. B., C. A. Napoli, and D. H. Hubbell (1976). Adsorption of bacteria to roots as related to host specificity in the *Rhizobium* clover symbiosis. *Appl. Environ. Microbiol.*, **32:**166–171.

Dupler, M. and R. Baker (1983). Survival of *Pseudomonas putida*, a biological control agent, in soil. *Phytopathology*, **74:**195–200.

Elad, Y., I. Chet, and Y. Henis (1982). Degradation of plant pathogenic fungi by *Trichoderma harzianum. Can. J. Microbiol.*, **28:**719–725.

Elad, Y., Y. Hadar, I. Chet, and Y. Henis (1982). Prevention of reinfestation of methyl bromide-fumigated soil and improvement of disease control by *Trichoderman harzianum. Crop Protection*, **1:**199–211.

Elad, Y., R. Barak, and I. Chet (1983). Possible role of lectins in mycoparasitism. *J. Bacteriol.*, **154:**1431–1435.

Elliston, J., J. Kuc, and E. B. Williams (1975). A comparative study of the development of compatible, incompatible, and induced incompatible interactions between *Colletotrichum* spp. and *Phaseolus vulgaris*. *Phytopath. Z.*, **87:**289–303.

Garraway, M. D. and R. C. Evans (1984). *Fungal Nutrition and Physiology*. Wiley, New York, 401 pp.

Geels, F. P. and B. Schippers (1983). Selection of antagonistic fluorescent *Pseudomonas* spp. and their root colonization and persistence following treatment of seed potatoes. *Phytopath. Z.*, **108:**193–206.

Griffin, G. J. (1981). Physiology of conidium and chlamydospore germination in Fusarium. In *Fusarium Diseases, Biology and Taxonomy* (P. E. Nelson, T. A. Toussoun, and R. J. Cook, eds.). The Pennsylvania State University Press, University Park and London, pp. 331–339.

Guy, S. D. and R. Baker (1977). Inoculum potential in relation to biological control of Fusarium wilt of peas. *Phytopathology*, **67:**72–78.

Hadar, Y., G. E. Harman, A. G. Taylor, and J. M. Norton (1983). Effects of pregermination of pea and cucumber seeds and of seed treatment with *Enterobacter cloacae* on rots caused by *Pythium* spp. *Phytopathology*, **73:**1322–1325.

Harman, G. E., I. Chet, and R. Baker (1981). Factors affecting *Trichoderma hamatum* applied to seeds as a biocontrol agent. *Phytopathology*, **71:**569–572.

Howell, C. R. and R. D. Stipanovic (1979). Control of *Rhizoctonia solani* on cotton seedlings with *Pseudomonas fluorescens* and with an antibiotic produced by the bacterium. *Phytopathology*, **69:**480–482.

Howell, C. R. and R. D. Stipanovic (1980). Suppression of *Pythium ultimum*-induced damping-off of cotton seedlings by *Pseudomonas fluorescens* and its antibiotic, pyoluteorin. *Phytopathology*, **70:**712–715.

Howie, W. J. and R. J. Cook (1983). Effect of soil matric potential on colonization of wheat roots by a *Pseudomonas* sp. suppressive to take-all. *Phytopathology*, **73:**822.

Hubbard, J. P., G. E. Harman, and Y. Hadar (1983). Effect of soil-borne *Pseudomonas* spp. on the biological control agent, *Trichoderma hamatum*, on pea seeds. *Phytopathology*, **73:**655–659.

James, Jr., D. W., T. V. Suslow, and K. E. Steinbeck (1984). The relationship between short-term adherence and long-term colonization of roots by bacteria. *Phytopathology*, **74:**866.

Jasalavich, C. A. and A. J. Anderson (1981). Isolation from legume tissues of an agglutinin of saprophytic pseudomonads. *Can. J. Bot.*, **59:**264–271.

Jones, J., J. S. Ziegle, and T. V. Suslow (1984). Cloning and expression of *Serratia* chitinase genes in *E. coli* and *Pseudomonas* rhizobacteria. *Phytopathology*, **743:**863.

Kerr, A. (1980). Biological control of crown gall through production of agrocin 84. *Plant Dis.*, **64:**24–30.

Kloepper, J. W. (1983). Effect of seed piece inoculation with plant growth-promoting rhizobacteria on populations of *Erwinia carotovora* on potato roots and in daughter tubers. *Phytopathology*, **73:**217–219.

Kloepper, J. W. and M. N. Schroth (1981a). Development of a powder formulation of rhizobacteria for inoculation of potato seed pieces. *Phytopathology*, **71:**590–592.

Kloepper, J. W. and M. N. Schroth (1981b). Relationship of *in vitro* antibiosis of plant growth-promoting rhizobacteria to plant growth and the displacement of root microflora. *Phytopathology*, **71:**1020–1024.

Kloepper, J. W., J. Leong, M. Teintze, and M. N. Schroth (1980a). *Pseudomonas* siderophores: A mechanism explaining disease suppressive soils. *Curr. Microbiol.*, **4:**317–320.

Kloepper, J. W., M. N. Schroth, and T. D. Miller (1980b). Effects of rhizosphere colonization by plant growth-promoting rhizobacteria on potato plant development and yield. *Phytopathology*, **70:**1078–1082.

Kuc, J. (1982). Induced immunity to plant disease. *Bio. Sci.*, **32:**854–860.

Kuc, J. and S. Richmond (1977). Aspects of the protection of cucumber against *Colletotrichum lagenarium* by *Colletotrichum lagenarium. Phytopathology*, **67:**533–536.

Kuc, J., G. Shockley, and K. Kearney (1975). Protection of cucumber against *Colletotrichum lagenarium* by *Colletotrichum lagenarium. Physiol. Plant Pathol.*, **7:**195–199.

Lapwood, D. N. and T. F. Hering (1970). Soil moisture and the infection of young potato tubers by *Streptomyces scabies* (common scab). *Potato Res.*, **13:**296–304.

Lifshitz, R. and J. G. Hancock (1983). Saprophytic development of *Pythium ultimum* in soil as a function of water matric potential and temperature. *Phytopathology*, **73:**257–261.

Lifshitz, R., B. Sneh, and R. Baker (1984a). Soil suppressiveness to a pathogenic *Pythium* species. *Phytopathology*, **74:**1054–1061.

Lifshitz, R., M. E. Stanghellini, and R. B. Baker (1984b). A new species of *Pythium* isolated from soil in Colorado. *Mycotaxon*, **20:**373–379.

Lifshitz, R., S. Lifshitz, and R. Baker (1985). Decrease in incidence of Rhizoctonia preemergence damping-off by use of integrated chemical and biological controls. *Plant Dis.* **69:**431–434.

Lindsay, W. L. (1974). Role of chelation in micronutrient availability. In *The Plant Root and Its Environment* (R. W. Carson, ed.). University Press Virginia, Charlottesville, pp. 507–524.

Liu, S. and R. Baker (1980). Mechanism of biological control in soil suppressive to *Rhizoctonia solani. Phytopathology*, **70:**404–412.

McFadden, A. G. and J. C. Sutton (1975). Relationships of populations of *Trichoderma* spp. in soil to disease in maize. *Can. Plant Sci.*, **55:**579–586.

McIntyre, J. L., J. Kuc, and E. B. Williams (1973). Protection of pear against fire blight by bacteria and bacterial sonicates. *Phytopathology*, **63:**872–877.

McIntyre, J. L., J. Kuc, and E. B. Williams (1975). Protection of Bartlett pear against fire blight with deoxyribonucleic acid from virulent and avirulent *Erwinia amylovora. Physiol. Plant Pathol.*, **7:**153–170.

Moore, K. J. and R. J. Cook (1984). Increased take-all of wheat with direct drilling in the Pacific Northwest. *Phytopathology*, **74:**1044–1049.

Nair, N. G. and P. C. Fahy (1972). Bacteria antagonistic to *Pseudomonas tolaasii* and their control of brown blotch of the cultivated mushroom *Agaricus bisporus. J. Appl. Bact.*, **35:**349–442.

Neilands, J. B. (1973). Microbial iron transport compounds (siderophores). In *Inorganic Chemistry* (I. C. L. Eickhorn, ed.). Elsevier, Amsterdam, pp. 167–202.

Nelson, E. B., G. T. Nash, and G. E. Harman (1984). Lectin-mediated attachment of the biocontrol agent *Enterobacter cloacae* to fungal hyphae: II. Role in the biological control of *Pythium* seed rot. *Phytopathology*, **74:**807.

Papavizas, G. C., J. A. Lewis, and T. H. Abd-El Moity (1982). Evaluation of new biotypes of *Trichoderma harzianum* for tolerance of benomy and enhanced biocontrol capabilities. *Phytopathology*, **72:**126–132.

Randall, R. and A. W. Helton (1976). Effect of inoculation date on induction of resistance to *Cytospora* in Italian prune trees by *Cytospora cincta. Phytopathology*, **66:**206–207.

Rothrock, C. S. and D. Gottlieb (1984). Role of antibiosis in antagonism of *Streptomyces hygroscopieus* var. *geldanus* to *Rhizoctonia solani* in soil. *Can. J. Microbiol.*, **30:**1440–1447.

Rouse, D. I. and R. Baker (1978). Modeling and quantitative analyses of biological control mechanisms. *Phytopathology*, **68:**1297–1302.

Scher, F. M. and R. Baker (1980). Mechanisms of biological control in a *Fusarium*-suppressive soil. *Phytopathology*, **70:**412–417.

Scher, F. M., M. Dupler, and R. Baker (1984a). Effect of synthetic iron chelates on populations of *Pseudomonas putida* and *Fusarium oxysporum* in soil. *Can. J. Microbiol.*, **30:**1271–1275.

Scher, F. M. and R. Baker (1982). Effect of *Pseudomonas putida* and a synthetic iron chelator on induction of soil suppressiveness to Fusarium wilt pathogens. *Phytopathology*, **72:**1567–1573.

Scher, F. M., J. W. Kloepper, and C. A. Singleton (1984b). Chemotaxis of *Pseudomonas putida* to soybean and seed exudates *in vitro* and in soil. *Can. J. Plant Pathol.*, **6:**267.

Scher, F. M., J. S. Ziegle, and J. W. Kloepper (1984c). A method for assessing the root-colonizing capacity of bacteria on maize. *Can. J. Microbiol.*, **30:**151–157.

Scher, F. M., J. W. Kloepper, and C. A. Singleton (1985). Chemotaxis of *Pseudomonas putida* to soybean and seed exudates *in vitro* and in soil *Can. J. Microbiol.*, (in press).

Shipton, P. J. (1977). The behavior of soil-borne diseases with monoculture: Prospects for biological control. In *Seminar Papers*. Australian Nurseryman's Association, Ltd., Hobart, Australia, pp. 21–30.

Sivan, A., Y. Elad, and I. Chet (1984). Biological control effects of a new isolate of *Trichoderma harzianum* on *Pythium aphanidermatum. Phytopathology*, **74:**489–501.

Sneh, B. (1981). Use of rhizosphere chitinolytic bacteria for biological control of *Fusarium oxysporum* f. sp. *dianthi* in carnation. *Phytopath. Z.*, **100:**251–256.

Suslow, T. V. and M. N. Schroth (1982). Rhizobacteria of sugar beets: Effects of seed application and root colonization on yield. *Phytopathology*, **72:**199–206.

Suslow, T. V., J. W. Kloepper, M. N. Schroth, and T. J. Burr (1979). Beneficial bacteria enhance growth. *Calif. Agric.*, **33:**15–17.

Vidaver, A. K. (1981). Bacteriocin-producing bacteria for biological control of bacterial plant pathogens. In *Handbook of Pest Management in Agriculture*, Vol. II (D. Pimentel, ed.). CRC Press, Boca Raton, FL, pp. 329–334.

Weller, D. M. (1983). Colonization of wheat roots by a fluorescent pseudomonad suppressive to take-all. *Phytopathology*, **73:**1548–2553.

Weller, D. M. and R. J. Cook (1981). Pseudomonads from take-all conducive and suppressive soils. *Phytopathology*, **71:**264.

Wells, H. D., D. K. Bell, and C. A. Jaworski (1972). Efficacy of *Trichoderma harzianum* as a biological control for *Sclerotium rolfsii. Phytopathology*, **62:**442–447.

Wiese, M. Y. and A. V. Ravenscroft (1981). Cephalosporium stripe decline in a wheat monoculture. *Plant Dis. Rep.*, **62:**721–723.

Wilkinson, H. T. and R. J. Cook (1981). The effect of size and concentration of inoculum on the infection of wheat by *Gaemannomyces graminis* var. *tritici* in different soils. *Phytopathology*, **81:**265.

Wong, P. T. W. and R. Baker (1984). Suppression of wheat take-all and Ophiobolus patch by fluorescent pseudomonads from a Fusarium-suppressive soil. *Soil Biol. Biochem.*, **16:**397–403.

Wymore, L. A. and R. Baker (1982). Factors affecting cross-protection in control of Fusarium wilt of tomato. *Plant Dis.*, **66:**908–910.

Yarwood, C. E. (1954). Mechanisms of acquired immunity to a plant rust. *Proc. Natl. Acad. Sci. USA*, **40:** 374–377.

Zentmyer, G. A. (1961). Chemotaxis of zoospores for root exudates. *Science*, **133:**1595–1596.

Zhang, B. X., D. M. Weller, and R. J. Cook (1984). Mass screening of fluorescent pseudomonads for suppressiveness to take-all. *Phytopathology*, **73:**963.

Zogg, H. and J. A. Amiet (1980). Laboratory studies on decline with different foot and root rot pathogens of wheat. *Phytopathol. Z.*, **97:**193–213.

2

PLANT GROWTH CONTROL BY FLUORESCENT PSEUDOMONADS

B. SCHIPPERS

Willie Commelin Scholten Phytopathological Laboratory, Department of Plant Pathology, University of Utrecht, University of Amsterdam, Baarn, The Netherlands

B. LUGTENBERG

Department of Molecular Botany, University of Leiden, Leiden, The Netherlands

P. J. WEISBEEK

Department of Molecular Cell Biology, University of Utrecht, Utrecht, The Netherlands

1. RHIZOSPHERE MICROORGANISMS AND PLANT GROWTH

Symbiotic and free-living nitrogen-fixing microorganisms as well as mycorrhizal fungi have long been considered beneficial for plant growth. Rhizosphere saprophytes, however, are thought to compete with plants for nutrients and to consume nutrients from soil and root exudate (Rovira and Davey, 1974; Rovira et al., 1979) which otherwise could have been used for plant growth.

Recently, rhizosphere saprophytes are regarded as an essential part of the plant ecosystem in supplying such elements as N, P, and possibly Fe for plant growth. In this view, the release of assimilates in root exudates creates an energy-rich rhizosphere zone that enables saprophytes to mineralize nutrients from soil organic debris, which thus becomes available for immediate uptake by the roots. The final remnants of microbiological degraded organic debris improve soil aggregate formation and soil structure.

Root exudates also deliver energy for microbial interactions. The resulting microbial equilibrium sustains the activity of beneficial organisms. It restricts the activity of plant pathogenic populations and also of rhizosphere microorganisms, which become harmful for root development with increasing numbers or activity. The considerable quantity of plant assimilates thus released into the soil (Barber and Martin, 1976) is an investment made by plants for their own benefit.

The harmful rhizosphere microorganisms other than regular soilborne pathogens have been little explored. Their detrimental effects seem to increase as a consequence of certain cultural practices (e.g., high cropping frequency) and to cause severe yield reduction (Schippers et al., 1985). Recently, introduction to the rhizosphere of selected antagonistic fluorescent pseudomonads was shown to suppress such harmful microbial activity (Kloepper and Schroth, 1981; Geels and Schippers, 1983b). Remarkable increases in crop yield have been obtained in field experiments (Kloepper et al., 1980a; Geels and Schippers, 1983c; Geels et al., 1985). Much remains to be learned about the mechanisms involved, however, before reliable applications can be expected.

Thus far, fluorescent pseudomonads have obtained the most attention. This is not by chance. With their flexible metabolism directed toward the dominating substrates released by young parts of the roots, their short generation times, and their mobility, they are well equipped as primary colonizers of young root parts. Moreover, the great variety of antagonistic secondary metabolites, including siderophores, which are suspected to be responsible for natural suppression of several soilborne pathogenic fungi (Kloepper et al., 1980b; Scher and Baker, 1980, 1982), has attracted attention to fluorescent pseudomonads as potential plant growth control agents. This does not exclude the possible importance of other rhizosphere microorganisms in this respect.

2. PLANT GROWTH-INHIBITING RHIZOSPHERE MICROORGANISMS

In many crops considerable reductions in yield occur with increasing cropping frequency. This is convincingly demonstrated by detailed long-term rotational experiments in The Netherlands (Schippers et al., 1985). The yield reductions in potato and wheat in these experiments became more pronounced over the years and reached their maximum after approximately 16 yr. They are clearly related to the cropping frequency in the rotation and are most serious for potatoes (Schippers et al., 1985). The yield reductions fluctuate around 10% in the 1/4 potato frequency and around 15% in the 1/3 potato frequency, compared with yields in a 1/6 frequency of potatoes in the rotation. They even rise to 20–30% in a 1/2 frequency and in continuous (1/1) potato cropping, compared with yields in a 1/3 potato frequency. Yield reductions in winter wheat with 1/2 cereals fluctuate around 9%, compared with 1/3 cereals in the rotation. In continuous wheat cropping in which eyespot (*Pseudocercosporella herpotrichoides*) was controlled with benomyl, yield reductions are approximately 10% of yields in plots with 1/4 winter wheat. Detailed analyses of soil physical and chemical factors have demonstrated that they do not account for the observed reductions (Schippers et al., 1985). In potato, nematodes and fungal pathogens like *Rhizoctonia solani*, *Streptomyces spp.*, and *Verticillium dahliae* also do not or may only partially account for the yield losses (Schippers et al., 1985; Scholte et al., 1985).

The reductions in yields of potato and wheat were reproduced in pot experiments with soil from the rotational field plots (Geels and Schippers, 1983b; Scholte et al., 1985). Soil treatments with methyl bromide and pasteurization at 55°C for half an hour eliminated the "rotational effects" completely (Scholte et al., 1985). This strongly suggests that the reductions are of microbial origin.

Root systems of potato stem cuttings grown for 2 wk attained less than half the weight in soil continuously (1/1) cropped to potato, compared with those developed in soil with 1/6 frequency of potatoes in the rotation. Both soils were from the same experimental field and were supplemented with optimal amounts of nutrients. Roots in both soils showed no obvious signs of microbial injury during these 2 wk when observed with the light microscope (Bakker et al., in press, Fig. 2). These experiments also suggest that unknown microbial factors other than regular pathogens can significantly inhibit root growth and decrease yields. They also indicate that these harmful effects increase with increasing cropping frequency, which may be a general phenomenon. Other cultural practices, however, may change the microbial equilibrium as well so that it becomes less optimal for plant growth despite the absence of regular pathogens. Yield reductions in high frequency cropping bear resemblance to the phenomenon of soil sickness described for many crops (Bouhout, 1979; Hoestra, 1979).

Rhizosphere microorganisms harmful to root development have been discussed and are considered to be minor pathogens (Schippers et al., 1985; Salt, 1979; Suslow and Schroth, 1982). Whether this is correct is now doubtful. According to Salt (1979), although minor pathogens are unspecialized and facultative parasites, they do parasitize plant tissue. Although such root parasites (minor pathogens) may indeed be partly responsible for yield reductions in high-frequency cropping systems, our observations suggest that nonparasitic rhizosphere microorganisms may also be involved (Schippers et al., 1985; Bakker et al., in press). Their harmful effects may be caused by their metabolites (Fredericson and Elliott, 1985). They may only become toxic or inhibitory for root cell activity above certain concentrations in the rhizosphere. This could also be caused by a relative decrease in the degradation of these metabolites by other rhizosphere organisms.

The increase in harmful microbial influences with increasing cropping frequency may originate in the accumulation of root debris of the particular crop. This may in turn sustain the survival of the causal microorganisms to the next season or crop. The increasing amount of root debris may also contain a precursor of microbial metabolites harmful to root cell activity.

3. PLANT GROWTH-STIMULATING PSEUDOMONADS

Research on seed bacterization revived following reports of plant growth stimulation by fluorescent pseudomonads by Burr et al. (1978), as has recently been reviewed by Suslow (1982), Schroth and Hancock (1982), and Schippers (1983). Plant growth-promoting pseudomonads (PGPP) were obtained by selecting *Pseudomonas* colonies that strongly inhibited a variety of pathogenic and saprophytic rhizosphere fungi and bacteria, including *Erwinia carotovora* on modified low iron King's B medium (Burr et al., 1978; Geels and Schippers, 1983c).

The addition of relatively high concentrations of the iron chelator 8-*O*-hydroxy quinoline to the medium is recommended to improve and simplify the search for pseudomonads with growth-promoting properties (Geels et al., 1985).

Plant growth stimulation is hypothesized to be mainly due to the suppression of harmful rhizosphere microorganisms, since it correlates with a decrease in the number of rhizosphere microorganisms, some of which negatively affect root development when reintroduced in high numbers in the rhizosphere by tuber treatment (Suslow and Schroth, 1982). Moreover *Pseudomonas* mutants that had lost their *in vitro* antagonistic activity as well as their plant growth-promoting properties did not affect the rhizosphere microflora (Kloepper and Schroth, 1981).

The main mechanism of the antagonism seems to be competition with harmful microorganisms for iron by the release of siderophores, secondary metabolites with a strong affinity to Fe^{3+} (Fig. 1), since the addition of dissolved Fe^{3+} to the environment not only abolishes *in vitro* antagonism but also prohibits plant growth promotion (Kloepper et al., 1980a). Furthermore, mutants obtained by exposure to ultraviolet light or mutagenic chemicals (Kloepper and Schroth, 1981) or by transposon mutagenesis (Marugg et al., 1985), which then lost their ability to produce siderophores *in vitro*, also lost their plant growth-stimulating properties, although they colonized roots equally well (Bakker et al., in press; Kloepper and Schroth, 1981). The siderophore pseudobactin produced by the plant growth promoting *Pseudomonas* B10 could mimic the antagonism *in vitro* and also promote plant growth when added to King's medium B and soil, respectively. Both FeEDTA and Fe-pseudobactin added to soil had no effect on the plant (Kloepper et al., 1980a). Production of pseudobactin in the rhizosphere has not yet been directly established.

Plant growth-promoting pseudomonads colonized plant roots after tuber or seed bacterization, following an *in vitro* mutation that conferred resistance to rifampicin and naladixic acid without affecting their plant growth-promoting properties (Geels and Schippers, 1983a; Kloepper et al., 1980c). They could be reisolated from the roots of treated plants during the season until harvest, although their numbers gradually decreased (Geels et al., 1985; Kloepper et al., 1980c). Apart from approximating a lognormal distribution (Loper et al., 1984), little is known, however, about their distribution and localization on the roots in the field. Significant increases in yield by seed and tuber bacterization have been demonstrated in field experiments with sugar beet, potato, and other crops

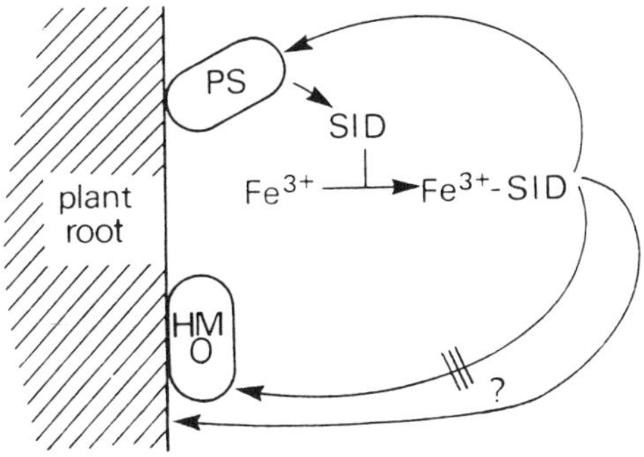

Fig. 1. Diagram of hypothesized mechanism of plant growth promotion: competition with harmful microorganisms (HMO) for iron (Fe^{3+}) by siderophores (S) released by plant growth promoting pseudomonads (PS).

(Burr et al., 1978; Geels and Schippers, 1983c; Geels et al., 1985; Kloepper et al., 1980c). In many cases, however, no significant increase was observed. This can be partly ascribed to the great variability in yields between the usually low number of replicates in field experiments. Most likely, inadequate colonization of those root parts responsible for tuber and seed development later in the season is the most important factor. The most impressive data on growth stimulation have, therefore, been obtained in pot and field experiments where root or tuber development were measured long before harvest (Burr et al., 1978; Geels et al., 1985; Kloepper et al., 1980c; Kloepper and Schroth, 1981). Impressive increases in final yields are therefore also obtained with crops that have a relatively short growing season and a relatively restricted root volume, such as radish (Geels et al., 1985; Kloepper and Schroth, 1978).

Rapid biotests are needed for fast screening of effective isolates and for unraveling the mechanisms of growth promotion. Upon treatment of rooted potato stem cuttings with bacteria, increases in root weight over 75% were obtained within 2 wk (Bakker et al., in press). Apparently, without tuber reserves to rely upon for first root development, root growth promotion by bacterial treatments becomes pronounced earlier.

Differences in numbers or activity of harmful rhizosphere microorganisms in soil may be another factor causing variations in growth promotion by bacterization. In our experiments significant growth promotion by bacterial treatments takes place *only* in soil with a history of high cropping frequency of the particular crop (Bakker et al., in press; Geels and Schippers, 1983b). Considerable increases in potato root development within 2 wk using stem cuttings also occurred only with soil that had been frequently cropped to potato. In such soil, root growth of nonbacterized stem cuttings was impressively inhibited compared to that in nonpotato soil from the same field (Fig. 2). Root development in nonpotato soil did not increase and was even slightly retarded upon bacterial treatments of rooted sprouts (Bakker et al., in press). These experiments make it very

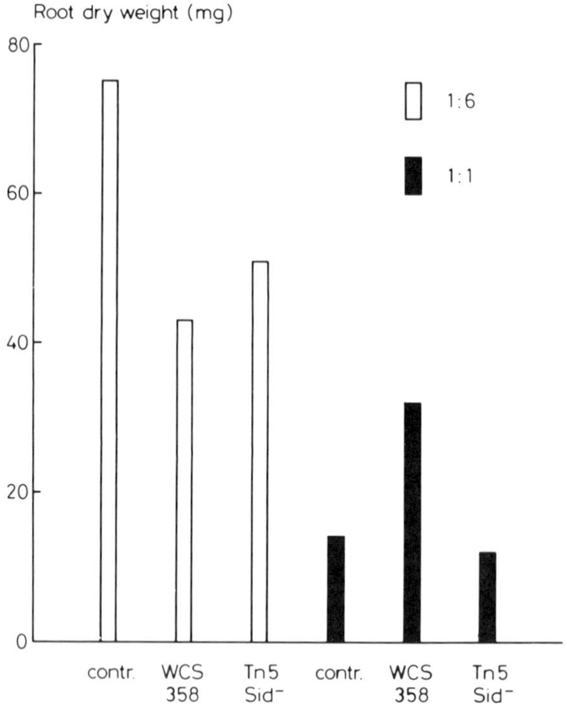

Fig. 2. Root dry weight after 1 wk of growth of rooted potato stem cuttings in soil from a field continuously cropped to potato (1/1) compared to that in soil from a field with 1/6 potato cropping frequency. The suppression of root development in 1/1 soil is partly compensated by treatment before planting of rooted sprouts with *P. putida* WCS 358, but not with its transposon mutant, which lost its ability to produce siderophores (Tn 5 Sid^-). Both WCS 358 and Tn 5 Sid^- inhibit root development in 1/6 soil similarly. This inhibition is thus not caused by WCS 358 siderophores (Bakker, unpublished).

likely that growth promotion by treatments with selected fluorescent pseudomonads occurs because the suppression of harmful microbial activity increases with cropping frequency.

In our experiments root growth inhibition in high frequency cropping soil, as well as *Pseudomonas*-induced growth promotion, are most pronounced in soils obtained immediately from the cropped field or shortly thereafter. It diminishes or may even disappear during storage and during fallow in the field, suggesting a decrease in the harmful microflora. The development of this microflora or its activity in the field may differ from year to year, or even from site to site in one field, depending on variations in abiotic factors, as is also known to occur for regular pathogens. This view is supported by the obvious fluctuations in yield depressions (rotational effects), as observed in long-term rotational field experiments (Schippers et al., 1985). Variations from year to year and from field to field in bacterization-induced yield increases may, thus, also be due to variations in rotational effects.

That imbalanced development of rhizosphere microbial populations can lead to inhibition of root development is demonstrated by tuber treatments with rhizosphere saprophytes (Suslow and Schroth, 1982). It is also demonstrated by a significant decrease in root development in soil with little or no history (1/6) of potato crops after treatments of stem cuttings with some *Pseudomonas* isolates that impressively stimulated root development in high-frequency potato soil (Bakker et al., unpublished; Fig. 2). This emphasizes that the results obtained by testing microorganisms for their deleterious or growth-promoting properties by introducing them in high numbers in the rhizosphere should be interpreted with care.

Although competition for iron seems to be a key factor in suppression of harmful rhizosphere microorganisms, other metabolites of pseudomonads may be involved as well (Howell and Stipanovic, 1979, 1980).

4. MICROBIAL COMPETITION FOR IRON

Microbial competition for iron is an important mechanism of antagonism in the rhizosphere. Iron plays a central role in the energy metabolism of aerobic and semianaerobic microorganisms and its availability as free ions is limited in natural environments. Although abundant in most soils, much of the iron is oxidized to highly insoluble ferric-oxyhydroxide polymers. The solubility constant of Fe(OH)3 is $10^{-38.7}$ and the free ferric-ion concentration is thus normally very low (Biedermann and Schindler, 1957). At dissolved iron concentrations of approximately 10^{-6} M Fe^{3+} or higher, which may occur at anaerobic conditions or at very low pH, enough iron is available to be directly transported into the microbial cells for growth (low iron affinity system). At lower concentrations of dissolved iron, which are more common in aerated soil, microorganisms have to compete for iron. A specific (high iron affinity) system is then induced to dissolve Fe^{3+} from the environment and transport it into the microbial cell. Such a system has been demonstrated for many aerobic and semianaerobic microorganisms (Neilands, 1981a,b, 1984). It includes the synthesis of secondary metabolites, called siderophores, which are water-soluble molecules with a high affinity for Fe^{3+}. They are too large (1000–1500) daltons to pass to the outer membrane through the usual pores. A specific high-affinity system is operative under these conditions. It contains special outer membrane proteins by which the iron can be bound and transported through the outer membrane, and subsequently through the cytoplasmic membrane, into the cell by an energy-dependent process.

Siderophores are molecules with bidentate ligands which, usually with oxygen atoms, contact the Fe^{3+} with its six octahedric directed bonds (Figs. 3 and 4). A variety of bidentate ligand systems of siderophores exists; hydroxamate and catechol systems are the most common (Neilands, 1981a, 1984).

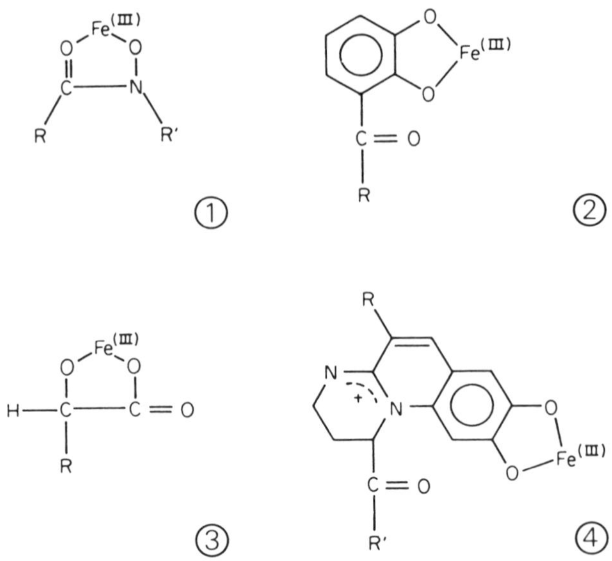

Fig. 3. Some bidentate ligand systems of siderophores. (1) Hydroxamate (ferrichrome and other hydroxamates). (2) Catechol (enterobactin and other catechols). (3) α-Hydroxy acid (pseudobactin, citrate-containing siderophores). (4) Fluorescent quinolin chromophore (pseudobactin, pyoverdine).

Hydroxamate siderophores, of which des-ferrichrome is the prototype, are formed by pro- and eukaryote microorganisms, especially fungi; the catechol type is formed by bacteria only (Neilands, 1984). A fluorescent quinolin chromophore is one of the bidentate ligands in siderophores of fluorescent pseudomonads (Figs. 3 and 4).

The affinity of siderophores for Fe^{3+} is expressed as a binding or stability constant log10 KFe (pH 7) (Lindsay, 1979). It is pH-dependent, varies widely among the various siderophores, and is thus an important ecological factor in the microbial competition for iron. In soil, the microbial competition for Fe^{3+} level is determined by soil pH (Scher and Baker, 1980; Lockwood and Schippers, 1984). The specificity of outer membrane receptor proteins of microorganisms for their own siderophores is an ecologically important mechanism in the competition for iron. A remarkable aspect is that some organisms possess only the capability to take up and process certain siderophore–iron complexes produced by other organisms, but do not possess the information to produce and secrete these siderophores themselves. For instance, the fungus produced des-ferrichrome is not synthesized by bacteria, although many bacteria (e.g., *Escherichia coli*) can use it. A similar, although less extreme, situation is found among the fluorescent pseudomonads. The major siderophore of *P. fluorescens* WCS 374 can be used by *P. putida* WCS 358 (both Tn5 mutants and wild type) and WCS 361. The reverse is not the case (Marugg et al., 1985; Bakker, unpublished). It is not yet clear whether this result is a consequence of the presence of independent synthesis systems or of minor modifications of basically the same process.

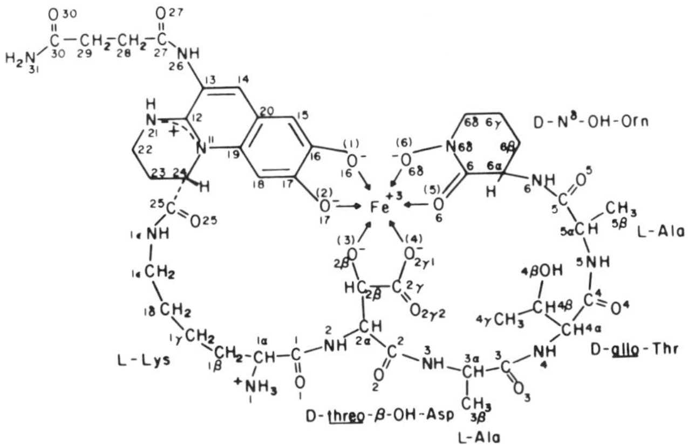

Fig. 4. The structure of ferric pseudobactin, an iron chelating siderophore produced by plant growth promoting rhizobacteria B10 (courtesy of J. Leong).

Optical and geometric isomers, and also different mechanisms to transfer iron from the iron chelate into the cell, contribute to the specificity of the siderophore-mediated iron transport systems (Monzyk and Crumbliss, 1983). According to Monzyk and Crumbliss (1983), the high stability constants of siderophore-Fe^{3+} complexes make it very likely that mechanisms of catalyzed iron-release function in regulating the biological availability of iron. Because of their presence in hydroxamate siderophore systems, they suggest hydroxamine acids to catalyze iron transfer from siderophores, thus delivering additional low-energy pathways for iron supply to the cell. In microbial competition for Fe^{3+} in nature, microorganisms with a siderophore system in which solubilization from the environment and uptake into the cell is fastest and least energy-demanding will probably be in a better position.

Environmental factors also influence microbial competition for iron. The synthesis and secretion of the siderophore molecules by the bacterial cell is, under laboratory conditions, controlled by Fe^{3+} concentrations, temperature, and pH. Only at Fe^{3+} concentrations lower than approximately $20 \mu M$ does the cell switch on its genetic capacity for the biosynthesis and secretion of the siderophore and for the biosynthesis of outer membrane receptor proteins. However, not all receptor proteins are completely absent at high Fe^{3+} concentrations, for example, Fe^{3+}-citrate transport systems can also be induced without Fe^{3+} limitation by the presence of citrate (von Braun, 1985).

The final concentrations of siderophore can reach levels of several milligrams per liter of culture. Moreover, siderophore biosynthesis continues long after the major cellular metabolic processes are shut down or severely decrease, for example, when cells have reached stationary phase conditions.

At 37°C, the growth of several fluorescent pseudomonads is completely normal. Their siderophore production and also their antagonistic properties, however, are completely inhibited at this temperature. This property may influence their applicability for growth stimulation in some situations. The pH may also differentially influence the production of the various siderophores and the stability of their ferri- and desferri- forms (Meyer and Abdallah, 1978; Philson and Llinas, 1982).

5. ANTAGONISTIC METABOLITES PRODUCED BY FLUORESCENT PSEUDOMONADS

Fluorescent pseudomonads include the species *P. fluorescens* (four biotypes), *P. putida* (two biotypes), *P. aeruginosa*, *P. chlororaphus*, *P. aureofaciens*, and *P. syringae*. These species are known to produce a variety of secondary metabolites with antagonistic characteristics, most of which are nitrogen-containing heterocyclic compounds. Among these are phenazines (pyocyanine and pyrrolnitrine-type compounds) and indole derivates (Leisinger and Margraff, 1966). Other secondary metabolites with antagonistic characteristics are unusual amino acids and peptides, including the iron-chelating agents, some of which act as iron carriers (siderophores). Some of the *Pseudomonas* secondary metabolites, however, may act as phytotoxins. The production of both strong siderophores and other antagonistic metabolites is of particular interest in the search for microbial agents for crop protection.

Pyoverdine and pseudobactin, water-soluble compounds with pronounced and strong siderophore properties produced by fluorescent pseudomonads, have been chemically characterized. Pseudobactin was described as the siderophore of plant growth promoting *P. fluorescens-putida* isolate B10 and claimed to be involved in the mechanism of plant growth promotion (Kloepper et al., 1980a). Pseudobactin is a linear hexapeptide with an *N*-hydroxy ornithine, a hydroxy acid of hydroxy aspartic acid, and a fluorescent *o*-dihydroxy aromatic group "quinoline" representing the three Fe^{3+}-chelating ligands (Teintze et al., 1981; Teintze and Leong, 1981; Figs. 3 and 4). This combination of Fe^{3+}-chelating ligands and the alternating D- and L- amino acids connecting them is considered to be responsible for its high log 10 KFe constant and its resistance to degradation by enzymes (Teintze et al., 1981).

Pyoverdine was already detected by Elliot in 1958 and was shown to have an important function in microbial iron transport by Meyer and Abdallah (1978). Philson and Llinas (1982a,b) isolated and characterized pyoverdine from cultures of an isolate of *P. fluorescens* from which the iron-chelating metabolite ferribactin had already been isolated and partially characterized by Maurer et al. (1968). Both pyoverdine and ferribactin carry a hydroxamate ligand (*N*-formyl *N*-hydroxy ornithine) and differ only slightly in their amino acid composition (Maurer et al., 1968). In pyoverdine, however, one of the amino acids, tyrosine, is replaced by a fluorescent chromophore *o*-dihydroxy quinoline, which considerably increases its affinity for Fe^{3+} relative to ferribactin. On the basis of their compositional analogy, Philson and Llinas (1982b) consider it plausible that ferribactin is a precursor of pyoverdine, which only accumulates after normal metabolism is blocked at late stages of growth. There is no evidence that ferribactin acts as an iron carrier.

Both pyoverdine and pseudobactin have strong optical absorption bands near 400 nm and two broad peaks at longer wave lengths that do not change between pH 3 and 10 (Philson and Llinas, 1982). Ferribactin, like other hydroxamate siderochromes, has one single broad absorption band with a maximum that is strongly affected by pH. Pyoverdine and pseudobactin are thus closely related, both containing two hydroxamate ligands and one catechol type ligand, but differing in their log KFe value.

All the fluorescent *Pseudomonas* isolates (Geels and Schippers, 1983a) tested produced a yellow-green fluorescent siderophore with an overall structure resembling those described for pyoverdine (Philson and Llinas, 1982b) and pseudobactin (Teintze et al., 1981). This does not exclude the presence and utilization of other siderophores in fluorescent pseudomonads.

The plant growth-promoting *P. putida* isolate WCS 358 (Geels and Schippers, 1983b) produces a siderophore that, based on fluorescence and absorption spectra at different pH's, has a very similar structure to pseudobactin (B10). The plant growth promoting *P. fluorescens* WCS 374 produces at least one fluorescent siderophore that, according to its absorption spectrum and retention values on anion exchange columns, very likely has a structure similar to that of pseudobactin (unpublished). In addition we have observed that the *Pseudomonas fluorescens* isolate WCS 374 contains a second nonfluorescing siderophore of considerable Fe^{3+}-acquisition efficiency next to its fluorescent siderophores (unpublished). Several of the nonfluorescing, noniron binding mutants of plant growth promoting pseudomonads have been tested for their antagonism toward other microorganisms and their ability to induce plant growth stimulation. Both properties were absent in these mutants (Kloepper and Schroth, 1981; Marugg et al., 1985; Bakker et al., in press).

6. PSEUDOMONADS AND SOILBORNE PLANT PATHOGENS

Fluorescent pseudomonads and their siderophores are involved in the natural microbial suppression of *Gaeumannomyces graminis* var. *tritici* (Ggt) (Kloepper et al., 1980b; Wong and Baker, 1984). The natural suppression could be nullified by the addition of FeEDTA to soil, suggesting that competition for iron plays a part in the mechanism of antagonism (Kloepper et al., 1980b; Scher and Baker, 1980, 1982; Schippers, unpublished). Seed treatments with selected isolates of fluorescent pseudomonads reduced the disease incidence of Ggt and increased grain yields in field experiments (Weller and Cook, 1983; Geels, unpublished). Isolates with the potential to produce, in addition to siderophores, metabolites antibiotic to Ggt were more effective than isolates with no or weak antibiotic production (unpublished). Whether competition for iron is the only mechanism involved in natural suppression of soilborne pathogens should therefore be doubted. Microorganisms other than pseudomonads are involved in the natural suppression of Ggt (Hornby, 1979). Their importance may very well depend on abiotic soil and climatic conditions.

Fluorescent pseudomonads also have potential as biocontrol agents of several other soilborne pathogens, for example, *Pythium* and *Rhizoctonia* in cotton seedlings (Howell and Stipanovic, 1979, 1980), *Erwinia carotovora* in potato (Kloepper, 1983), and *Fusarium oxysporum* in flax, radish, and cucumber (Scher and Baker, 1980, 1982).

7. MOLECULAR MICROBIOLOGICAL ASPECTS

The suppression of harmful microorganisms on plant roots and promotion of plant growth by fluorescent *Pseudomonas spp.* might be improved and extended to new areas of application when the factors involved are understood at the molecular level. For comparative examination of these factors with different isolates and in the search for new isolates with better properties, it would be helpful to characterize isolates at the molecular level. It was therefore attempted to classify 27 antagonistic fluorescent *Pseudomonas* isolates obtained from potato roots (Geels and Schippers, 1983a), using SDS-PAGE (sodium dodecyl sulfate polyacrylamide gel electrophoresis) patterns of outer membrane proteins and lipopolysaccharides as the criteria (Fig. 5). Most isolates tested appeared to be distinct strains using these criteria. The only isolates that could not be distinguished from each other were those derived from a single plant grown in a pot (de Weger et al., 1985).

Because of their high specificity toward Fe^{3+}-siderophore complexes, patterns of ferric-ion limitation induced proteins may be a more appropriate indicator to distinguish the isolates with respect to their siderophore-dependent antagonistic properties than patterns of outer membrane proteins. Among 21 distinct isolates, 16 different patterns of ferric-ion regulated high molecular weight membrane proteins were easily distinguished. This suggests that most of these isolates differ in their siderophore-dependent antagonistic properties (de Weger et al., 1985).

In *E. coli*, genes involved in the synthesis and uptake of siderophores are either found on the chromosome or on plasmids (von Braun, 1985; Neilands, 1982). The possibility that the siderophore-dependent antagonistic properties of all the distinct isolates are present on the same plasmid was checked. Small plasmids appeared to be absent and megaplasmids were only present in a minority of the isolates (de Weger et al., 1985).

Knowledge of the genetics of the siderophore system may reveal possibilities to improve antagonistic and growth-promoting properties of *Pseudomonas* isolates. The genetic basis of siderophore synthesis has been approached via mutation and selection and via the isolation of the genes at various steps in the biosynthetic pathway. Both chemical- and transposon-mediated mutagenesis have been applied. The advantage of using mutagenic chemicals is that any isolate can be approached, irrespective of the availability of genetic transformation or

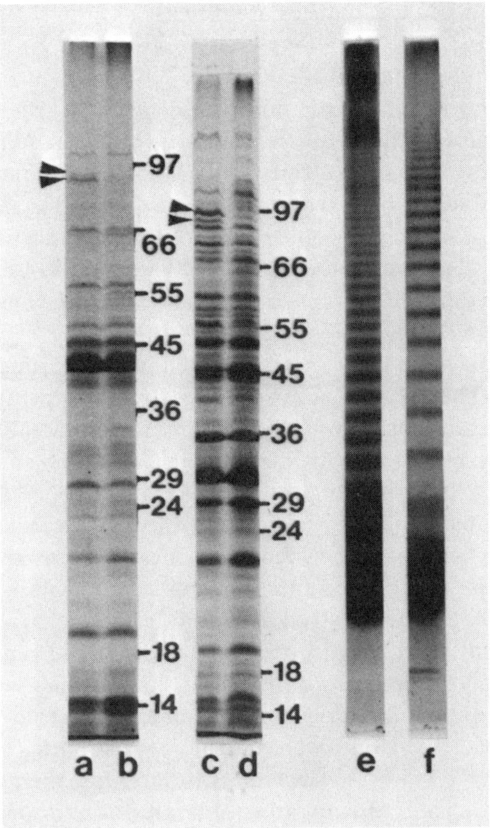

Fig. 5. SDS-PAGE of membrane proteins (*a–d*; 11% gel; stained with Fast Green) and lipopolysaccharides (*e* and *f*; 13% gel; silver stained) of fluorescent *Pseudomonas* isolates WCS 358 (*a*, *b*, and *e*) and WCS 374 (*c*, *d*, and *f*). The molecular weights of standard proteins are indicated in kilodaltons. Closed arrows indicate the positions of proteins that appear upon Fe^{3+} starvation.

conjugation methods. The major disadvantage is in general that many additional (>10) chemical damages may be introduced to the chromosome of the selected mutant. This can interfere with the interpretation of the physiological and ecological effects of the mutants. A genetically cleaner technique is the use of transposons; Tn5 is a very effective transposon for use in *Pseudomonas* (Haas, 1983; Marugg et al., 1985). Chemical mutagenesis has produced a large number of nonfluorescing mutants that were all unable to produce an effective siderophore (Moores et al., 1984). Studies with cosmid gene banks (Moores et al., 1984) revealed that for isolate B10 at least 12 genes in four gene clusters are required for complementation of all the mutants. The results indicate the presence of at least 12 separate transcription units. In view of the size of the cosmid inserts (22 kb) and the clustering of the genes, this may well be a considerable underestimate of the number of genes involved. An important question is whether the isolated DNA sequences are directly used in siderophore biosynthesis or whether more general pathways such as those involved with the biosynthesis of certain amino acids are also involved.

By using transposon Tn5 as the mutagenic agent (Marugg et al., 1985) mutants defective in fluorescence, iron-sequestering ability, and amino-acid biosynthesis have been obtained from strain WCS 358. The mutants could be divided into six groups (Table 1). The largest group (Flu^- Sid^-) is very likely comparable to the mutants isolated by Moores et al. (1984). In addition, nonfluorescent mutants that are still able to sequester iron efficiently (Flu^- Sid^+), and mutants with either intermediate fluorescence ($Flu^{\pm}$) and/or intermediate iron binding ($Sid^{\pm}$) were distinguished.

These mutants have all been complemented with a cosmid-gene bank in the plasmid pLAFR1. These experiments produced five independent complementation groups. The exact position of the Tn5 insert within each mutant chromosome has been determined by hybridization and restriction analyses; the results are presented in Fig. 6. From the position of the inserts it could be deduced that those mutants complemented by cosmids in this cluster fall into at least three different transcription units, of which the largest is approximately 14 kb. The size of this transcript points toward the presence of multiple genes on this mRNA and therefore to an operon-like structure for at least this part of the biosynthetic pathway. These results indicate the existence of at least 13 genes that are directly involved in siderophore production. None of the mutants was defective in the uptake and intracellular utilization of the siderophore–iron complex normally present in wild type cells. Nonfluorescent, noniron binding mutants of antagonistic and growth-promoting *Pseudomonas* isolates lost their antagonistic activity toward other microorganisms and their ability to promote plant growth (Bakker et al., in press; Marugg et al., 1985).

TABLE 1. Siderophore-Defective Mutants of *Pseudomonas putida* WCS 358

Class	*Phenotype*[a]	*Number of Mutants*	*Fluorescent*	*Maximum Growth Concentration of Bipyridyl (M)*	*In Vitro Antibiosis*
I	Flu^- Sid^+	3	—	800	+
II	Flu^- Sid^-	18		200	—
III	Flu^- $Sid^{\pm}$	1	—	400	+
IV	$Flu^{\pm}$ $Sid^{\pm}$	1	±	600	+
V	$Flu^{\pm}$ Sid^+	4	±	800	+
VI	Flu^+ $Sid^{\pm}$	1	+	400	+

[a]Flu^-, no fluorescence; $Flu^{\pm}$, moderate fluorescence; Flu^+, normal (wt) fluorescence; Sid^-, growth in presence of max 200 μM bipyridyl; $Sid^{\pm}$, growth in presence of 400–600 μM bipyridyl; Sid^+, growth in presence of max 800 μM bipyridyl. Fluorescence was determined with irradiation of UV light (>360 nm). In all cases mutants were grown on KB medium.

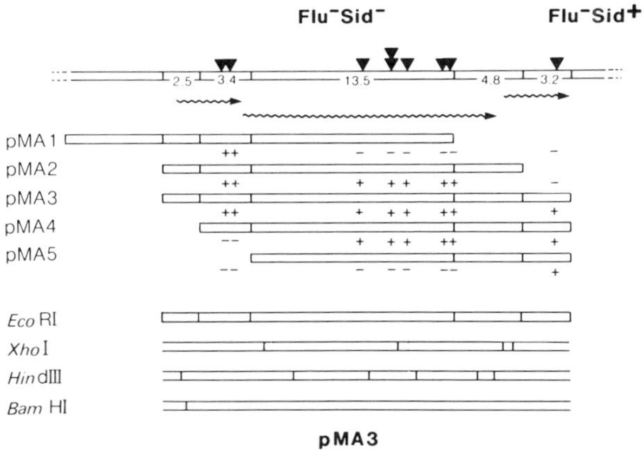

Fig. 6. Physical map of the cosmid inserts of group A, visualizing their regions of overlap and the complementation pattern of this group with eight Flu^- Sid^- mutants and one Flu^- Sid^+ mutant of *Pseudomonas putida* WCS 358. The +/− symbols indicate complementation yes/no. The fragment sizes are given in kilobases.

Because failures to increase plant growth and yield in the field by bacterization with growth promoting pseudomonads are likely to be due to inadequate colonization of the roots, the study of *Pseudomonas*-plant interactions deserves serious attention. To be optimally beneficial for the plant, the *Pseudomonas* cells should be present on or very close to the plant root surface.

The bacterium could accomplish this by chemotaxis toward molecules present in root exudates, by adhesion to the root tissue, or by a combination of these two mechanisms.

The mechanisms of interaction are presently being investigated for the Dutch isolates (Geels and Schippers, 1983a), which are all highly motile. Mutants with no flagella have recently been isolated for these studies (de Weger, unpublished), and the physical adhesion of *Pseudomonas* cells to sterile potato roots has been observed under laboratory conditions (de Weger, unpublished). We intend to use this system for the isolation of adhesive-defective mutants. Whether this mechanism of adhesion indeed plays a role under field conditions will then be studied. Fimbriae are the organelles that are usually responsible for adhesion (Gaastra and de Graaf, 1982). A preliminary search for fimbriae on our *Pseudomonas* isolates was negative (de Weger, unpublished). Adhesion of bacteria by other mechanisms has also been reported, for example, by extracellular polysaccharides (Ramphal and Pier, 1985) and by an additional protein layer outside the outer membrane (Udey and Fryer, 1978).

Despite some detailed studies (Dazzo and Gardiol, 1984; Dazzo et al., 1985; Matthijse, 1984), knowledge of the interaction between bacteria and plants at the molecular level is scarce. Concepts used to explain the interaction between bacteria and other cells or surfaces are usually based on knowledge of surface structures derived from studies on organisms grown independently of each other. Recent work has shown that at least some bacterium–plant responses can be induced by molecular signaling between the two partners. One of the virulence genes of the pathogen *Agrobacterium tumefaciens* is induced by a root exudate factor (Okker et al., 1984). In the nodule-forming bacterium *Rhizobium* the situation might be even more complex in that a root exudate factor is required for the synthesis of a bacterial factor, which in turn causes an altered growth pattern of the plant root (van Brussel et al., 1985). These examples show that studies of bacteria and plants in their natural environment, or under conditions closely resembling this environment, can contribute information essential for understanding their interactions at the molecular level.

So far, mutants resistant to antibiotics have been used to label certain strains, thus allowing the study of root colonization by a particular isolate throughout the season in the field (Geels et al., 1985). Since the most frequently used antibiotics, nalidixic acid and rifampicin, alter the structures of enzymes essential for the growth of the bacterial cell, it is conceivable that such mutants have a disadvantage under natural conditions. Mutants labeled with an antibiotic resistance marker derived from a transposon are more promising because this resistance marker is added to the existing genes. By screening for mutant strains that survive well under field conditions it is likely that mutants can be isolated with an insertion in a gene that is not important for survival under the conditions studied.

8. PSEUDOMONADS, SIDEROPHORES, AND IRON NUTRITION OF PLANTS

Because the plant appears to benefit from introduction into its rhizosphere of pseudomonads capable of producing siderophores with a strong affinity to Fe^{3+}, it is likely that the plant is somehow able to use the Fe^{3+} of the Fe^{3+}-siderophore complexes. The alternative possibility is that certain plants have developed their own mechanisms to cope with iron deficiency and deal in some other way with microorganisms competing for iron.

Many plants react to iron deficiency by extrusion of H^+-ions (Römheld and Kramer, 1983; Römheld and Marschner, 1983). Furthermore, probably as a consequence of rhizosphere acidification, reducing and chelating compounds may diffuse out of the roots (Barrett-Lennard et al., 1983; Bienfait et al., 1983; Olsen et al., 1981; Römheld and Marschner, 1983). Both effects result in an increased solubility of iron. Moreover, the roots may develop a highly active reduction capacity for ferric chelates at the plasma membrane of the epidermis cells (Bienfait et al., 1983). Because the ferrous ion is taken up much faster than the ferric chelates tested (Chaney et al., 1972), these responses together result in an increased efficiency of the root to obtain iron. Grasses, however, seem to follow a strategy more in line with microorganisms; ferric-chelating compounds like mugineic acid (Mino et al., 1983) and avenic acid (Fushiya et al., 1980) were found in exudates from iron-deficient barley and oats, respectively.

The question arises whether microbial siderophores interfere with the iron nutrition of plants. Some evidence has been presented that hydroxamate-type siderophores (HS) like desferrioxamine B, desferrichrome, desferrichrome C, desferrichrome A, and HS siderophores of *Boletus* spp. can function as iron sources for higher plants such as tomato (Orlando and Neilands, 1982; Powell and Szaniszlo, 1982; Stutz, 1964).

Using hydroxamate auxotroph bacteria Powell et al. (1980, 1983) demonstrated that hydroxamate siderophore concentrations in bulk soil were sufficient to guarantee iron transport to plant roots as calculated by O'Conner et al. (1971). Moreover, concentrations of HS siderophores in the rhizosphere soil of ectomycorrhizal *Pinus* seedlings were 11–56 times as high as in bulk soil (Powell and Szaniszlo, 1982). The possible depletion of Fe–HS in the rhizosphere is caused by iron uptake by the plants and the low mobility of HS due to binding to organic material and clay particles (Cline et al., 1983; Powell et al., 1980). This may be compensated for by the accumulation or production of HS by rhizosphere microorganisms (Powell and Szaniszlo, 1982). Cline et al. (1983) demonstrated the ability of HS to chelate Fe^{3+} in acid as well as alkaline soils. Other ions (Mn, Zn, Mg, Al, Cu, Ca) could not compete successfully for Fe^{3+} with the HS ligand. Other natural iron-chelating organic acids, because of their lower stability constants, were ineffective iron chelators at alkaline soil pH under optimal oxidized redox conditions for iron-chelating organic acids. These observations also reinforce the possible role of Fe^{3+}–HS as source of iron in ecosystems. Whether the pyoverdine/pseudobactin-type siderophore of fluorescent pseudomonads can be utilized as an iron source by plants remains to be decided. A positive role for microbial siderophores in iron nutrition of higher plants, however, is still not evident. Thus desferrichrome significantly reduced the amount of iron taken up from $Fe(OH)^3$ (Reid et al., 1984). A similar effect was shown by desferrioxamine on iron utilization by sorghum, whereas under comparable conditions with sunflower a stimulation of growth and chlorophyll synthesis was found (Cline et al., 1984).

The introduction into the rhizosphere of microorganisms producing siderophores that compete with the plant for iron might thus have a negative effect on the iron nutrition of plants. However, bacterization of potato roots with pyoverdine/pseudobactin-type siderophore producing pseudomonads increased plant growth and root development, at least in high-frequency potato soil (Bakker et al., in press; Geels and Schippers, 1983b). In soil without a history of potato, and thus without yield-depressing and root growth-inhibiting microbial activity, however, a significant decrease in root development was observed following bacterization (Bakker et al., in press, Fig. 2). The final outcome of bacterization with fluorescent siderophore producing pseudomonads on plant growth in the rhizosphere might therefore depend on concentration and quality of siderophores produced *in situ* and on the equilibrium between plant growth inhibiting and stimulating rhizosphere microorganisms in the soil. More information on the relationship between siderophore producing microorganisms and plant roots is therefore needed.

ACKNOWLEDGMENT

We wish to thank Dr. K. A. Seifert for correcting the English in this chapter. We also acknowledge Dr. H. F. Bienfait for his suggestions and critical reading of the section on the iron nutrition of plants.

REFERENCES

Bakker, P. A. H. M., A. W. Bakker, J. D. Marugg, P. J. Weisbeek, and B. Schippers (1986). A bioassay to study the mechanisms of potato growth promotion by *Pseudomonas* spp. in short rotations (submitted for publication).

Barber, D. A. and J. K. Martin (1976). The release of organic substances by cereal roots into soil. *New Phytol.*, **76:**69–80.

Barrett-Lennard, E. C., H. Marschner, and V. Römheld (1983). Mechanism of short-term Fe III reduction by roots. *Plant Physiol.*, **73:**893–898.

Biedermann, G. and P. Schindler (1957). On the solubility product of precipitated iron (III) hydroxyde. *Acta Chem. Scand.*, **11:**735.

Bienfait, H. F., R. J. Bino, A. M. van der Bliek, J. F. Duivenvoorden, and J. M. Fontaine (1983). Characterization of ferric reducing activity in roots of Fe-deficient *Phaseolus vulgaris. Physiol. Plant*, **59:**196–202.

Bouhout, D. (1979). Un test biologique à deux niveaux por l'etude des fatigues de sol. Application à l'etude des necroses des racines, *Ann. Phytopathol.*, **11:**95–109.

Burr, T. J., M. N. Schroth, and T. Suslow (1978). Increased potato yields by treatments of seedpieces with specific strains of *Pseudomonas fluorescens* and *P. putida. Phytopathology*, **68:**1377–1383.

Chaney, R. L., J. C. Brown, and L. O. Tiffin (1972). Obligatory reduction of ferric chelates in iron uptake by soybeans. *Plant Physiol.*, **50:**208–213.

Cline, G. R., P. E. Powell, P. J. Szaniszlo, and C. P. Reid (1983). Comparison of the abilities of hydroxamic and other natural organic acids to chelate iron and other ions in soil. *Soil. Sci.*, **136:**145–157.

Cline, G. R., C. P. P. Reid, P. E. Powell, and P. J. Szaniszlo (1984). Effects of a hydroxamatic siderophore on iron adsorption by sunflower and sorghum. *Plant Physiol.*, **76:**36.

Dazzo, F. B. and A. E. Gardiol (1984). Host specificity in *Rhizobium*-legum interactions. In *Genes Involved in Microbe-Plant Interactions* (D. P. S. Verma and T. H. Hohn, eds.). Springer-Verlag, New York, pp. 3–31.

Dazzo, F. B., J. W. Kijne, K. Haahtela, and T. Korhonen (1985). Fimbriae, lectins and agglutinins of nitrogen-fixing bacteria. In *Microbial Lectins and Agglutinins* (D. Mirelman, ed.). Wiley, New York (in press).

Eliott, R. P. (1958). Some properties of pyoverdine, the water-soluble pigment of the pseudomonads. *Appl. Microbiol.*, **6:**241–246.

Fredericson, J. K. and L. F. Elliott (1985). Effects on winter wheat seedling growth by toxin-producing rhizobacteria. *Plant Soil*, **83:**399–409.

Fushiya, S., Y. Sato, S. Nozoe, K. Momoto, T. Takemoto, and S. Takagi (1980). Avenic Acid A, a new amino-acid possessing an iron chelating activity. *Tetrahedron Lett.*, **21:**3071–3072.

Gaastra, W. and F. K. de Graaf (1982). Host-specific fimbrial adhesions of noninvasive enterotoxigenic *Escherichia coli* strains. *Microbiol. Rev.*, **46:**129–161.

Geels, F. P. and B. Schippers (1983a). Selection of antagonistic fluorescent *Pseudomonas* spp. and their root colonization and persistence following treatment of seed potatoes. *Phytopathol. Z.*, **108:**193–214.

Geels, F. P. and B. Schippers (1983b). Reduction of yield depressions in high frequency potato cropping soil and seed tuber treatments with antagonistic fluorescent *Pseudomonas* spp., *Phytopathol. Z.*, **108:**207–214.

Geels, F. P. and B. Schippers (1983c). Reduction of yield depressions in high frequency potato cropping by fluorescent *Pseudomonas* spp., *Les Collogues de l'INRA*, **18:**231–238.

Geels, F. P., J. L. Lamers, O. Hoekstra, and B. Schippers (1985). Potato plant response to seed tuber bacterization in the field in various rotations. *Neth. J. Pl. Pathol.*, (submitted for publication).

Geels, F. P., E. D. L. Schmidt, and B. Schippers (1985). The use of 8-hydroxy quinoline for the isolation and pre-qualification of plant growth-stimulating rhizosphere pseudomonads. *Biol. Fert. Soils*, **1:**167–173.

Haas, D. (1983). Genetic aspects of biodegradation by pseudomonads. *Experientia*, **39:**1199–1213.

Hoestra, H. (1979). Self-induced disease. In *Plant Disease, An Advanced Treatise*, Vol. IV (J. G. Horsfall and E. B. Cowling, eds.). Academic, New York, pp. 331–342.

Hornby, D. (1979). Take-all decline: a theorist's paradise. In *Soil-Borne Plant Pathogens* (B. Schippers and W. Gams, eds.). Academic, New York, pp. 133–156.

Howell, C. R. and R. D. Stipanovic (1979). Control of *Rhizoctonia solani* on cotton seedlings with *Pseudomonas fluorescens* and with an antibiotic produced by the bacterium. *Phytopathology*, **69:**480–482.

Howell, C. R. and R. D. Stipanovic (1980). Suppression of *Pythium ultimum* induced damping-off of cotton seedlings by *Pseudomonas fluorescens* and its antibiotic, pyoluteorin. *Phytopathology*, **70:**712–715.

Kloepper, J. W. (1983). Effect of seed piece inoculation with PGPR on populations of *Erwinia carotovora* on potato roots and in daughter tubers. *Phytopathology*, **73:**217–219.

Kloepper, J. W. and M. N. Schroth (1978). Plant growth promoting rhizobacteria on radishes. In *Proceedings of the IV International Conference on Plant Pathogenic Bacteria*, Vol. 2. Angers, France, pp. 879–882.

Kloepper, J. W. and M. N. Schroth (1981). Relationship of *in vitro* antibiosis of PGPR to plant growth and the displacement of root microflora. *Phytopathology*, **71:**1020–1024.

Kloepper, J. W., J. Leong, M. Teintze, and M. N. Schroth (1980a). Enhanced plant growth by siderophores produced by plant growth promoting rhizobacteria. *Nature*, **286:**885–886.

Kloepper, J. W., J. Leong, M. Teintze, and M. N. Schroth (1980b). *Pseudomonas* siderophores: a mechanism explaining disease-suppressive soils. *Curr. Microbiol.*, **4:**317–320.

Kloepper, J. W., M. N. Schroth, and T. D. Miller (1980c). Effects of rhizosphere colonization by plant growth-promoting rhizobacteria on potato development and yield. *Phytopathology*, **70:**1078–1082.

Leisinger, T. and R. Margraff (1966). Secondary metabolites of the fluorescent pseudomonads. *J. Appl. Bact.*, **29:**470–489.

Lindsay, W. L. (1979). *Chemical Equilibria in Soils*. Wiley, New York.

Lockwood, J. L. and B. Schippers (1984). Evaluation of siderophores as a factor in soil mycostasis. *Trans. Br. Mycol. Soc.*, **82:**589–594.

Loper, J. E., T. V. Suslow, and M. N. Schroth (1984). Lognormal distribution of bacterial populations in the rhizosphere. *Phytopathology*, **74:**1454–1460.

Marugg, J. D., M. van Spanje, W. P. M. Hoekstra, B. Schippers, and P. J. Weisbeek (1985). Isolation and analysis of genes involved in siderophore-biosynthesis in the plant growth-stimulating *Pseudomonas putida* strain WCS 358. *J. Bacteriol*, **164:**563–570.

Matthijse, A. G. (1984). Interaction of *Agrobacterium tumefaciens* with the plant cell surface. In *Genes Involved in Microbe-Plant Interactions* (D. P. S. Verma and Th. Hohn, eds.), Springer-Verlag, New York, pp. 33–54.

Maurer, B., A. Müller, W. Keller-Schierlein, and H. Zähner (1968). Ferribactin, a siderochrome from *Pseudomonas fluorescens* migula, *Arch. Mikrob.*, **60:**326–339.

Meyer, J. M. and M. A. Abdallah (1978). The fluorescent pigment of *Pseudomonas fluorescens*: biosynthesis, purification and physicochemical Properties. *J. Gen. Microbiol.*, **107**:319–328.

Mino, Y., T. Ishida, N. Ota, M. Inone, K. Nomoto, T. Takemoto, H. Tanaka, and Y. Sugiura (1983). Mugineic acid-iron (III) complex and its structurally analogous cobalt (III) complex: characterization and implication for absorption and transport of iron in gramineus plants. *J. Am. Chem. Soc.*, **105**:4671–4676.

Monzyk, B. and A. L. Crumbliss (1983). Factors that influence siderophore-mediated iron bioavailability: catalysis of interligand Fe transfer from ferrioxamine B to EDTA by hydroxamic acids. *J. Inorg. Biochem.*, **19**:19–39.

Moores, J. C., M. Magazin, G. S. Ditta, and J. Leong (1984). Cloning of genes involved in the biosynthesis of pseudobactin, a high-affinity iron transport agent of a plant growth-promoting *Pseudomonas* strain. *J. Bacteriol.*, **157**:53–58.

Neilands, J. B. (1981a). Microbial iron compounds. *Annu. Rev. Biochem.*, **50**:715–731.

Neilands, J. B. (1981b). Iron absorption and transport in microorganisms, *Annu. Rev. Nutr.*, **1**:27–46.

Neilands, J. B. (1982). Microbial envelope proteins related to iron. *Annu. Rev. Microbiol.*, **36**:285–309.

Neilands, J. B. (1984). Siderophores of bacteria and fungi. *Microbiol. Sci.*, **1**:9–14.

O'Connor, G. A., W. L. Lindsay, and S. R. Olsen (1971). Diffusion of iron and iron chelates in soil. *Soil Sci. Soc. Am. Proc.*, **35**:407–410.

Okker, R. J. H., H. Spaink, J. Hille, T. A. N. van Brussel, B. Lugtenberg, and R. A. Schilperoort (1984). Plant-inducible virulence promotor of the *Agrobacterium tumefaciens* tiplasmid. *Nature*, **312**:564–566.

Olsen, R. A., J. H. Bennett, D. Blume, and J. C. Brown (1981). Chemical aspects of the Fe stress response mechanism in tomatoes. *J. Plant. Nutr.*, **3**:905–921.

Orlando, J. A. and J. B. Neilands (1982). Ferrichrome compounds as a source of iron for higher plants. In *Chemistry and Biology of Hydroxamic Acids* (H. Kehl, ed.). S. Karger, Basel, pp. 123–129.

Philson, S. B. and M. Llinas (1982a). Siderochromes from *Pseudomonas fluorescens*. I. Isolation and characterization. *J. Biol. Chem.*, **257**:8081–8090.

Philson, S. B. and M. Llinas (1982b). Siderochromes from *Pseudomonas fluorescens*. II. Structural homology as revealed by NMR spectroscopy. *J. Biol. Chem.*, **257**:8086–8090.

Powell, P. E. and P. J. Szaniszlo (1982). Hydroxamate siderophores in the iron nutrition of plants. *J. Plant Nutr.*, **5**:653–673.

Powell, P. E., G. R. Cline, C. P. P. Reid, and P. J. Szaniszlo (1980). Occurrence of hydroxamate siderophore iron chelators in soils. *Nature*, **287**:833–834.

Powell, P. E., P. J. Szaniszlo, and C. P. P. Reid (1983). Confirmation of occurrence of hydroxamate siderophores in soil by a novel *Escherichia coli* bioassay. *Appl. Environ. Microbiol.*, **46**:1080–1083.

Ramphal, R. and G. B. Pier (1985). Role of *Pseudomonas aeruginosa* mucoid exopolysaccharide in adherence to tracheal cells. *Infect. Immun.*, **47**:1–4.

Reid, C. P. P., D. E. Crowley, H. J. Kim, P. E. Powell, and P. J. Szaniszlo (1984). Utilization of iron by oat when applied as ferrated synthetic chelate or as ferrated hydroxamate siderophore. *J. Plant Nutr.*, **7**:437–447.

Römheld, V. and D. Kramer (1983). Relationship between proton efflue and rhizodermal transfer cells induced by iron deficiency. *Z. Pflanzen-physiol.*, **113**:73–83.

Römheld, V. and H. Marschner (1983). Mechanism of iron uptake by peanut plants. *Plant Physiol.*, **71**:949–954.

Römheld, V., C. Müller, and H. Marschner (1984). Localization and capacity of proton pumps in roots of intact sunflower plants. *Plant Physiol.*, **76**:603–606.

Rovira, A. D. and C. B. Davey (1974). Biology of the rhizosphere. In *The Plant Root and its Environment* (E. W. Carson, ed.). University Press Virginia, Charlottesville, pp. 153–204.

Rovira, A. D., R. C. Foster, and J. K. Martin (1979). Note on terminology: origin, nature and nomenclature of the organic material in the rhizosphere. In *The Soil–Root Interface* (J. L. Harley, ed.). Academic, New York, pp. 1–4.

Salt, G. E. (1979). The increasing interest in "minor pathogens." In *Soil-Borne Plant Pathogens* (B. Schippers and W. Gams, eds.). Academic, New York, pp. 289–312.

Scher, F. M. and R. Baker (1980). Mechanisms of biological control in a *Fusarium*-suppressive soil. *Phytopathology*, **70:**412–417.

Scher, F. M. and R. Baker (1982). Effect of *Pseudomonas putida* and a synthetic iron chelator on induction of soil suppressiveness to *Fusarium* wilt pathogens. *Phytopathology*, **72:**1567–1573.

Schippers, B. (1983). Prospects of biological control of plant pathogens with fluorescent *Pseudomonas* spp. In *Proceedings of the 10th International Congress on Plant Protection, Brighton*, The Lavenham Press Ltd., Lavenham, Suffolk, pp. 767–771.

Schippers, B., F. P. Geels, O. Hoekstra, J. G. Lamers, C. A. A. A. Maenhout, and K. Scholte (1985). Yield depressions in narrow rotations caused by unknown microbial factors and their suppression by selected pseudomonads. In *Ecology and Management of Soil-Borne Plant Pathogens* (C. A. Parker, K. J. Moore, P. T. W. Wong, A. E. Rovira, and J. F. Kolmorgen, eds.). The American Phytological Society, St Paul, MN, pp. 127–130.

Scholte, K., J. W. Veenbaans-Rijks, and R. E. Labruyère (1985). Potato growing in short rotations and the effect of *Streptomyces* spp., *Colletotrichum coccoides*, *Fusarium tabacinum* and *Verticillium dahliae* on plant growth and tuber yield. *Potato Res.*, **28:**331–348.

Schroth, M. N. and J. G. Hancock (1982). Disease-suppressive soil and root-colonizing bacteria. *Science*, **216:**1376–1381.

Stutz, E. (1964). Aufnahme von ferrioxamin B durch tomatenpflanzen, *Experimentia*, **20:**430–431.

Suslow, T. V. (1982). Role of root-colonizing bacteria in plant growth. In *Phytopathogenic Prokaryotes*, Vol. I (M. S. Mount and G. H. Lacy, eds.). Academic, New York, pp. 187–223.

Suslow, T. V. and M. N. Schroth (1982). Role of deleterious rhizobacteria as minor pathogens in reducing crop growth. *Phytopathology*, **72:**111–115.

Teintze, M. and J. Leong (1981). Structure of Pseudobactin A, a second siderophore from plant growth promoting *Pseudomonas* B10. *Biochemistry*, **20:**6457–6462.

Teintze, M., M. B. Hossain, C. L. Barnes, J. Leong, and D. van der Helm (1981). Structure of ferric pseudobactin, a siderophore from a plant growth promoting *Pseudomonas. Biochemistry*, **20:**6446–6457.

Udey, L. R. and J. L. Fryer (1978). Immunization of fish with bacterins of *Aeromonas salmonicida. Mar. Fish. Rev.*, **40:**12–17.

van Brussel, A. A. N., S. Zaat, H. Canter-Cremers, C. Wijffelman, E. Pees, T. Tak, and B. Lugtenberg (1985). Role of plant root exudate and symplasmid-localized nodulation genes in the synthesis of thick and short roots of common vetch. J. Bacteriol. **165:**517–522.

von Braun, V. (1985). The unusual features of the iron transport systems of *Escherichia coli. Trends Biochem. Sci.*, **10:**75–78.

Weger de, L. A., R. van Boxtel, B. van der Burg, R. Gruters, F. P. Geels, B. Schippers, and B. Lugtenberg (1985). Outer membrane proteins of plant growth-stimulating antagonistic root-colonizing *Pseudomonas* spp. J. Bacteriol. **165:**585–594.

Weller, D. M. and R. J. Cook (1983). Suppression of take-all of wheat by seed treatments with fluorescent pseudomonads. *Phytopathology*, **73:**463–469.

Wong, P. T. W. and R. Baker (1984). Suppression of wheat take-all and *Ophiobolus* patch by fluorescent pseudomonas from a *Fusarium*-suppressive soil. *Soil Biol. Biochem.*, **16:**397–403.

3

MANAGEMENT OF TAKE-ALL IN CONSECUTIVE CROPS OF WHEAT OR BARLEY

R. JAMES COOK
DAVID M. WELLER
USDA–ARS, Root Disease and Biological Control Research Unit, Pullman, Washington

Take-all caused by *Gaeumannomyces graminis* var. *tritici* is probably the most important root disease of wheat worldwide (Asher and Shipton, 1981). Barley is affected less than wheat, but wheat after barley can be as severely diseased as wheat after wheat. Infections cause the roots to become black and result in nutrient-deficiency symptoms in the leaves, owing to the inability of the roots to provide nutrients for the tops. Infections of the tiller bases result in premature blighting of the plants during or shortly after heading, due to the restriction of water movement to the tops. The pathogen exists between host crops as a saprophyte in the fragments of roots and stems of previous host plants, but it exhausts these food reserves relatively quickly or is displaced in the crop residue by other microorganisms. One or two years away from wheat, barley, or other grass hosts is usually sufficient to control take-all, but the disease continues to be important because of the need in many cereal-growing areas of the world to grow two, three, or even several consecutive crops of wheat or barley before introducing a break crop or fallow. Unfortunately, adequate resistance to take-all does not exist in *Triticum* and *Hordeum* species (Scott, 1981). Thus it has become necessary to find methods of take-all control that will permit the continuous cropping of susceptible cereals.

We use soil fumigation (methyl bromide, chloropicrin, or 1,3-dichloropropene plus chloropicrin) routinely in our field experiments to produce a take-all "free" check, but the cost of fumigation is prohibitive for commercial wheat and barley production. Triadimenol applied to seeds (Bockus, 1983) and benomyl applied in-furrow (Bateman, 1980, 1981) will limit take-all, but triadimenol can be phytotoxic to wheat seedlings and the resultant yield responses, if any, are sometimes only a fraction of the potential yield if take-all was controlled completely. For example, in a trial at Mount Vernon, Washington, in 1984, winter wheat, sown in a field cropped to winter wheat the previous year and with take-all as the only significant root disease, produced 3.6 mt/ha without seed treatment and 5.0 mt/ha with a triadimenol seed treatment. However, wheat sown at the same site with the same amount of rainfall and fertilizer produced about 8.0 mt/ha following a 1-yr break with a nonhost crop. A 40% yield increase with triadimenol is impressive, but it still represents only two-thirds of the healthy crop.

In England, cross protection by *Phialophora graminicola* against *G. graminis* var. *tritici* occurs naturally in the initial wheat crops after a grass ley (Scott, 1970; Deacon, 1976; Speakman and Lewis, 1978). Without such protection, wheat can be damaged severely by take-all when sown immediately after grasses that serve as hosts of this pathogen (Sprague, 1950; Yarham, 1981). Attempts have been made in Australia to introduce *Phialophora* spp. or similar avirulent fungi related to *G. graminis* var. *tritici* to control take-all (Wong and Southwell, 1980). This approach has potential but is risky with *P. graminicola* in view of recent evidence (R. W. Smiley, personal communication) that it is pathogenic to wheat as temperatures approach 29–30°C. Other groups, including our own, have attempted to use the biological control responsible for take-all decline (Cook and Rovira, 1976; Hornby, 1979; Shipton, 1975, 1977). In an experimental plot of irrigated wheat at Lind, Washington (soil pH of about 7.0), take-all was severe in the seventh–tenth consecutive wheat crops and yields averaged only 40–50% of those in fumigated subplots, but it declined by the fifteenth year of continuous wheat (soil pH had also declined by about 1 unit) and yields averaged 8.0 and 7.4 mt/ha, respectively, with and without soil fumigation. Commercial fields in central Washington state, with soil pH values of 5.8–6.5, yield between 90 and 95% of the fumigated check in the fourth–sixth years of consecutive irrigated wheat because of take-all decline.

Maximizing or accelerating the benefits of take-all decline require the use of tillage and depend on soil pH, the choice of planting date, whether spring wheat or winter wheat is grown, and the method of fertilization and irrigation. This chapter summarizes the approaches to controlling take-all in consecutive cereals and the scientific explanation or theory for each, with emphasis on our research and experiences in the Pacific Northwest, USA. The approaches are broadly grouped under management of the inoculum potential of the pathogen in soil; management of take-all decline; biological control with fluorescent pseudomonads; and management with soil fertility and irrigation.

1. MANAGEMENT OF INOCULUM POTENTIAL IN SOIL

1.1. The Concept of Inoculum Potential Applied to *Gaeumannomyces graminis* var. *tritici*

Inoculum potential is used in this chapter sensu Garrett (1970) as an estimate of the energy available for infection at the surface of the host. Like any other concept of potential energy, the values are relative and a standard is selected arbitrarily for reference purposes. Moreover, the emphasis is on the number and condition of hyphal tips of the pathogen poised in the infection court (Garrett, 1970) and not on total biomass of the pathogen in the soil, although presumably these two parameters were related.

Ideally, a measure of the inoculum potential should include a count of the number of hyphal tips that reach the root surface and the value then adjusted upward or downward by a factor that reflects the physiological state of the tips. However, two less direct methods provide values that may be almost as accurate. Wildermuth and Rovira (1977) estimated the total length of pigmented hyphae of *G. graminis* var. *tritici* on the surface of the root. Assuming that the number of hyphal tips is proportional to total length of hyphae, the values for total length of hyphae are as useful as those for total number of tips since all values are relative. The only factor missing is the adjustment value for physiological state of the hyphae. Using another method, Wilkinson et al. (1985a) infested soil with a known mass and particle size of inoculum and counted the number of lesions on roots of seedlings 21 days after planting in the soil. They assumed that the only important hyphal tips are those that cause lesions, and that energetic hyphae cause more lesions than the same number of weak hyphae. This method is most accurate with particle sizes and concentrations that produce discreet lesions.

Brown and Hornby (1971) observed that fragments of infested tiller bases were better than infested roots as a source of inoculum of *G. graminis* var. *tritici*. Wilkinson et al. (1985a) confirmed this by comparing the number of infections produced per unit mass of infested particles. Tiller bases naturally infected in the field, roots infected in the greenhouse, and oat grains colonized axenially in the laboratory were each fragmented in a Waring blender and then sieved to provide particles 0.09 mm and smaller, 0.09–0.15, 0.15–0.25, 0.25–0.5, 0.5–1.0, and 1.0–2.0 mm. The different particle sizes were blended into each of two fumigated soils (Shano and Ritzville silt loams) at 10 concentrations between 0.01 and 10 mg/g soil. The minimum particle size of tiller bases and oat grains required for infection was 0.15–0.25 mm in the Shano silt loam and 0.25–0.5 mm in the Ritzville silt loam. The threshold size was the same for tiller bases and oat grain particles, but the oat grains produced a greater number of infections per unit mass of material than did infested tiller bases. Infested root particles were essentially noninfectious in the Shano silt loam at sizes smaller than 0.5–1.0 mm and even the largest particles tested (1–2 mm) were noninfectious in the Ritzville silt loam. Any cultural practice that can limit *G. graminis* var. *tritici* to root infection and prevent the pathogen from establishing in the tillers should lower the inoculum potential of the pathogen and lead to less take-all the following year.

Of the infested crop residue sieved from soil from Washington wheat fields, only inoculum particles larger than 0.5–1.0 mm produced lesions; this material was mainly infested remnants of tiller bases (Wilkinson et al., 1985a). MacNish (1973) reported for soils in South Australia that sieving the soil to 0.5 mm reduced the incidence of take-all on seedlings then grown in the soil, but did not eliminate all infectious material from the soil. These findings confirm reports of Hornby (1969, 1975) that the infectious particles recovered from wheat field soils in England were 0.42 mm and larger. Significantly, at all particle sizes above the threshold size, the number of infections is proportional to total mass of infested material (Wilkinson et al., 1985a).

The energy available to *G. graminis* var. *tritici* in the infested fragments of crop residue affects the distance the pathogen can grow through soil to reach and infect the root. Wilkinson et al. (1985b) developed a statistical model to estimate this distance. As might be expected, the smaller the particle the closer to the root it must be to cause an infection. The pathogen was estimated to grow about 10 mm through soil from particles 1–2 mm in diameter. Gilligan (1980) similarly estimated that *G. graminis* var. *tritici* established axenically in millet seeds (about 2 mm in diameter) can grow 10–12 mm through soil to reach and infect a root. Particles smaller than the threshold or minimum size probably have insufficient energy to support infection at any distance away from the root in natural soil. That the number of infections is proportional to total mass of infested particles above the threshold size can be explained on the basis that, with increasing particle size, fewer particles make up a given mass, but each particle supports more hyphae and the hyphae can grow further through soil to infect roots. Smaller particles must be closer to the root to infect, but there are more of them per unit mass of infested material. With time, and depending on the soil and cultural practices, the infested debris will become smaller and the inoculum potential of the pathogen will decrease. This is important since no combination of cultural practices aimed at take-all control in consecutive crops of wheat is likely to succeed if the inoculum potential is still high when the next crop is sown.

1.2. Suppression of *G. graminis* var. *tritici* by Competitors in the Food Base

Like other cereal root- and foot-rot fungi that survive in the residue of tiller bases colonized through parasitism, *G. graminis* var. *tritici* is dependent for survival on its ability to digest the cellulose in the residue. Garrett (1966) compared *G. graminis* var. *tritici* with other cereal foot-rot fungi for ability to break down cellulose relative to the need for sugars for respiration and growth as estimated by linear growth rate. The relationship between these two parameters was defined as the "cellulose adequacy index." A value less than 1 indicates that the need for sugars is greater than the supply produced by enzymatic breakdown of cellulose by hyphae growing in the substrate. Of five fungi compared, *G. graminis* var. *tritici* had the lowest cellulose adequacy index at 0.59 and *Cochliobolus sativus* had the highest at 2.1–3.5 (Garrett, 1966, 1976). Any limitation on the rate of metabolism of a fungus in a substratum such as straw in soil will weaken its ability to maintain active possession of the substratum (Bruehl, 1975) and thereby increase its vulnerability to displacement by other microorganisms. In addition, commensal bacteria contiguous to the hyphae of *G. graminis* var. *tritici* may use a portion of the sugars released by enzymes from the hyphae in the residue (Garrett, 1983); the concentration of sugars may then not reach the level necessary for catabolite repression so that cellulolysis continues unrestricted and the foodbase is exhausted even more rapidly (Garrett, 1985). The net result of competition from saprophytic fungi and bacteria in the straw is a more rapid reduction in inoculum potential, an acceleration in the successions of microorganisms responsible for decomposition of the residue, and a shortening of the survival period of *G. graminis* var. *tritici*. On the other hand, any soil condition unfavorable to microbial decomposition of the infested residue during saprophytic existence of the pathogen, such as dry soil conditions (MacNish and Dodman, 1973), could effectively preserve the inoculum and require longer rotations to control this disease. In the Columbia Basin of Washington, where essentially no precipitation occurs during the summer and early fall, it is essential for take-all control in consecutive crops of wheat to irrigate two or three times during the period of soil preparation between the harvest of one crop (July) and sowing of the next (August) to facilitate the microbial decomposition of infested crop residue.

Nitrogen that diffuses into the residue from the soil solution greatly enhances the extension of hyphae and formation of new hyphal branches of *G. graminis* var. *tritici* (Garrett, 1940, 1976). Cellulolysis is then also enhanced since cellulases are produced most abundantly from young hyphae. Thus it is important in management of the inoculum potential of *G. graminis* var. *tritici*, especially during the period between harvest and sowing, to keep the supply of available nitrogen low or almost nil in the top 15–20 cm of soil where most inoculum of this pathogen resides, and to fertilize each crop at sowing by applying the nitrogen below the seed.

Most of the important competitors of *G. graminis* var. *tritici* probably become established in the substrate with this fungus as secondary colonists of the infected roots and tiller bases while the plant is still alive. Those that wait to colonize the residue after the plant is dead must overcome the advantages of *G. graminis* var. *tritici* as well as the secondary colonists already in the residue. Garrett (1985) points out the significance of these secondary colonists of infected tissues by reinterpreting some of his own earlier data (Garrett, 1938). *Gaeumannomyces graminis* var. *tritici* survived in 50% of axenically colonized straws for about 9.9 and 7 months, respectively, in field plots in a sandy soil of pH 7.2 near Bridgham in Norfolk, England and in a clay loam of pH 6.8 on the Rothamsted Experimental Station. However, the longevity of the fungus in naturally infested straws buried in the two soils was about the same (7.5 and 7 months, respectively). Straws anexically colonized by the pathogen apparently were subject to intense competition from soil microorganisms, with the competition being greater in the loam than in the sandy soil. However, in naturally infested tiller bases longevity of the pathogen was determined by the secondary colonists established in the residue before burial and not by competition from microorganisms in soils at the respective sites; thus longevity in naturally infested tiller bases was about the same in the two soils. Cook et al. (1986) observed significantly more lesions per dose of inoculum in pasteurized soil (60°C/30 min) than in natural soil, if the inoculum was axenically colonized oat grains but not if it was naturally infested tiller bases. Pasteurization of the soil eliminated potential colonists of the residue (fast-growing fungi such as *Trichoderma* and *Fusarium* spp.), but probably the important competitors were already in the particles of naturally infested tiller bases. Biological control of *G. graminis* var. *tritici* aimed at reducing its inoculum potential must begin, therefore, with the use of practices that favor the most antagonistic secondary colonists beginning prior harvest.

1.3. A Model to Explain General Suppression sensu Gerlagh

Gerlagh (1968) proposed that in addition to the *specific suppression* responsible for take-all decline (see the following section), there is a background *general suppression* of soils that limits the amount of disease produced by *G. graminis* var. *tritici*. General suppression is thought to be microbial in origin but not related to wheat monoculture.

Alabouvette and Cook (unpublished data) obtained evidence for the generally suppressive nature of natural soils by showing that two to three times more lesions developed with a given dose of axenically colonized oat-grain inoculum in some soils than in other soils. As found by Cook et al. (1986), the differences among soils could not be explained by whether or not the soil came from fields where take-all had declined. They tested the hypothesis that the differences in infection efficiency were the result of differences in competition for root exudates in the rhizosphere during prepenetration growth of the pathogen. However, neither the total microbial biomass of the soil nor responsiveness of the biomass to added glucose as estimated, respectively, by the methods of Jenkinson (1966) and Anderson and Domsch (1973) related to the degree of suppression of infection efficiency. However, the values for infection efficiency were greater for all soils pasteurized with 60°C moist heat for 30 min before adding the inoculum and planting wheat compared to the respective untreated soils. They concluded that the general suppressiveness of all soils to *G. graminis* var. *tritici* did not result from competition for nutrients in the soil solution by the total microbial biomass active during prepenetration growth of the pathogen.

Tang and Cook (unpublished) subsequently demonstrated that the important competition occurs in the foodbase. A benomyl-resistant strain of the pathogen in axenically colonized oat-grain particles impregnated with benomyl produced significantly more lesions per dose of inoculum than the same strain in the same size particles not protected by benomyl. The increase in infection efficiency obtained with benomyl approached that obtained with pasteurization of the soil.

General suppression of take-all may, therefore, be the result of competition from the microorganisms for the foodbase (Cook, 1985), with some soils supporting more intense competition than others, depending on conditions for growth of the competitors in the residue. When an axenically colonized substrate is used, suppression results from saprophytic colonization of the residue after burial in soil, but where the residue is naturally infected material, it may result primarily from microorganisms that enter the residue during the parasitic phase of the pathogen and then carry over with the pathogen in the residue. Take-all is almost totally suppressed in essentially all natural soils at soil temperatures above 16–20°C (Henry, 1932), which can also be explained on the basis of a greater competition in the foodbase in warmer soils. Any practice or soil condition that helps intensify competition with *G. graminis* var. *tritici* in its foodbase enhances the biological control of this fungus.

1.4. Tillage

Scott (1969) demonstrated that the number of white heads caused by take-all is less in plots rotovated immediately after harvest than if delayed until near the time of sowing. In the Pacific Northwest, take-all is significantly less severe where the soil is tilled several times prior to sowing than if it is left undisturbed and the crop is direct drilled (Moore and Cook, 1984), even in fields that have undergone take-all decline. For example, in our plot of irrigated wheat at Lind, where the yield of the fifteenth crop was about 95% of that in subplots of fumigated soil, take-all was devastating in the direct-drilled strips but was still suppressed in tilled strips at the site in the eighteenth crop (Fig. 1). Take-all is also more severe in Australia on direct-drilled wheat compared to wheat sown in a conventionally prepared seedbed (Rovira and Venn, 1985; Neate, 1984). Ferraz (1979) noted from results in pot trials in England that take-all was more severe with direct drilling than simulated plowing, especially when planting was delayed somewhat. In contrast, Slope (1966) and Brooks and Dawson (1968) observed in field trials in England that the disease is greater when wheat is sown following some tillage than if direct-drilled, and Yarham and Hirst (1975) noted no apparent affect of tillage on take-all. The reasons for these differences are not understood (Yarham, 1981), but should be investigated for new clues to control this disease.

At least two mechanisms of biological control are thought to be favored by tillage and the resultant fragmentation of infested crop residue. Tillage results in a temporary flush of microbial activity, because of improved soil aeration and/or exposure of new surfaces on the crop residue to colonization by soil microorganisms. Each burst of fungal and bacterial growth is thought to enhance or accelerate displacement of *G. graminis* var. *tritici* in the residue by saprophytic competitors, thereby lowering the inoculum potential and increasing the minimum threshold particle size required for infection (Scott, 1969; Hornby, 1975; Ferraz, 1979; Moore and Cook, 1984). In addition, mixing infested residue into the soil forces the pathogen to begin its parasitic phase by infecting roots, where it is then subject to antagonism by the root-colonizing microorganisms associated

Fig. 1. Winter wheat growing on a site in the 18th consecutive crop, with alternate strips of tilled (each side) and not tilled (center). The foreground was fumigated with methyl bromide. The lack of vigorous wheat in the no-till strips is due mainly to severe take-all caused by *Gaeumannomyces graminis* var. *tritici*.

with take-all decline (see the section following). If infection by the fungus is confined exclusively to roots, it will also then be limited to this tissue during its saprophytic existence in soil; as pointed out above, infested roots are less effective than infested tiller bases as sources of inoculum for infection of the next crop. There is no more ideal situation for *G. graminis* var. *tritici* than to exist undisturbed in large unfractured crowns (tiller bases joined together) of wheat in the top 2–3 cm of soil where it can then grow directly into the tiller bases of the next wheat crop when direct drilled (Ferraz, 1979; MacNish, 1973; Moore and Cook, 1984). We have no method for take-all control other than soil fumigation that can be used for continuous direct-drilled wheat. If rotation is not used, some form of tillage is essential.

1.5. Delayed Sowing

Observations in Ireland (Cunningham, 1966a) and England (Scott, 1969; Yarham, 1981) make abundantly clear that early tillage followed by late sowing can greatly limit or at least retard the onset of take-all. The longer the interval between tillage and sowing in consecutive cereal crops, the less the amount of take-all. In the Pacific Northwest, wheat sown in early October shows significantly more take-all than wheat sown in late October (Taylor et al., 1983). The additional 2–3 weeks allows saprophytes to establish in the residue, and the roots of late-seeded wheat are exposed to infection for less time than wheat seeded 2–3 weeks earlier (Taylor et al., 1983).

Surprisingly, if spring wheat is sown the following March or April, take-all is commonly severe under Pacific Northwest conditions, even in fields where take-all severity on winter wheat has declined. This is an anomaly if a long exposure of infested residue to soil saprophytes and a short exposure of the crop to the pathogen are important to take-all management. Spring wheat is planted when the soil is nearly frozen (1–5°C) and then warms gradually, whereas winter wheat is planted while soils are relatively warm (10–15°C) and then cool gradually. We think the pathogen somehow escapes antagonists less able to grow at temperatures below about 5°C and prevalent during the early stages of spring wheat growth. Bockus (1983) has shown that early infections have the greatest effect on yield and certainly for spring wheat the infections occur early and greatly limit yield.

Management of the inoculum potential by tillage and delayed seeding are important, but are insufficient by themselves to control take-all. Methods to protect plants from surviving inoculum are also needed.

2. MANAGEMENT OF TAKE-ALL DECLINE AND EVIDENCE FOR AN INVOLVEMENT OF PSEUDOMONADS

Take-all decline (TAD), a spontaneous diminution in the severity of take-all and concomitant increase in yield with monoculture of wheat (Shipton, 1975), was first noted more than 50 years ago in England by Glynne (1935) and in the United States by Fellows and Ficke (1934). It was first demonstrated experimentally in pots by Gerlagh (1968), and has been documented in field studies by many workers around the world (Shipton, 1972; Lester and Shipton, 1967; Pope and Jackson, 1973; Shipton et al., 1973; Gerlagh, 1968; Vojinović, 1972a and b, 1973; Lemaire and Coppenet, 1968; Zogg and Fossati, 1966). In Shangtung Province, People's Republic of China, peasant farmers refer to take-all as "Quan shi bing ban jia," which means that it moves its house to another field after one or two severe outbreaks in a field (Tang, personal communication).

Take-all decline results from suppression of the pathogen in the soil, referred to by Gerlagh (1968) as *specific suppression*. This distinction by Gerlagh (1968) between specific and general suppression provided a basis for experimental studies aimed at distinguishing between suppressiveness responsible for take-all decline and the suppression that occurs in all nonsterile soils, especially at temperatures above 16–20°C. Specific suppression (1) develops with wheat monoculture and following one or more outbreaks of take-all; (2) may develop with the addition of mycelium of *G. graminis* var. *tritici* to soil; (3) is eliminated by treating the soil with moist heat (60°C for 30 min), methyl bromide, or chloropicrin; and (4) is transferable. Cook et al. (1986) found no evidence that the antagonists in soil following take-all decline had any greater effect than those from known conducive soils on either the minimum (threshold) size of axenically colonized oat-grain particles required for infection or the infection efficiency of the particles in soils. There was also no evidence that the antagonists responsible for specific suppression affect the distance the fungus can grow to infect roots (Wilkinson et al., 1985b). Instead, the suppressive organisms are thought to operate primarily on the rhizoplane and in the root (Cook and Rovira, 1976). Specific suppression is most effective at soil temperatures between about 5°C and 16–20°C (Cook and Rovira, 1976). Cook and Baker (1983) proposed that the suppressive microorganisms responsible for take-all decline effectively diminish or eliminate a significant portion of the temperature niche open to this pathogen. The severe take-all on spring wheat in the Pacific Northwest is thought to result from an escape by the pathogen at soil temperatures below about 5°C from the antagonists responsible for take-all decline.

The mechanisms proposed to explain take-all decline can be grouped into two general categories (Hornby, 1979; Rovira and Wildermuth, 1981): the result of microbiological changes in the soil or rhizosphere suppressive to the pathogen (Gerlagh, 1968; Vojinović, 1970, 1972a and b, 1973; Shipton, 1975; Zogg and Jäggi, 1974; Cook and Rovira, 1976; Smiley, 1978a and b, 1979; Wildermuth and Rovira, 1977; Rovira and Wildermuth, 1981; Wildermuth et al., 1979); and the result of changes within the pathogen itself and responsible for loss of virulence (Lapierre et al., 1970; Lemaire et al., 1970, 1971). *Gaeumannomyces graminis* var. *tritici* is commonly infected with dsRNA mycovirus-like agents (Buck et al., 1981) and the fungus usually loses virulence during maintenance in pure culture (Chambers, 1970; Naiki and Cook, 1983a). However, there is no evidence that mycovirus infections occur at any greater frequency within the population of *G. graminis* var. *tritici* after than before take-all decline (Buck et al., 1981), or that changes in virulence within the population of *G. graminis* var. *tritici* can account for take-all decline (Asher, 1980; Cook and Naiki, 1982). Among the antagonists, giant vampyrellid amoebae perforate pigmented hyphae of *G. graminis* var. *tritici* in suppressive soils (Homma et al., 1979; Chakraborty and Warcup, 1983), but their effect is more likely on the longevity of the pathogen during saprophytic survival than as a deterent to parasitism of roots as seems to occur in soils following take-all decline. All evidence points to a build-up of antagonistic microorganisms (bacteria) that inhibit the take-all fungus on the root (Cook and Rovira, 1976; Smiley, 1978b, 1979) or in the pieces of inocula (Wildermuth et al., 1979; Rovira and Wildermuth, 1981; Vojinović, 1973). In this section, evidence is presented for the role of fluorescent pseudomonads in the specific suppression associated with take-all decline.

2.1. Evidence for the Role of Fluorescent Pseudomonads

Cook and Rovira (1976) selected over 100 bacteria at random from roots of wheat grown in take-all suppressive (specific, i.e., from a field where take-all had declined) or conducive soils, grew them in a dilute nutrient medium in autoclaved soil, and reintroduced them as soil cultures into a conducive soil. Only eight strains provided protection equal to that obtained by adding the same quantity of a known suppressive soil into the conducive soil; all were pseudomonads, and seven were fluorescent. Smiley (1979) studied *Pseudomonas* spp. from the rhizoplane of wheat grown in a Wimmera grey clay from a field cropped to wheat since 1916 (take-all had declined) and in a Rosedale sandy clay loam from a field with a variable cropping history. In the absence of *G. graminis* var. *tritici*, the rhizoplane *Pseudomonas* population from the Wimmera grey clay was significantly higher than in the Rosedale sandy clay loam, but the populations were the same when the take-all pathogen was present. However, the proportion of the population of *Pseudomonas* spp. that was inhibitory to *G. graminis* var. *tritici in vitro* on potato dextrose agar was greater from roots grown in the Wimmera grey clay than from roots grown in the Rosedale sandy clay loam. Smiley (1979) demonstrated that severity of take-all in the Wimmera grey clay was inversely correlated with an antagonism rating for pseudomonads (decimal percentage of antagonists × zone of inhibition in millimeters).

Weller and Cook (1981) compared the fluorescent pseudomonads from roots of wheat growing in Ritzville and Puget silt loams (conducive soils from fields in which a variety of crops had been grown) to those from a Shano silt loam (suppressive soil from a field cropped to wheat for over 20 yr and in which take-all had declined). The soils were diluted 1 : 10 (w/w) with the same fumigated Ritzville silt loam (to minimize differences in physical/chemical properties between soils) and then amended with 1.0% oat-grain inoculum of *G. graminis* var. *tritici*. After two successive croppings, the wheat was protected against take-all in the suppressive but not in the conducive soils. The population of total bacteria was the same on roots from all soils but about 10 times more fluorescent pseudomonads were associated with the roots in the suppressive than in the conducive soils. Equally important, the proportion of the strains of fluorescent pseudomonads inhibitory to *G. graminis* var. *tritici in vitro* were significantly greater in the suppressive soil (Weller and Cook, 1981). In a second study (Weller, in Cook and Baker, 1983), the population of fluorescent pseudomonads on roots of wheat grown in two suppressive soils was the same as that on roots from two conducive soils, but again the number of isolates inhibitory to *G. graminis* var. *tritici in vitro* were three- to sixfold greater (significant) on roots in the suppressive compared to the conducive soils. Subsequent studies have demonstrated that a significantly greater portion of strains isolated from roots in suppressive soils controlled take-all *in vivo* as compared to isolates from conducive soils (Weller et al., 1985b) (Table 1). Kloepper et al. (1980b) demonstrated that fluorescent pseudomonads had at least a partial role in take-all decline by converting a conducive soil to a suppressive soil by adding *Pseudomonas* strain B10 or its siderophore, pseudobactin. Conversely, adding FeEDTA (exogenous iron) to a suppressive soil eliminated the suppressiveness of the soil. They concluded that the iron eliminated the ability of the pseudomonads and other microorganisms to suppress disease by shutting down siderophore production.

Many of the studies designed to determine the identity of the suppressive agent have used results from seedlings from greenhouse experiments. These experiments assume that specific suppression associated with take-all decline op-

TABLE 1. Proportion of Fluorescent Pseudomonads from Either Suppressive or Conducive Soils that Suppressed Take-All When Tested in Shano Silt Loam (Fumigated) and Puget Silt Loam Amended with Either 0.45 or 0.15% (w/w) *Gaeumannomyces graminis* var. *tritici* (Ggt) Oat Inoculum

		Effective Strains Per Source[a]		
Test Soil	*Ggt*[a] *(w/w)*	*Suppressive Soil (%)*	*Conducive Soil (%)*	*Probability*
Shano silt loam	0.45%	84a[b]	44b	0.01
(Fumigated)	0.15%	79a	72a	0.05
Puget silt loam	0.45%	34a	25a	0.05
(Nonfumigated)	0.15%	74a	50b	0.05

[a]Wheat from seeds treated with effective strains had significantly ($P = 0.05$) less disease than wheat from untreated seed. Data represent tests with 38 strains from suppressive soil and 32 from conducive soils. Ggt was added to soil as Ggt-colonized oat kernels that were pulverized and sieved; sizes 0.25–0.50 mm were used as the inoculum source.

[b]Values in the same line followed by the same letter (for the same test soil and inoculum concentration) are not significantly different using chi-square test as the indicated level.

[c]Data from Weller et al., 1985b

erates in greenhouse tests as in the field. Brown (1981) and Hornby (1979) indicated, however, that results in the greenhouse may have little bearing on the phenomenon occurring in the field. Indeed, many kinds of suppressiveness that have little or no relationship to the suppressiveness responsible for take-all decline may be produced in a greenhouse pot, which is why the pot assay was developed (Shipton et al., 1973; Cook and Rovira, 1976). Soils from fields where take-all has declined are suppressive when diluted 1:100 or 1:10 with fumigated soil, whereas soils from fields in a rotation are not. Only soils in which wheat has been grown continuously and in which take-all has declined were suppressive when tested in the pot assay. Brown (1981) found no correlation between the population of pseudomonads and the severity of disease in the field. In our work, it is the proportion of strains inhibitory to *G. graminis* var. *tritici* on roots that relates to suppressiveness and not the total numbers of pseudomonads.

No single strain of *Pseudomonas*, when introduced alone, suppresses take-all as consistently or to the same degree as occurs in soil from fields where take-all has declined. This indicates the complexity of the phenomenon and that several strains of pseudomonads and possibly other microorganisms operate collectively in specific suppression. Further, the naturally occurring suppressive agents no doubt operate throughout the life history of the fungus. On the other hand, the elimination of specific suppression by 60°C moist heat treatment for 30 min points to the significance of heat-sensitive organisms such as pseudomonads and other nonsporing bacteria.

2.2. Model to Explain Take-All Decline

Rovira and Wildermuth (1981) proposed, and we concur in part, that take-all decline develops because of a gradual build-up of inhibitory bacteria (of which pseudomonads are a component) in infested root debris with several years of monoculture, and that these bacteria then retard the growth of *G. graminis* var. *tritici* from the debris toward the root or cause lysis of the hyphae. These suppressive bacteria are thought to build up initially in lesions on roots of living plants as secondary colonists of the infected tissues and then carry over with the pathogen in crop residue where they limit the ability of the pathogen to attack the next crop. Pope and Jackson (1973) similarly noted that ability of *G. graminis* var. *tritici* to grow toward the root was more limited in soils that had undergone take-all decline than in conducive soils. As an extension of the Rovira and Wildermuth (1981) model, we propose that the fluorescent pseudomonads, operating against a background of general suppression (Gerlagh, 1968), suppress *G. graminis* var. *tritici* during parasitism of roots (in lesions) as well as during saprophytism in the plant debris, and in the rhizosphere during growth required to reach and infect new roots (Fig. 2).

It is known that bacterial populations are greater on roots with take-all than on healthy roots (Vojinović, 1972a,b; Brown, 1981; Weller, 1983). Both scanning and transmission electron microscopic studies indicate that bacteria proliferate in root lesions caused by *G. graminis* var. *tritici* (Rovira and Wildermuth, 1981). Lesions caused by the take-all fungus are presumably a rich source of nutrients for bacteria such as fluorescent pseudomonads. Gradually, the most antagonistic and aggressive strains could dominate as colonists of the lesions, which can explain the qualitative shift toward antibiotic- and siderophore-producing types following wheat monoculture and one or more severe outbreaks of take-all. The selection pressure provided by the environment of leaky root lesions (Cook and Baker, 1983) may explain why a greater proportion of the *Pseudomonas* population from wheat roots is inhibitory to *G. graminis* var. *tritici in vitro* when isolated from roots growing in suppressive soils in the presence as compared to the absence of the pathogen (Weller, in Cook and Baker, 1983; Smiley, 1979); why more inhibitory fluorescent pseudomonads occur on roots of wheat grown in the presence of *G. graminis* var. *tritici* in suppressive than in conducive soil (Weller, in Cook and Baker, 1983); and why the strains from suppressive soils, as a group, are more suppressive to take-all *in vivo* than strains from roots grown in conducive soils (Weller et al., 1985a,b).

Direct evidence that *P. fluorescens* is successful as a root colonist in the presence of *G. graminis* var. *tritici* and that it affects the pathogen during its parasitic as well as saprophytic existence was obtained by Weller (1983) using strain 2-79RN_{10}, which is resistant to rifampin and nalidixic acid. The original strain 2-79 was isolated from wheat roots grown in soil from a field at Lind, Washington, in which wheat had been grown for 11 consecutive years and take-all was beginning to decline. The strain is highly inhibitory to *G. graminis* var. *tritici in*

FIRST YEAR

Occasional infected plant; inoculum from infected grass weeds or rare fragments of infested crop residue

Successions of secondary colonists compete with Ggt in lesions and during saprophytic survival in crop residue

SECOND YEAR

Take-all in patches; inoculum mainly from infested remains of plants infected first year

Successions of secondary colonists compete with Ggt in lesions and during saprophytic survival in crop residue. The more intense the competition within lesions, the greater the selection pressure in favor of antibiotic- and siderophore-producing bacteria

THIRD YEAR

Take-all widespread, but many plants only mildly diseased because of pathogen suppression by inhibitory bacteria and other secondaries established as cohabitants in infested residue

Inhibitory bacteria become more dominant as secondary colonists of lesions; possibly they accompany Ggt beginning with the earliest stages of pathogenesis

FOURTH AND SUBSEQUENT YEARS

Most or all plants infected but disease is mild; pathogen is suppressed by its cohabitants in the crop residue, especially inhibitory bacteria that establish early in lesions

A fluctuating equilibrium develops between inhibitory effects of bacteria versus sufficient infected tissue to support populations of the bacteria

Fig. 2. Proposed sequence of events and microbiological processes that result in take-all decline.

vitro and is suppressive of take-all in both the greenhouse and field when applied as a seed treatment (Weller and Cook, 1983). The ratio of the population of strain 2-79RN_{10} from roots with and without *G. graminis* var. *tritici* was consistently greater than the ratio of populations of indigenous fluorescent pseudomonads and of total aerobic bacteria in the same tissue (Weller, 1983). Specific nutrients released from lesions, parasitism of the hyphae of *G. graminis* var. *tritici*, and/or production of a more favorable physical/chemical environment for 2-79RN_{10} in the lesions may account for this result. The same conditions that favored 2-79RN_{10} may also favor the build-up of other take-all suppressive bacteria during wheat monoculture and outbreaks of take-all and eventually lead to take-all decline.

In another study (Wilkinson et al., 1982), *P. fluorescens* strains 2-79 and 13-79 (from the same source as 2-79 and also inhibitory to *G. graminis* var. *tritici in vitro*) and two fluorescent pseudomonads from a conducive soil and lacking *in vitro* inhibition of *G. graminis* var. *tritici* each were (1) coated on wheat seeds, (2) added directly to soil, or (3) infiltrated into *G. graminis* var. *tritici* oat-grain inoculum. The seed treatment helped establish the bacteria on the roots where their main effect would be expressed during the parasitic phase. Infiltration of the bacteria into oat grains concentrated their inhibitory effect mainly during the saprophytic phase of the pathogen. The bacteria in soil (about 10^6 CFU/g) was intended to assess their suppressive ability when in the bulk soil. Only strains 2-79 and 13-79 were suppressive to take-all *in vivo*. Suppression was strongest when the bacteria were infiltrated into the particles, intermediate when introduced with seed, and virtually nonexistent when cells were mixed with the soil. Both these results and those of Weller (1983) point to the importance of the inhibitory bacterial activity on roots and in lesions during parasitism as well as in the inoculum and rhizosphere during saprophytism and prepenetration growth of the pathogen. Further, this study demonstrates the importance of the location and the quality of strains necessary for suppression to occur.

3. BIOLOGICAL CONTROL OF TAKE-ALL WITH ROOT-COLONIZING PSEUDOMONADS

The evidence that take-all decline is due to root-colonizing pseudomonads has prompted research on their application as either a seed or soil treatment for take-all suppression where decline has not yet begun. Although no practical commercial biological control has emerged to date, the progress is encouraging. Even in the unlikely case that pseudomonads are not the agents responsible for take-all decline, recent work with other diseases on other crops (Burr et al., 1978; Howell and Stipanovic, 1979, 1980; Geels and Schippers, 1983; Kloepper et al., 1980c; Suslow and Schroth, 1982b; Scher and Baker, 1980, 1982; Howie and Echandi, 1983; Sneh et al., 1984; Colyer and Mount, 1984; Xu and Gross, 1986a and b) as well as take-all (Weller and Cook, 1983; Wong and Baker, 1984; Kloepper et al., 1980b; Vraný et al., 1981) lends strong support to our decision to proceed with a screening and testing program aimed at finding effective *Pseudomonas* strains. Moreover, fluorescent pseudomonads possess characteristics that make them ideal for biological control (Weller, 1985). They are well adapted to the rhizosphere and rhizoplane (Bowen and Rovira, 1976; Rovira and Davey, 1974) where they are needed; they are able to utilize a large number of organic substrates (Stolp and Gadkari, 1981) including root exudates (Rovira and Davey, 1974; Sands and Rovira, 1971); they have a fast growth rate in the rhizosphere relative to other bacteria (Rouatt and Katznelson, 1961; Bowen and

Rovira, 1976); they produce a wide variety of antibiotics (Leisinger and Margraff, 1979) and siderophores (Kloepper et al., 1980a; Misaghi et al., 1982; Teintze and Leong, 1981; Teintze et al., 1981; Vandenbergh et al, 1983) that can inhibit the growth of the take-all fungus and other pathogens; and they can be successfully introduced into the rhizosphere and rhizoplane of a variety of crops via the planting material (Burr et al., 1978; Kloepper et al., 1980c; Suslow and Schroth, 1982b; Schroth and Hancock, 1981, 1982; Weller and Cook, 1983).

3.1. Selection of Effective Strains

There are no *in vitro* tests that will predict which strains from a natural population will control take-all. In work on plant growth-promoting pseudomonads, an attempt to separate effective strains from other root-colonizing pseudomonads using 100 physiological and nutritional tests was unsuccessful (Suslow, 1982). Further, only about 2–5% of the rhizobacteria tested produced a positive plant-growth response (Schroth and Hancock, 1981; Suslow and Schroth, 1982b). Nevertheless, in the case of take-all, some isolation and testing procedures are available that can increase the quantity and quality of effective strains (Weller et al., 1985b).

The first critical step is the isolation of candidate bacteria. Fluorescent pseudomonads should be selected from the environment in which they will be expected to function. For example, the bacteria used to control a root disease should be adapted to the root and have a high rhizosphere competence. Because of the greater proportion of effective strains that can be recovered (Weller et al., 1985b), we now isolate routinely from roots of wheat or barley infected with *G. graminis* var. *tritici* and growing in soil from a field in which take-all has declined.

The second step involves screening for the ability of the strains to inhibit *G. graminis* var. *tritici in vitro* through the production of antibiotics and/or siderophores. Antibiosis cannot be used as a sole determinant in the selection process (Schroth and Hancock, 1981; Baker, 1969; Wong and Baker, 1984), but our results continue to indicate that some kind of inhibitory activity is important to take-all suppression *in vivo* (Wilkinson et al., 1982; Weller et al., 1985a). Research on the mechanisms of suppression of other root diseases has also indicated the important role of antibiotics (Howell and Stipanovic, 1979, 1980; Xu and Gross, 1986) and/or siderophores (Kloepper et al., 1980a; Kloepper and Schroth, 1981; Scher and Baker, 1982; Sneh et al., 1984; Wong and Baker, 1984). Wong and Baker (1984) concluded that antibiotic production was not a valuable criterion for selecting potential antagonists against *G. graminis* var. *tritici*. However, Weller et al. (1985a) found that nine fluorescent pseudomonads isolated from roots of plants grown in suppressive soils and highly inhibitory to *G. graminis* var. *tritici* on both King medium B (King et al., 1954) and potato dextrose agar were significantly more suppressive to take-all *in vivo* as a group than were 9 fluorescent pseudomonads from a conducive soil that showed little or no *in vitro* inhibition of *G. graminis* var. *tritici* (Fig. 3). Further, when suppressive *P. fluorescens* strains 2-79 and R1a-80 were treated with nitrosoquanidine and eight mutants selected for reduced or no inhibition of *G. graminis* var. *tritici in vitro*, the mutants colonized wheat roots like the parents but were less suppressive than the parents to take-all *in vivo*. In another study, the suppressive ability of *P. fluorescens* strains R1a-80 and 2-79 and *P. putida* strain L30b-80 was unchanged, reduced, and completely eliminated, respectively, by the addition of FeEDTA to the soil, suggesting that strain R1a-80 suppressed take-all by antibiotic production, L30b-80 by siderophore production,

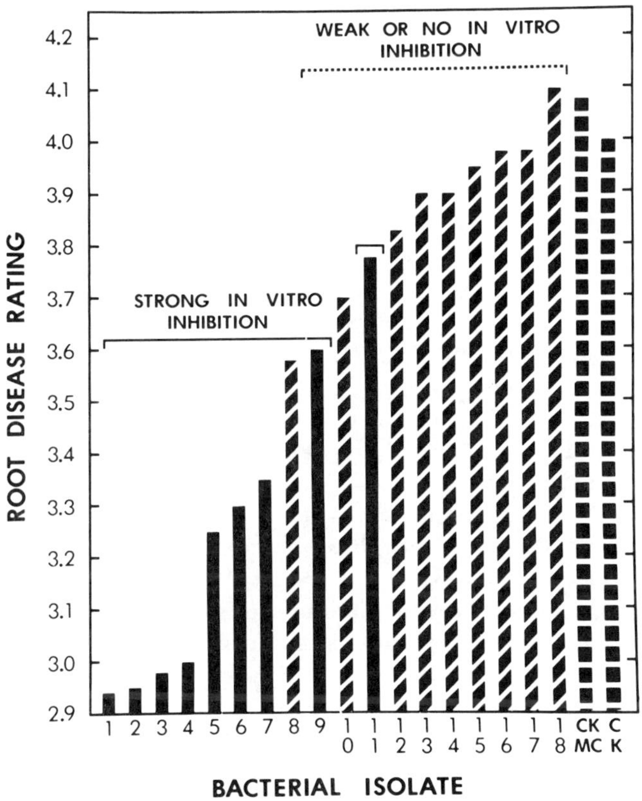

Fig. 3. Relationship of source of fluorescent pseudomonads and their ability to inhibit *G. graminis* var. *tritici in vitro* and suppress take-all *in vivo*. Solid black bars represent strains from wheat roots grown in soil from a field where take-all had declined (suppressive soil). Broken bars represent strains from wheat roots grown in soil from a field in which alfalfa had previously been grown (conducive soil). Solid brackets indicte isolates strongly inhibitory to *G. graminis* var. *tritici* on potato dextrose agar and King's medium B and broken-line brackets indicate isolates weakly inhibitory or noninhibitory on these media. *In vivo* tests were conducted with the respective strains applied to the wheat seed at 10^8 CFU/seed and tested in fumigated soil with 0.45% inoculum of *G. graminis* var. *tritici*. The checks (CK and CK MC) received no bacterial treatments and a 1.0% solution of methylcellulose, respectively. Disease severity was evaluated on a scale of 0–5, where 0 = no disease evident and 5 = plants dead or nearly so.

and 2-79 by both kinds of compounds (Weller et al., 1985a). The main antibiotic produced by Strain 2-79 has been identified as a dimer of phenazine carboxylic acid (Gurusiddaiah et al., 1986), and further, T*n 5* mutants of 2-79, no longer able to produce this antibiotic, are greatly reduced in ability to suppress take-all *in vivo* (L. S. Thomashow, unpublished). Although effective strains that are noninhibitory *in vitro* may be missed by selecting for antibiotic and siderophore producers, nevertheless, the risk is justified and worth the benefits (Weller et al., 1985b).

An *in vivo* greenhouse test was developed as a third step to further reduce the number of candidate strains to be tested in the field (Weller et al., 1985b). The assay uses plastic tapered tubes (2.5 cm diameter × 16.5 cm length), each with a hole in the bottom (Ray Leach Cone-tainer, Canby, Oregon) supported in a hanging position in plastic racks, 200 tubes per rack. Each tube is filled with a 6.5-cm-thick column of sterile vermiculite followed by 5.0 g of test soil. *Gaeumannomyces graminis* var. *tritici* is blended into the soil as particles (0.25–0.5 mm) of colonized oat kernels (Wilkinson et al., 1985a) at 0.15 and 0.45% w/w. Two bacteria-treated seeds (10^8 CFU/seed) are placed on the soil and then covered with a 1.5-cm-thick topping of vermiculite. The tubes are incubated at 15- 18°C in a dark/light cycle of 12 hr. Each cone is watered with 5 ml of diluted Hoagland's solution (macroelements only) twice a week, and after 3–4 wk the seedlings are washed and evaluated for disease severity.

Of 121 strains tested in the cone assay, a significantly greater percentage were suppressive to take-all in fumigated soil at both inoculum concentrations than in natural soil. Further, 71% of the strains suppressed take-all in an untreated conducive natural soil with 0.15% oat-kernel inoculum, but only 49% of the strains were effective in the same soil with 0.45% oat inoculum. By adjusting the inoculum potential of the pathogen, isolates with a maximum level of suppressive activity can be identified while those that are only weakly suppressive can be eliminated (Weller et al., 1985b).

The final step is to test the candidate strains in experimental field plots. Over 20 different strains have been tested in experiments conducted throughout Washington since 1979 with several showing some activity. Strain 2-79 alone or combined with strain 13-79 has received the most attention. The studies have been conducted at locations where take-all is a problem, for example, in the high rainfall area of western Washington, the irrigated Columbia Basin in central Washington, and the moderate rainfall region of eastern Washington. The bacteria are grown on plates of KMB for 2–3 days, scraped into a 1.5% solution of methylcellulose, and then mixed as a slurry with wheat seed. Each seed generally contains about 10^8 CFU at the time of sowing. The bacterial population remains stable if the seed is frozen, declines slowly over several weeks at 5°C, and declines precipitously as the temperature is increased up to 25°C (Weller and Cook, 1983). For small-scale field tests, three or four rows of treated wheat (each 3-5 m long) are placed next to an equal number of rows of untreated wheat. The take-all fungus is applied to the seed furrow as colonized whole oat-grain inoculum at approximately 4–5 g per 3-m length of row to produce uniform take-all.

Occasionally the severity of the disease may be so great on the seedlings as to totally swamp the effects of the bacteria. Rovira (unpublished) tested another method whereby millet seed colonized by *G. graminis* var. *tritici* is mixed by rototilling into the top few centimeters of the soil. Unfortunately, even with the best methods of infesting the soil, take-all may still develop unevenly through the field, particularly when seeded on a slope. The pairing of treated wheat with nontreated wheat greatly increases the ability to detect effective strains when disease develops unevenly in the field.

The use of soil fumigation with methyl bromide or chloropicrin followed by the reintroduction of *G. graminis* var. *tritici* allows take-all to develop more uniformly throughout the field and the beneficial effects of the bacteria are more easily detected. This method also demonstrates that the enhanced growth of *Pseudomonas*-treated wheat is due to take-all control rather than inhibition of other root pathogens or direct stimulation of the plant. The improved growth of potatoes and sugarbeets by the PGPR apparently results from the suppression of a complex of deleterious rhizoplane microorganisms and not a single root pathogen (Schroth and Hancock, 1981, 1982; Suslow and Schroth, 1982a). At least 10 species or varieties of *Pythium* attack wheat roots in Washington (Chamswarng and Cook, 1985) and also are subject to control by *Pseudomonas* spp. (Weller and Graham, 1984). Knowing which pathogen is controlled by the pseudomonad improves the ability to predict when a growth response should be expected and helps to explain failures.

3.2. Biological Control with Pseudomonads in the Field

In small-scale field plots where *G. graminis* var. *tritici* was introduced into natural soil, both spring and winter wheat treated with 2-79, or 2-79 + 13-79 had less take-all and 5–27% greater yield than nontreated wheat in six of eight tests. The greatest success occurred in western Washington where soils are highly conducive to take-all and rainfall averages about 100 cm per year. Yield increases up to 147% as compared to the untreated check have been observed with tests in fumigated soils reinfested with the pathogen (Weller and Cook, 1983). A greater growth response in fumigated than in natural soil is not unexpected since, although the disease is more severe because of the generally more conducive nature of the treated soil, the fluorescent pseudomonads likewise have less competition from indigenous rhizosphere colonists and proliferate in treated soil (Ridge, 1976). Furthermore, Weller et al. (1985b) demonstrated that a greater percentage of strains suppressed take-all in fumigated soil than in natural soil.

A combination of isolates 2-79 and 13-79 has been superior to either isolate alone in about 50% of the tests (Weller and Cook, 1983). Sivasithamparam and Parker (1978) reported no control of take-all by individual strains, but significant control by a mixture. It seems reasonable that a mixture of strains might better simulate the natural suppressive microflora, especially that responsible for take-all decline. Using organisms with different but complementary mechanisms of inhibiting *G. graminis* var. *tritici*, such as a strain having a siderophore with a high affinity for iron with one having the ability to produce a potent antibiotic, might provide the best approach to designing a mixture.

Since 1982, large-scale pilot tests have been conducted in Washington with 2-79 or 2-79 + 13-79 applied on seed sown in commercial fields with commercial techniques in plots up to 0.5 acres per replication. All fields were selected because they had a history of take-all or were expected to have a problem. Of five field tests completed by 1984, yields were 10–21% greater in three trials and 5% greater in one trial, and showed no difference in the fifth. The latter site was in wheat for the fourth consecutive year when the test was conducted and take-all decline had occurred; yields of the untreated check were therefore at a level expected for a relatively healthy crop and were not improved by the bacterial treatment.

4. ROOT COLONIZATION BY SUPPRESSIVE FLUORESCENT PSEUDOMONADS

4.1. Mechanisms of Colonization

The ability of introduced pseudomonads to become distributed along the root and maintain a significant population is important in suppression of take-all as well as other root diseases (Schroth and Hancock, 1981, 1982; Loper et al., 1985; Suslow, 1982). All results to date suggest that take-all suppressive fluorescent pseudomonads introduced on the seed become distributed along the root axis primarily by passive carriage of bacteria with the elongating root (Weller, 1983, 1984; Howie, 1985). Initiation of colonization depends on the bacteria becoming established near the root tips. As the roots elongate, bacteria are carried downward with the root tips while others are left behind. Motility mediated by flagella can be ruled out as an explanation for movement over distances greater than a few millimeters, since three nonmotile mutants that lacked flagella were equal to their motile parents both as root colonists and in their ability to suppress take-all (Howie, 1985). Movement in percolating water can be ruled out on the basis that bacteria were detected on roots down to 7–9 cm in soil at matric potentials down to −4.0 bars in a system where water was not added at the soil surface and where the only movement of water would have been toward the root in response to the gradient formed as water was absorbed by the roots (Howie, 1985).

Howie (1985) has suggested that the phrase "rhizosphere colonization by the bacteria" is misleading because it suggests that the primary process of colonization is due totally to the bacteria. He suggests that a phrase such as "spread of bacteria introduced onto seed and their multiplication in the rhizosphere" would be better. By this model, the important components in the successful establishment of an introduced bacteria on roots relative to indigenous bacteria would occur after the cells become distributed along the root. Whether the cells carried downward are specifically bound to the root is unknown. In any case, the key to success is ability to multiply on the root and not be preempted or displaced by indigenous organisms once dispersed along the root. A strain such as 2-79 has these characteristics (Weller, 1984).

Motility and percolating water probably assist in the movement of cells over shorter distances, for example, a few millimeters. Soby and Bergman (1983) demonstrated that a motile strain of *R. melilotii* spread through autoclaved soil at a matric potential maintained between -2 and -8 mbars, whereas a flagella-minus strain did not spread. Leben (1983) studied colonization by *Pseudomonas syringae* pv. *lachrymans* of roots of cucumber grown on agar and in soil and concluded that active spread of the bacteria was responsible for their movement along the root. Chao et al. (1986) reported that *Enterobacter cloacae* and *P. fluorescens*, applied to pea seed, were able to colonize the upper half of the root system in nontreated soil, but percolating water enhanced the downward movement and the depth of penetration was influenced by the amount of water added. Madsen and Alexander (1982) reported that the downward transport of *P. putida* through the soil was enhanced by both the addition of water and the growth of bean roots when the bacteria was applied to the top layer of soil. In a natural and artificial soil, Parke, Rovira, and Bowen (unpublished) observed strain 2-79 on roots in distinct microcolonies, but once water was added the bacteria became dispersed over the root.

4.2. Population Dynamics in the Field

Pseudomonas fluorescens strain 2-79RN_{10} has been used as a model to study colonization, multiplication, and survival of take-all suppressive pseudomonads on roots in the field when introduced on seed of either winter or spring wheat. Bacteria were isolated from a bulked sample of roots (Weller, 1983). On winter wheat at Pullman, during the first month after planting, 2-79RN_{10} was present at over 10^6 CFU/0.1g (2.5 cm) of seminal root. It made up more than 90% of the total indigenous fluorescent pseudomonads and up to 50% of the total aerobic bacteria. During the fall and winter, the population of 2-79RN_{10} on the roots, whether infected or not infected with *G. graminis* var. *tritici*, declined to a minimum level in March to 2.8×10^3 CFU/0.1 g root. With the onset of spring, the population of 2-79RN_{10} increased nearly tenfold on infected roots and then remained fairly steady until harvest, and take-all was suppressed. In contrast, the population of 2-79RN_{10} on roots of the plants without take-all continued to decline. The populations on wheat roots were comparable to those of PGPR on roots of potatoes (Kloepper et al., 1980c; Loper et al., 1984) and sugarbeets (Suslow and Schroth, 1982b).

In a second study, the population density was determined on individual roots. A population gradient developed along the wheat root, with the highest number of 2-79RN_{10} at the base of the root (near the seed) and a significant linear decline in the population at progressively greater distances from the seed and toward the tip (Weller, 1984). The population profile along the root was described by a series of curves, each representing the changes in the population of 2-79RN_{10} on specific sections of the root over time. In sections 0–3 cm below the seed, 10^5–10^6 CFU/cm root were detected from the onset and no increase in the population was evident in later samples. However, in sections further along the root, where initial populations were near 10^3 CFU/cm of root, the population increased up to 100-fold, with average apparent doubling times of 15–67 hr. Bowen and Rovira (1976) reported doubling times of *Pseudomonas* spp. on roots of *Pinus radiata* of 5.2 hr.

The growth curves of 2-79RN$_{10}$ on roots are similar to those published for indigenous (Bowen, 1979; Bowen and Rovira, 1976; Foster and Bowen, 1982) and introduced pseudomonads (Bennett and Lynch, 1981a,b; Loper et al., 1985). Bowen (1979) describes the growth of bacteria on roots as occurring in an initial rapid growth phase that can be regarded as an "intensity" factor, indicating the richness or suitability of the substrate for microbial growth, followed by a slow phase, regarded as a capacity factor, in which substrate supply balances maintenance requirements. Bowen and Rovira (1976) earlier referred to the "capacity" factor as the total supply of energy-yielding materials from roots, which determines the total biomass on the root. By this concept, the maximum population obtained on a root is determined by the quantity and quality of the nutrients available. Bennett and Lynch (1981b) studied the growth of a *Pseudomonas* sp., a *Mycoplana* sp., and a *Curtobacterium* sp. introduced into the rhizosphere of gnotobiotic barley plants. The final populations on the root were independent of the initial amount of bacteria added into the sterile sand in which the roots were growing. They proposed the term "colonization potential" to describe the maximum population of a bacterium per unit mass of root. In contrast, Weller (1984) and Loper et al. (1985) found that in natural soil the rhizosphere populations of introduced fluorescent pseudomonads were dependent on initial inoculum load. This difference between growth in an aseptic versus septic system demonstrates how the population of introduced bacteria are influenced by competition from indigenous rhizosphere microflora (Loper et al., 1985).

The effects of competition also explain why, despite multiplication of strain 2-79RN$_{10}$, sections more distant from the seed had lower maximum populations than sections close to the seed (Weller, 1984). Sections close to the seed received a larger initial dose of 2-79RN$_{10}$, which would allow the introduced bacteria to dominate this portion of the root in advance of colonization by the indigenous bacteria. In contrast, sections more distant from the seed would have a lower initial density of the introduced strain, which would limit the chances for domination of the root by this strain.

The pattern of root colonization by 2-79RN$_{10}$ seems to follow closely the growth of *G. graminis* var. *tritici* on the roots (Weller, 1983), thus contributing to the success of the bacteria in the suppression of take-all. Take-all generally begins on the seminal roots on both fall- and spring-sown wheat. Brown, thick-walled macrohyphae (runner hyphae) of *G. graminis* var. *tritici* grow along the surface of the roots, and from these, thin-walled microhyphae (infection hyphae) penetrate the roots. Strain 2-79RN$_{10}$ becomes established in high numbers along the entire length of the seminal roots, including the root tip. Protection of the root near the tip is especially important since this portion of the root is more susceptible to infection by the take-all fungus than older root tissue (Deacon and Henry, 1980). The ability of 2-79RN$_{10}$ to multiply on the root affords it the ability to protect the young tissue behind the tip even though the bacterial population there is initially low. By their ability to proliferate in lesions, the introduced bacteria then retard the spread of the pathogen. Those hyphae that continue to grow also encounter resistance from bacteria of 2-79RN$_{10}$ (up to 10^6 CFU/0.1 g tissue) that are present on the subcrown internode. Finally, infection of crown roots, whether by inoculum in the soil or hyphae from other parts of the plant, is suppressed by bacteria that colonize these roots from colonies located at the base of the tillers and subcrown internode.

4.3. Influence of Rhizosphere Matric Potential on Root Colonization

Loper et al. (1985) stated that the dynamics of the rhizosphere populations of introduced bacteria must be governed by many interacting processes that influence instantaneous population size. These processes might include bacterial growth, survival, death, emigration and immigration, as influenced by the chemical, biological, and physical environment of the rhizosphere. Howie (1985) studied both the effect of matric potential and rhizosphere pH on root colonization by strain 2-79RN$_{10}$ in a Shano silt loam, a Quincy loamy fine sand, and a Thatuna silt loam collected, respectively, from wheat fields near Quincy, Pasco, and Pullman, Washington. The effect of water potential was studied over a range from 0 to −4.0 bars by sampling the bacterial populations on segments of root 1–3 and 7–9 cm below the treated seed. The populations of the introduced bacteria were greatest after 10 days on both segments at −0.3 bars in the Shano silt loam and −0.7 bars in the Thatuna silt loam and the Quincy loamy fine sand. The introduced bacteria were detected on segments at 7–9 cm below the seed at −4.0 bars, but whether the numbers represented cells carried to this distance with no multiplication was not determined. Based on extrapolations, Howie predicted that downward spread by 2-79 could occur down to −7.0 bars. He suggested that the poor colonization in wet soil is due to poor aeration, whereas the relatively poor colonization in dry soils (−4.0 to −8.0 bars) results from a direct restriction of low matric potential on growth of the bacteria. The optimal of −0.3–0.7 bars might be where oxygen and turgor potential for the cells both are maximal. In contrast to the activity of 2-79RN$_{10}$, the total bacterial populations did not show the same trends and the populations generally varied little at each matric potential. Parke et al. (1984) demonstrated that in a natural South Australian soil colonization was greatest at −1.4 bars and lowest at −0.14 bars, but in the same soil, irradiated, it was greatest at −0.53 bars.

4.4. Effect of Rhizosphere pH on Rhizosphere Colonization and Take-All Suppression

The populations of 2-79RN$_{10}$ in both a Shano silt loam and the Thatuna silt loam were highest at rhizosphere pH (pHr) 6 and 6.5, whether or not inoculum of *G. graminis* var. *tritici* was present (Howie, 1985). The population of total fluorescent pseudomonads also was highest at pH 6.0–6.5, but only when the soil was also infested with *G. graminis* var. *tritici*. Of interest was that in the Thatuna silt loam, the population of 2-79RN_{10} dropped precipitously at pH 7. Moreover, take-all was suppressed significantly by 2-79RN_{10} in either soil at pHr 6.0 or 6.5, but less so at pHr 5.1 or 7.2. Thus suppression overlaps with the range of pH in which colonization of 2-79 is maximal. Whether the pHr effect is due to lower population or less activity of the introduced antagonists is not known. That a pHr above 7 is not optimal for growth of 2-79 and suppression by the pseudomonads does not occur is further support for the role of pseudomonads in take-all decline, since take-all decline is slow to occur or may not occur in alkaline soils in the Pacific Northwest (Cook and Baker, 1983).

5. MANAGEMENT OF TAKE-ALL WITH SOIL FERTILITY AND IRRIGATION

5.1. Manipulation of Soil pH and Form of Nitrogen

Take-all is usually more severe in calcareous soils (pH 6.0–7.0 and above) and in acid soils (less than pH 5.5) that have been limed (Cook, 1981). The effect of liming is the result of elevated soil pH and not disease enhancement by calcium in the lime (Ponchet, 1962; Reis et al., 1983).

Garrett (1937) suggested that since hyphal growth of *G. graminis* var. *tritici* is restricted by high concentrations of carbon dioxide, and since the equilibrium relationship between carbon dioxide and carbonate-bicarbonate ions shifts toward the gas as the pH is lowered, take-all suppression in acid soils may result from inhibition of hyphal growth by carbon dioxide. Subsequent findings indicate that concentrations of carbon dioxide sufficient to inhibit *G. graminis* var. *tritici* are unlikely to occur in soil (Ferraz, 1973; Garrett, 1981). Reis et al. (1983) suggested that wheat may be more resistant to take-all in acid than in neutral or alkaline soil because of a greater availability of trace nutrients to the plant in acid soil. However, although the disease-suppressing effects of an adequate and balanced supply of trace nutrients to the plant can be striking, this explanation is inadequate to account for the highly suppressive effect of acid soil on the disease.

Certain isolates of *G. graminis* var. *tritici* produce a low molecular weight substance in agar at pH 3.5–5.0 that inhibits their own growth as well as the growth of nonproducer strains and even other fungi (Romanos et al., 1980; Naiki and Cook, 1983b; McGinty et al., 1984). Conceivably these strains produce the inhibitor and thereby limit their own growth in crop residue in acid soils. However, strains able to grow at pH 3.5–5.0 would almost certainly become dominant with the selection pressure of a low pH, although there is no evidence for such a shift in the population of *G. graminis* var. *tritici*, as the soil becomes more acid. Ability to produce a small amount of inhibitor actually may be advantageous to the pathogen as a means to limit competition from other microorganisms in the foodbase (Naiki and Cook, 1983b; McGinty et al., 1984). Understanding the genetic control of self-inhibition may offer a means to control of take-all in the future, but is not likely to explain why take-all is suppressed in acid soil.

Different mechanisms of take-all suppression may be operative according to the degree of acidity. Smiley and Cook (1973) suggested that pathogen growth is inhibited by acidity per se at about pH 5.0 and below and by antagonistic soil microorganisms between pH 5.0 and 7.0. However, except for those strains unable to grow because of inhibitor production at low pH, *G. graminis* var. *tritici* can grow remarkably well down to pH 4.0 (Sivasithamparam and Parker, 1981). It seems likely, therefore, that inhibition is also of microbial origin below pH 5.0, but perhaps different microorganisms are operative at the different pH values.

Possibly take-all suppression is the result of competition mainly from fungi in the foodbase in soil at pH values between approximately 4.5–5.0 and 6.0, and from this competition and also inhibition by antagonistic pseudomonads on roots and in the lesions at pH values between 5.5–6.0 and about 7.0. The latter kind of inhibition may depend on wheat monoculture and the occurrence of one or two outbreaks of take-all first, to produce the qualitative shift in make-up of the *Pseudomonas* population on wheat roots in favor of the types inhibitory to the take-all fungus. Until this qualitative shift and take-all decline has occurred, take-all is likely to be severe at pH values down to 5.5–6.0, but usually not much lower. This is consistent with Gerlagh's (1968) concept that specific suppression operates against a background of general suppression. At still higher pH (e.g., 7.0 and above), neither form of suppression seems adequate, at least under Pacific Northwest conditions, and therefore the disease is slow to decline or may not decline for several years of monoculture.

In the Pacific Northwest, some soils that were once pH 7.0–7.5 are now at pH 5.0–5.5 owing to the use of nitrogen fertilizer for nearly 40 yr. Suppression of take-all has been facilitated in these soils because of their lower pH. Besides the long-term depression of soil pH by ammonium fertilizer, take-all can be suppressed in neutral or slightly alkaline soils by acidification of the rhizosphere with ammonium (Smiley and Cook, 1973). The rhizosphere can be 0.5–1.0 pH units lower or higher than that of bulk soil pH on plants fertilized with largely ammonium and nitrate, respectively. This slight drop in rhizosphere pH may be of little consequence in soils at pH 7.0 and above, but can be significant if the bulk soil pH is already naturally acid. It is therefore possible to limit take-all severity by keeping the nitrogen in the ammonium form (Huber et al., 1968; Smiley and Cook, 1973; Trolldenier, 1981; MacNish, 1980). The results of Christensen and Brett (1985) indicate that take-all severity is negatively correlated with the length of time after spring fertilization that the ratio of ammonium : nitrate was about 3 : 1 or greater in favor of ammonium. This may be accomplished by the use of a nitrification inhibitor such as N-serve or chloride (Halsey and Powelson, 1981; Christensen et al., 1981) applied with the ammonium, provided the soil is not limed.

5.2. Management of Take-All with Fertility and Irrigation

Wheat (and probably barley, although fewer studies have been conducted with this cereal) becomes highly susceptible to take-all if deficient in certain macro and/or micro nutrients (Huber, 1981). The importance of phosphorus has long been known (Garrett, 1948; Schaffnit, 1930; Stumbo et al., 1942; and Slope et al., 1978) but the significance of K, Mg, and trace nutrients was shown only recently (Reis et al., 1982). Plants deficient in any of these nutrients become so susceptible that they die in the seedling stage with an inoculum potential that normally permits heading and at least some yield on the wheat supplied with these nutrients. Comparable deficiencies in Ca, S, and N have little or no effect on the relative susceptibility of tissues to *G. graminis* var. *tritici* (Reis et al., 1982), except that high nitrogen promotes root growth and helps the plant outgrow or tolerate the disease (Garrett, 1948). Cook and Baker (1983) proposed that deficiencies of P, K, Mg, and trace nutrients result in greater severity of take-all because of their critical role in basal cell functions and maintenance, in contrast to Ca, S, and N, which are involved more in formation of new tissues. An ample supply of nitrogen in particular is essential to take-all management in consecutive crops of wheat (Cunningham, 1966b; Taylor et al., 1983) as a means to sustain rapid plant growth and hence ability to tolerate or outgrow the disease.

When wheat is grown in monoculture in the Pacific Northwest, phosphorus is supplied with nitrogen in the seed furrow or below and to one side of the seed at the time of planting (Taylor et al., 1983). This placement is preferred to broadcast application because it makes the phosphorus more accessible to the plant. Damage to the roots reduces the efficiency of nutrient uptake and thereby increases susceptibility of the plant to the disease; concentrating phosphorus in the rooting zone helps avoid this problem. More work is needed on trace nutrient management for suppression of take-all and especially on the benefits of foliarly applied nutrients.

Take-all is favored by moist soil conditions (Cook, 1981). In the Pacific Northwest, the disease is limited to areas that receive 450–500 mm or more of precipitation or precipitation plus irrigation per year. Since the pathogen operates mainly in the surface 20–30 cm of soil, conditions that promote drainage and drying of this soil layer can limit take-all significantly. Besides the direct effects of low soil and plant water potentials on the growth of this fungus is dry soil, fungi such as *Fusarium*, *Trichoderma*, *Penicillium*, and others able to grow at these lower water potentials are likely to thrive as secondary colonists and further suppress *G. graminis* var. *tritici*. Unfortunately for take-all control, the trend in irrigation equipment has been toward the center pivots that apply 1–2 cm of water as frequently as every 6 hr. This environment may be ideal for root-colonizing bacteria. Where rill or furrow irrigation is used; some take-all control can be achieved by filling the soil profile with large aliquots of water and then allowing 7–10 days between irrigations to permit the surface soil to dry out.

6. PROSPECTS FOR MANAGEMENT OF TAKE-ALL WITH AVIRULENCE AND HYPOVIRULENCE DETERMINANTS

Considerable interest has emerged in recent years in the use of a determinant for avirulence or hypovirulence in a pathogen as an agent of biological control of that pathogen. Such agents include the dsRNA mycoviruses associated with hypovirulence in *Endothia parasitica* and shown to be transferred through hyphal anastomosis between infected and healthy strains within vegetatively compatible groups of *E. parasitica* (Anagnostakis, 1982). A dsRNA mycoviruslike agent also occurs in a hypovirulent strain of *Rhizoctonia solani* and is transmissible by hyphal anastomosis between a culture of the infected strain and a culture of the same strain that had been cured of the infection (Castanho et al., 1978). In Japan, certain weakly virulent or avirulent strains of *R. solani* harbor a DNA plasmid (Hashiba et al., 1984). The exact role and effect of dsRNA mycoviruslike agents and DNA plasmids in fungi and how these genetic determinants interact with or depend for expression on nuclear genes in the host fungus is still not understood. Nevertheless, any determinant for avirulence or hypovirulence in a pathogen should be investigated for its potential as an agent or source of genes for biological control.

The existence of virus particles in *G. graminis* var. *tritici* was first reported by Lemaire et al. (1970) and Lapierre et al. (1970). At least three sizes of particles have now been distinguished in mycelium of this fungus, namely 29, 35, and 40 nm (Rawlinson and Buck, 1981). Particles 29 nm in diameter contain a single dsRNA component, but the other two sizes contain up to four dsRNA components each (Buck et al., 1981; McFadden et al., 1983; Buck, 1984). Particles 35 nm in diameter have been divided into two groups according to whether the dsRNA components are 1.94–1.53 kbp or 2.33–2.02 kbp, respectively. Virus infections are common in this pathogen, including apparently healthy mycelium (Rawlinson et al., 1973). Frick and Lister (1978) showed by serology that 20 of 22 isolates obtained from wheat in Indiana contained at least one, and some more than one, dsRNA component. As for other fungi, transmission and spread of mycoviruses in *G. graminis* var. *tritici* may depend on hyphal fusion between vegetatively compatible strains. Ascospores produced by infected cultures are usually virus-free (Rawlinson et al., 1973). The epidemiological significance of mycoviruses in this fungus remains unknown, but obviously deserves continued study and consideration.

The suggestion (Lemaire et al., 1970) that take-all decline results from mycovirus spread and a gradual loss of virulence in the population of *G. graminis* var. *tritici* has not been supported on the basis that (1) mycovirus-infected strains occur at about the same frequency in the pathogen population whether or not take-all has declined (Rawlinson et al., 1973) and (2) virulence in the pathogen population does not decline significantly if at all during monoculture of the host and take-all decline (Asher, 1980; Cook and Naiki, 1982). Mycovirus infections have been associated with loss of virulence and aggressiveness during vegetative culture of this fungus on agar media (Tivoli et al., 1974), but whether the mycovirus causes loss of virulence is still unknown. However, the introduction into field plots in France of inoculum of a strain of *G. graminis* var. *tritici* that had lost virulence resulted in less take-all and 10–30% greater yield for up to 4 yr (Lemaire et al., 1979; Brun et al., 1976). A special technique was developed to mass produce the "hypovirulent" strain and apply it as a seed treatment to protect wheat against virulent inoculum of the pathogen in soil (Lemaire, personal communication). The current status and future prospects of this approach are uncertain, but the concept is novel. Even a small response to such a treatment, if real, is worthy of further investigation because of the potential for improvement.

If known, the genetic determinant for production of the *Q* factor (Romanos et al., 1980) responsible for self-inhibition (Romanos et al., 1980; McGinty et al., 1984; Naiki and Cook, 1983b) and possibly an inability of isolates to express virulence (Naiki and Cook, 1983b) might also be subject to manipulation for biological control purposes. Approximately 10–15% of isolates obtained from nature and maintained for a short period in culture are positive for inhibitor production and self-inhibition at pH 4.0–4.5 (McGinty et al., 1984). However, this percentage may eventually increase during prolonged culture of the isolates on agar media (Naiki and Cook, 1983b). The kind of media also affects expression of inhibitor production (McGinty et al., 1984). The phenomenon cannot be explained on the basis of dsRNA mycovirus infections (Rommanos et al., 1980; McGinty et al., 1984; Naiki and Cook, 1983b). Naiki and Cook (1983b) suggest that possibly the fungus is heterokanyotic, and that during prolonged vegetative culture, the nuclear makeup in each cell shifts so that the gene(s) for inhibitor production occurs in higher frequency. However, whether this fungus occurs as a heterokaryon in nature is still not clear. Alternatively, Naiki and Cook (1983b) suggest that inhibitor production is controlled by a cytoplasmic factor, perhaps a plasmid, since ascosporic cultures converted from normal, virulent "nonproducer" types to weakly virulent or avirulent producer types at nearly the same rate as did mycelial cultures obtained directly from diseased plants. If more were known about the conditions required for inhibitor production, or expression of the gene(s) responsible for inhibitor production, a method for controlling the fungus might then emerge. Alternatively, the gene(s) that control expression of this phenotype might be isolated and used to develop a superior antagonist or resistant host. This area of research is only in its infancy, but offers unlimited opportunities for new unconventional approaches to control of plant pathogens. Perhaps the take-all fungus can emerge as one of the first controlled by this molecular approach.

REFERENCES

Anagnostakis, S. L. (1982). Biological control of chesnut blight. *Science*, **215:**466–471.

Anderson, J. P. E. and K. H. Domsch (1973). Quantification of bacterial and fungal contributions to soil respiration. *Arch. Mikrobiol.*, **93:**113–127.

Asher, M. J. C. (1980). Variation in pathogenicity and cultural characters in *Gaeumannomyces graminis* var. *tritici*. *Trans. Br. Mycol. Soc.*, **75:**213–220.

Asher, M. J. C. and P. J. Shipton, Eds. (1981). *Biology and Control of Take-All.* Academic, New York, 538 pp.

Baker, R. (1968). Mechanisms of biological control of soil-borne pathogens. *Annu. Rev. Phytopathol.*, **6:**263–294.

Bateman, G. L. (1980). Prospects for fungicidal control of take-all of wheat. *Ann. Appl. Biol.*, **96:**275–282.

Bateman, G. L. (1981). Effects of soil application of benomyl against take-all (*Gaeumannomyces graminis*) and footrot diseases of wheat. *Z. Pflanzenkn. Pflanzenschultz*, **88:**249–255.

Bennett, R. A. and J. M. Lynch (1981a). Colonization potential of bacteria in the rhizosphere. *Curr. Microbiol.*, **6:**137–138.

Bennett, R. A. and J. M. Lynch (1981b). Bacterial growth and development in the rhizosphere of gnotobiotic cereal plants. *J. Gen. Microbiol.*, **125:**95–102.

Bockus, W. W. (1983). Effects of fall infection by *Gaeumannomyces graminis* var. *tritic* and triadimenol seed treatment on severity of take-all in winter wheat. *Phytopathology*, **73:**540–543.

Bowen, G. D. (1979). Integrated and experimental approaches to the study of growth of organisms around roots. In *Soil-Borne Plant Pathogens* (B. Schippers and W. Gams, eds.). Academic, London, pp. 209–227.

Bowen, G. D. and A. D. Rovira (1976). Microbial colonization of plant roots. *Annu. Rev. Phytopathol.*, **14:**121–144.

Brooks, D. H. and M. G. Dawson (1968). Influence of direct-drilling of winter wheat on incidence of take-all and eyespot. *Ann. Appl. Biol.*, **61:**57–64.

Brown, M. E. (1981). Microbiology of roots infected with the take-all fungus (*Gaeumannomyces graminis* var. *tritici*) in phased sequence of winter wheat. *Soil. Biol. Biochem.*, **13:**285–291.

Brown, M. E. and D. Hornby (1971). Behavior of *Ophiobolus graminis* on slides buried in soil in the presence of wheat seedlings. *Trans. Br. Mycol. Soc.* **56:**95–103.

Bruehl, G. W. (1975). Systems and mechanisms of residue possession by pioneer fungal colonists. In *Biology and Control of Soil-Borne Plant Pathogens* (G. W. Bruehl, ed.). American Phytopathology Society, St. Paul, MN, pp. 77–83.

Brun, H., B. Tivoli, and Carpentier (1976). Influence des caracteristiques des sols sur la manifestation de maladies d'origine tellurique et la mise en oeuvre de méthodes de lutte biologique. *Sci. Agron. Rennes*, **1976:**37–48.

Buck, K. W. (1984). A new double-stranded RNA virus from *Gaeumannomyces graminis. J. Gen. Virol.*, **65:**987–990.

Buck, K. W., M. R. Almond, J. J. P. McFadden, M. A. Romanos, and C. J. Rawlinson (1981). Properties of thirteen viruses and virus variants obtained from eight isolates of the wheat take-all fungus, *Gaeumannomyces graminis* var. *tritici. J. Gen. Virol.*, **53:**235–245.

Burr, T. J., M. N. Schroth, and T. Suslow (1978). Increased potato yields by treatment of seedpieces with specific strains of *Pseudomonas fluorescens* and *P. putida. Phytopathology*, **68:**1377–1383.

Castanho, B., E. E. Butler, and R. J. Shepherd (1978). The association of double-stranded RNA with Rhizoctonia decline. *Phytopathology*, **68:**1515–1519.

Chakraborty, S. and H. H. Warcup (1983). Soil amoebae and saprophytic survival of *Gaeumannomyces graminis* var. *tritici* in a suppressive pasture soil. *Soil Biol. Biochem.*, **15:**181–185.

Chambers, S. C. (1970). Pathogenic variation in *Ophiobolus graminis. Aust. J. Biol. Sci.*, **23:**1099–1103.

Chao, W. L., E. B. Nelson, G. E. Harman, and H. C. Hoch (1986). Colonization of the rhizosphere by biological control agents applied to seeds. *Phytopathology*, **76:**60–65.

Chamswarng, C. and R. J. Cook (1985). Identification and comparative pathogenicity of *Pythium* species from wheat roots and wheat-field soils in the Pacific Northwest. *Phytopathology*, **75**:821-827.

Christensen, N. W. and M. Brett (1985). Chloride and liming effects on soil nitrogen form and take-all of wheat. *Agron. J.*, **77**:157-163.

Christensen, N. W., R. G. Taylor, T. L. Jackson, and B. L. Mitchell (1981). Chloride effects on water potentials and yield of winter wheat infected with take-all root rot. *Agron. J.*, **73**:1053-1058.

Colyer, P. D. and M. S. Mount (1984). Bacterization of potatoes with *Pseudomonas putida* and its influence on postharvest soft rot diseases. *Plant Dis.*, **68**:703-706.

Cook, R. J. (1981). The effects of soil reaction and physical conditions. In *Biology and Control of Take-All* (M. J. C. Asher and P. J. Shipton, eds.). Academic, New York, pp. 343-352.

Cook, R. J. (1985). Biological control of plant pathogens, with special reference to the take-all fungus in suppressive soils. *Taiwan Plant Prot. Bull.* **27**:179-198.

Cook, R. J. and K. F. Baker (1983). *The Nature and Practice of Biological Control of Plant Pathogens.* American Phytopathological Society, St. Paul, MN.

Cook, R. J. and T. Naiki (1982). Virulence of *Gaeumannomyces graminis* var. *tritici* from fields under short-term and long-term wheat cultivation in the Pacific Northwest, USA. *Plant Pathol.*, **31**:201-207.

Cook, R. J. and A. D. Rovira (1976). The role of bacteria in the biological control of *Gaeumannomyces graminis* by suppressive soils. *Soil Biol. Biochem.*, **8**:269-273.

Cook, R. J., H. T. Wilkinson, and J. R. Alldredge (1986). Evidence that microorganisms in suppressive soil associated with wheat take-all decline do not limit the number of lesions produced by *Gaeumannomyces graminis* var. *tritici. Phytopathology*, **76**:342-345.

Cunningham, P. C. (1966a). Effects of ploughing and sowing dates on behavior of soil-borne pathogens of spring wheat. *Ir. J. Agric. Res.*, **5**:215-236.

Cunningham, P. C. (1966b). Influence of nitrogen on the incidence of foot and root rots of spring-sown cereals. *Ir. J. Agric. Res.*, **5**:63-78.

Deacon, J. W. (1976). Biological control of the take-all fungus, *Gaeumannomyces graminis*, by *Phialophora radicicola* and similar fungi. *Soil Biol. Biochem.*, **8**:275-283.

Deacon, J. W. and C. M. Henry (1980). Age of wheat and barley roots and infection by *Gaeumannomyces graminis* var. *tritici. Soil Biol. Biochem.*, **12**:113-118.

Fellows, H. and C. H. Ficke (1934). Report on wheat take-all in cereal and forage crop disease investigations. *Rep. Kans. Exp. Stn.*, **7**:1932-1934.

Ferraz, J. F. P. (1973). Influence of the soil atmosphere on spread of *Ophiobolus graminis* along wheat roots. *Trans. Br. Mycol. Soc.*, **61**:237-249.

Ferraz, J. F. P. (1979). Effect of direct-drilling on infection of wheat plants by *Gaeumannomyces graminis* var. *tritici. Agron. Lusitan.*, **39**:343-360.

Foster, R. C. and G. D. Bowen (1982). Plant surfaces and bacterial growth: the rhizosphere and rhizoplane. In *Phytopathogenic Prokaryotes* (M. S. Mount and G. H. Lacy, ed.). Academic, London, pp. 159-185.

Frick, L. J. and R. M. Lister (1978). Serotype variability in virus-like particles from *Gaeumannomyces graminis. Virology*, **85**:504-517.

Garrett, S. D. (1937). Soil conditions and the take-all disease of wheat. II. The relation between soil reaction and soil aeration. *Ann. Appl. Biol.*, **24**:747-751.

Garrett, S. D. (1938). Soil conditions and the take-all disease of wheat. III. Decomposition of the resting mycelium of *Ophiobolus graminis* in infected wheat stubble buried in the soil. *Ann. Appl. Biol.*, **25**:742-766.

Garrett, S. D. (1940). Soil conditions and the take-all disease of wheat. V. Further experiments on the survival of *Ophiobolus graminis* in infected wheat stubble buried in the soil. *Ann. Appl. Biol.*, **27**:199-204.

Garrett, S. D. (1948). Soil conditions and the take-all of wheat. IX. Interaction between host plant nutrition, disease escape, and disease resistance. *Ann. Appl. Biol.*, **35:**14–17.

Garrett, S. D. (1966). Cellulose-decomposing ability of some cereal foot-rot fungi in relation to their saprophytic survival. *Trans. Br. Mycol. Soc.*, **49:**57–68.

Garrett, S. D. (1970). *Pathogenic Root-Infecting Fungi*. Cambridge University Press, London, 294 pp.

Garrett, S. D. (1976). Influence of nitrogen on cellulolysis rate and saprophytic survival in soil of some cereal foot-rot fungi. *Soil Biol. Biochem.*, **8:**229–234.

Garrett, S. D. (1981). Introduction. In *Biology and Control of Take-All* (M. J. C. Asher and P. J. Shipton, eds.). Academic, New York, pp. 1–11.

Garrett, S. D. (1983). Weight losses of unsterilized filter-paper caused by colonies of cereal foot-rot fungi. *Trans. Br. Mycol. Soc.*, **81:**421–423.

Garrett, S. D. (1985). Effect of soil texture on microbial abbreviation of saprophytic survival by the take-all fungus of wheat. *Proc. Indian Acad. Sci. (Plant Sci.)*, **94:**85–90.

Geels, F. P. and B. Schippers (1983). Reduction of yield depressions in high frequency potato cropping soil after seed tuber treatments with antagonistic fluorescent *Pseudomonas* spp. *Phytopathol. Z.*, **108:**207–214.

Gerlagh, M. (1968). Introduction of *Ophiobolus graminis* into new polders and its decline. *Neth. J. Plant Pathol.*, **74:**(Suppl. 2):1–97.

Gilligan, C. A. (1980). Dynamics of root colonization by the take-all fungus, *Gaeumannomyces graminis. Soil. Biol. Biochem.*, **12:**507–512.

Glynne, M. D. (1935). Incidence of take-all on wheat and barley on experimental plots at Woburn. *Ann. Appl. Biol.*, **22,** 225–235.

Gurusiddaiah, S., D. M. Weller, A. Sarkar, and R. J. Cook (1986). Characterization of an antibiotic produced by a strain of *Pseudomonas fluorescens* inhibitory to *Gaeumannomyces graminis* var. *tritici* and *Pythium* spp. *Antimicrob. Agents Chemother.* **29:**488–495.

Halsey, M. E. and R. L. Powelson (1981). "Bacterization" seed treatment and banded fertilizers for suppression of take-all root rot of wheat. *Oregon Agric. Exp. Sta. Spec. Rep.*, **663:**41–46.

Hashiba, T., Y. Homma, M. Hyakumachi, and I. Matsuda (1984). Isolation of a DNA plasmid in the fungus *Rhizoctonia solani. J. Gen. Microb.*, **130:**2067–2070.

Henry, A. W. (1932). Influence of soil temperature and soil sterilization on the reaction of wheat seedlings to *Ophiobolus graminis* Sacc. *Can. J. Res.*, **7:**198–203.

Homma, Y., J. W. Sitton, R. J. Cook, and K. M. Old (1979). Perforation and destruction of pigmented hyphae of *Gaeumannomyces graminis* by vampyrellid amoebae from Pacific Northwest wheat field soils. *Phytopathology*, **69:**1118–1122.

Hornby, D. (1969). Method of investigating populations of the take-all fungus (*Ophiobolus graminis*) in soil. *Ann. Appl. Biol.*, **64:**503–513.

Hornby, D. (1975). Inoculum of the take-all fungus: Nature, measurement distribution and survival. *EPPO Bull.*, **5:**319–333.

Hornby, D. (1979). Take-all decline: A theorist's paradise. In *Soil-Borne Plant Pathogens* (B. Schippers and W. Gams, eds.). Academic, London, pp. 133–156.

Howell, C. R. and R. D. Stipanovic (1979). Control of *Rhizoctonia solani* on cotton seedlings with *Pseudomonas fluorescens* and with an antibiotic produced by the bacterium. *Phytopathology*, **69:**480–482.

Howell, C. R. and R. D. Stipanovic (1980). Suppression of *Pythium ultimum*-induced damping-off of cotton seedlings by *Pseudomonas fluorescens* and its antibiotic, pyoluteorin. *Phytopathology*, **70:**712–715.

Howie, W. J. (1985). Factors affecting colonization of wheat roots and suppression of take-all by pseudomonads antagonistic to *Gaeumannomyces graminis* var. *tritici.* Ph.D. Thesis, Washington State University, Pullman, 82 pp.

Howie, W. J. and E. Echandi (1983). Rhizobacteria: influence of cultivar and soil type on plant growth and yield of potatoes. *Soil Biol. Biochem.*, **15:**127–132.

Huber, D. M. (1981). The role of nutrients and chemicals. In *Biology and Control of Take-All* (M. J. C. Asher and P. J. Shipton, eds.). Academic, New York, pp. 317–341.

Huber, D. M., C. G. Painter, H. C. McKay, and D. L. Peterson (1968). Effect of nitrogen fertilization on take-all of winter wheat. *Phytopathology*, **58:**1470–1472.

Jenkinson, D. S. (1966). Studies on the decomposition of plant material in soil. II. Partial sterilization of soil and the soil biomass. *J. Soil. Sci.*, **17:**280–302.

King, E. O., M. K. Ward, and D. E. Raney (1954). Two simple media for the demonstration of pyocyanin and fluorescein. *J. Lab. Clin. Med.*, **44:**301–307.

Kloepper, J. W. and M. N. Schroth (1981). Relationship of in vitro antibiosis of plant growth-promoting rhizobacteria to plant growth and the displacement of root microflora. *Phytopathology*, **71:**1020–1024.

Kloepper, J. W., J. Leong, M. Teintze, and M. N. Schroth (1980a). Enhanced plant growth by siderophores produced by plant growth-promoting rhizobacteria. *Nature*, **286:**885–886.

Kloepper, J. W., J. Leong, M. Teintze, and M. N. Schroth (1980b). *Pseudomonas* siderophores: A mechanism explaining disease suppressive soils. *Curr. Microbiol.*, **4:**317–320.

Kloepper, J. W., M. N. Schroth, and T. D. Miller (1980c). Effects of rhizosphere colonization by plant growth-promoting rhizobacteria on potato plant development and yield. *Phytopathology*, **70:**1078–1082.

Lapierre, H., J. M. Lamaire, B. Jouan, and G. Molin (1970). Mise en évidence de particules virales associées á une perte de pathogenicité chez le Piétin-échaudage des céréales, *Ophiobolus graminis* Sacc. C. R. Hebd. Séanc. *Acad. Sci. Paris D*, **271:**1833–1836.

Leben, C. (1983). Association of *Pseudomonas syringae* pv. *lachrymans* and other bacterial pathogens with roots. *Phytopathology*, **73:**577–581.

Leisinger, T. and R. Margraff (1979). Secondary metabolites of the fluorescent pseudomonads. *Microbiol. Rev.*, **43:**422–442.

Lemaire, J. M. and M. Coppenet (1968). Influence de la succession céréalière sur les fluctuations de la gravité du Piétin-échaudage (*Ophiobolus graminis* Sacc.) *Ann. Épiphyt.*, **19:**589–599.

Lemaire, J. M., H. Lapierre, B. Jouan, and G. Bertrand (1970). Découverte de particules virales chez certaines souches *d'Ophiobolus graminis*, agent du Piétin-échaudage des céréales: consequences agronomique prévisibles. *C. R. Hebd. Seanc. Acad. Agric. Fr.*, **56:**1134–1138.

Lemaire, J. M., B. Jouan, B. Perraton, and M. Sailly (1971). Perspectives de lutte biologique contre les parasites des céréales d'origine tellurique en particulier *Ophiobolus graminis* Sacc. *Sci. Agron. Rennes*, **1971:**1–8.

Lemaire, J. M., F. Carpentier, J. F. Dalle, G. Doussinault, B. Perraton, and C. Moule (1979). Lutte biologique contre le Piétin-échaudage des céréales. Modifications physiologiques chez le Blé inoculé par une souche atténuée d'*Ophiobolus graminis*. 1. Précocité accrue aux premiers stades du Blé. *C. R. Acad. Agric. Fr.*, **65:**766–772.

Lester, E. and P. J. Shipton (1967). A technique for studying inhibition of the parasitic activity of *Ophiobolus graminis* (Sacc.) in field soils. *Plant Pathol.*, **16:**121–123.

Loper, J. E., T. V. Suslow, and M. N. Schroth (1984). Log normal distribution of bacterial populations in the rhizosphere. *Phytopathology*, **74:**1454–1460.

Loper, J. E., C. Haack, and M. N. Schroth (1985). Population dynamics of soil pseudomonads in the rhizosphere of potato (*Solanum tuberosum* L.). *Appl. Environ. Microbiol.*, **49:**416–422.

MacNish, G. C. (1973). Effect of mixing and sieving on incidence of *Gaeumannomyces graminis* var. *tritici* in field soil. *Aust. J. Biol. Sci.*, **26:**1277–1283.

MacNish, G. C. (1980). The use of ammonium nitrogen to reduce take-all (*Gaeumannomyces graminis* var. *tritici*) in wheat. *APPS Natl. Pathol. Conf.*, **4:**27.

MacNish, G. C. and R. L. Dodman (1973). Survival of *Gaeumannomyces graminis* var. *tritici* in the field. *Aust. J. Biol. Sci.*, **26:**1309–1317.

Madsen, E. L. and M. Alexander (1982). Transport of *Rhizobium* and *Pseudomonas* through soil. *Soil. Sci. Soc. Am. J.*, **46:**557–560.

McGinty, R. M., J. J. P. McFadden, C. J. Rawlinson, and K. W. Buck (1984). Widespread inhibitor production in culture by isolates of *Gaeumannomyces graminis* var. *tritici. Trans. Br. Mycol. Soc.*, **82:**429–434.

McFadden, J. J. P., K. W. Buck, and C. J. Rawlinson (1983). Infrequent transmission of double-stranded RNA virus particles but absence of DNA proviruses in single ascospore cultures of *Gaeumannomyces* graminis. *J. Gen. Virol.*, **64:**927–937.

Misaghi, I. J., L. J. Stowell, R. G. Grogan, and L. C. Spearman (1982). Fungistatic activity of water-soluble fluorescent pigments of fluorescent pseudomonads. *Phytopathology*, **72:**33–36.

Moore, K. J. and R. J. Cook (1984). Increased take-all of wheat with direct drilling in the Pacific Northwest. *Phytopathology*, **74:**1044–1049.

Naiki, T. and R. J. Cook (1983a). Factors in loss of pathogenicity in *Gaeumannomyces graminis* var. *tritici. Phytopathology*, **73:**1652–1656.

Naiki, T. and R. J. Cook (1983b). Relationship between production of a self-inhibitor and inability of *Gaeumannomyces graminis* var. *tritici* to cause take-all. *Phytopathology*, **73:**1657–1660.

Neate, S. M. (1984). Minimum cultivation and root disease of wheat. Ph.D. Thesis, The University of Adelaide, Adelaide, Australia.

Parke, J. L., A. D. Rovira, and G. D. Bowen (1984). Soil matric potential affects colonization of wheat roots by a pseudomonad suppressive to take-all. *Phytopathology*, **84:**806 (Abstract).

Ponchet, J. (1962). Etude des facteurs qui conditionnent le developpement du pietin-echaudage: *Linocarpon cariceti* B. *Br. Ann. Epiphyt.*, **13:**151–165.

Pope, A. M. S. and R. M. Jackson (1973). Effects of wheatfield soil on inocula of *Gaeumannomyces graminis* (Sacc.) Arx and Olivier var. *tritici* J. Walker in relation to take-all decline. *Soil Biol. Biochem.*, **5:**881–890.

Rawlinson, C. J. and K. W. Buck (1981). Viruses in *Gaeumannomyces* and *Phialophora* spp. In *Biology and Control of Take-All* (M. J. C. Asher and P. J. Shipton, eds.). Academic, New York, pp. 151–172.

Rawlinson, C. J., D. Hornby, V. Pearson, and J. M. Carpenter (1973). Virus-like particles in the take-all fungus, *Gaeumannomyces graminis. Ann. Appl. Biol.*, **74:**197–209.

Reis, E. M., R. J. Cook, and B. L. McNeal (1982). Effect of mineral nutrition on take-all of wheat. *Phytopathology*, **72:**224–229.

Reis, E. M., R. J. Cook, and B. L. McNeal (1983). Elevated pH and associated reduced trace-nutrient availability as factors contributing to increased take-all of wheat upon soil liming. *Phytopathology*, **73:**411–413.

Ridge, E. H. (1976). Studies on soil fumigation–II effects on bacteria. *Soil Biol. Biochem.*, **8:**249–253.

Romanos, M. A., C. J. Rawlinson, M. R. Almond, and K. W. Buck (1980). Production of fungal growth inhibitors by isolates of *Gaeumannomyces graminis* var. *tritici. Trans. Br. Mycol. Soc.*, **74:**79–88.

Rouatt, J. W. and H. Katznelson (1961). A study of the bacteria on the root surface and in the rhizosphere soil of crop plants. *J. Appl. Bact.*, **24:**164–171.

Rovira, A. D. and C. B. Davey (1974). Biology of the rhizosphere. In *The Plant Root and its Environment* (E. W. Carson, ed.). University Press of Virginia, Charlottesville, pp. 153–204.

Rovira, A. D. and N. R. Venn (1985). Effect of rotation and tillage on take-all and Rhizoctonia root rot in wheat. In *Ecology and Management of Soilborne Plant Pathogens* (C. A. Parker, A. D. Rovira, K. J. Moore, P. T. W. Wong, and J. F. Kollmorgen eds.), pp. 255–258.

Rovira, A. D. and G. B. Wildermuth (1981). The nature and mechanisms of suppression. In *Biology and Control of Take-All* (M. J. C. Asher and P. J. Shipton, eds.). Academic, New York, pp. 385–415.

Sands, D. and A. D. Rovira (1971). *Pseudomonas fluorescens* biotype G, the dominant fluorescent pseudomonad in South Australian soils and wheat rhizospheres. *J. Appl. Bact.*, **34:**261–275.

Schaffnit, E. (1930). Ertragseinbussen im Getreidebau durch Fusskrankheiten. *Mitt. Deutsch. Landwirtsch. Ges.*, **45:**247–251.

Scher, F. M. and R. Baker (1980). Mechanism of biological control in a Fusarium-suppressive soil. *Phytopathology*, **70:**412–417.

Scher, F. M. and R. Baker (1982). Effect of *Pseudomonas putida* and a synthetic iron chelator on induction of soil suppressiveness to Fusarium wilt pathogens. *Phytopathology*, **72:**1567–1573.

Schroth, M. N. and J. G. Hancock (1981). Selected topics in biological control. *Annu. Rev. Microbiol.*, **35:**453–476.

Schroth, M. N. and J. G. Hancock (1982). Disease-suppressive soil and root-colonizing bacteria. *Science*, **216:**1376–1381.

Scott, P. R. (1969). Control of survival of *Ophiobolus graminis* between consecutive crops of winter wheat. *Ann. Appl. Biol.*, **63:**37–43.

Scott, P. R. (1970). *Phialophora radicicola*, an avirulent parasite of wheat and grass roots. *Trans. Br. Mycol. Soc.*, **55:**163–167.

Scott, P. R. (1981). Variation in host susceptibility. In *Biology and Control of Take-All* (M. J. C. Asher and P. J. Shipton, eds.). Academic, New York, pp. 219–236.

Shipton, P. J. (1972). Take-all in spring-sown cereals under continuous cultivation: disease progress and decline in relation to crop succession and nitrogen. *Ann. Appl. Biol.*, **71:**33–46.

Shipton, P. J. (1975). Take-all decline during cereal monoculture. In *Biology and Control of Soil-Borne Plant Pathogens* (G. W. Bruehl, ed.). American Phytopathology Society, St. Paul, MN, pp. 137–144.

Shipton, P. J. (1977). Monoculture and soilborne plant pathogens. *Ann. Rev. Phytopathol.*, **15:**387–407.

Shipton, P. J., R. J. Cook, and J. W. Sitton (1973). Occurrence and transfer of a biological factor in soil that suppresses take-all of wheat in eastern Washington. *Phytopathology*, **63:**511–517.

Sivasithamparam, K. and C. A. Parker (1978). Effect of certain isolates of bacteria and actinomycetes on *Gaeumannomyces graminis* var. *tritici* and take-all of wheat. *Aust. J. Bot.*, **26:**773–782.

Sivasithamparam, K. and C. A. Parker (1981). Physiology and nutrition in culture. In *Biology and Control of Take-All* (M. J. C. Asher and P. J. Shipton, eds.). Academic, New York, pp. 125–150.

Slope, D. B. (1966). Effect of paraquat on take-all. *Rep. Rothamsted Exp. Sta.*, **1965:**125.

Slope, D. B., G. A. Salt, E. W. Broom, and R. J. Gutteridge (1978). Occurrence of *Phialophora radicicola* var. *graminicola* and *Gaeumannomyces graminis* var. *tritici* on roots of wheat in field crops. *Ann. Appl. Biol.*, **88:**239–246.

Smiley, R. W. (1978a). Antagonists of *Gaeumannomyces graminis* from the rhizoplane of wheat in soils fertilized with ammonium- or nitrate-nitrogen. *Soil Biol. Biochem.*, **10:**169–174.

Smiley, R. W. (1978b). Colonization of wheat roots by *Gaeumannomyces graminis* inhibited by specific soils, microorganisms and ammonium-nitrogen. *Soil Biol. Biochem.*, **10:**175–179.

Smiley, R. W. (1979). Wheat-rhizoplane pseudomonads as antagonists of *Gaeumannomyces graminis*. *Soil Biol. Biochem.*, **11:**371–376.

Smiley, R. W. and R. J. Cook (1973). Relationship between take-all of wheat and rhizosphere pH in soils fertilized with ammonium vs. nitrate-nitrogen. *Phytopathology*, **63:**882–890.

Sneh, B., M. Dupler, Y. Elad, and R. Baker (1984). Chlamydospore germination of *Fusarium oxysporum* f. sp. *cucumerinum* as affected by fluorescent and lytic bacteria from Fusarium-suppressive soil. *Phytopathology*, **74:**1115–1124.

Soby, S. and K. Bergman (1983). Motility and chemotaxis of *Rhizobium mililoti* in soil. *Appl. Environ. Microbiol.*, **46:**995–998.

Speakman, J. B. and B. G. Lewis (1978). Limitation of *Gaeumannomyces graminis* by wheat root responses to *Phialophora radicicola*. *New Phytol.*, **80:**373–380.

Sprague, R. (1950). *Diseases of Cereals and Grasses in North America*. Ronald Press, New York, 538 pp.

Stolp, H. and P. Gadkari (1981). Nonpathogenic members of the genus *Pseudomonas*. In *The Prokaryotes: A Handbook on Habitats, Isolation, and Identification of Bacteria*, Vol. I (M. P. Starr, H. Stolp, H. G. Trüper, A. Balows, and H. G. Schlegel eds.). Springer-Verlag, Berlin, pp. 719–741.

Stumbo, C. R., P. L. Gainey, and F. E. Clark (1942). Microbiological and nutritional factors in the take-all disease of wheat. *J. Agric. Res.*, **64:**653–665.

Suslow, T. V. (1982). Role of root-colonizing bacteria in plant growth. In *Phytopathogenic Prokaryotes*, Vol. I. (M. S. Mount and G. H. Lacy eds.). Academic, London, pp. 187–223.

Suslow, T. V. and M. N. Schroth (1982a). Role of deleterious rhizobacteria as minor pathogens in reducing crop growth. *Phytopathology*, **72:**111–115.

Suslow, T. V. and M. N. Schroth (1982b). Rhizobacteria of sugar beets: Effects of seed application and root colonization on yield. *Phytopathology*, **72:**199–206.

Taylor, R. G., T. L. Jackson, R. L. Powelson, and N. W. Christensen (1983). Chloride, nitrogen form, lime, and planting date effects on take-all root rot of winter wheat. *Plant Dis.*, **67:**1116–1120.

Teintze, M. and J. Leong (1981). Structure of pseudobactin A, a second siderophore from plant growth promoting *Pseudomonas* B10. *Biochemistry*, **20:**6457–6462.

Teintze, M., M. B. Hossain, C. L. Barnes, J. Leong, and D. van der Helm (1981). Structure of ferric pseudobactin, a siderophore from a plant growth promoting *Pseudomonas. Biochemistry*, **20:**6446–6457.

Tivoli, B., J. M. Lemaire, and B. Jouan (1974). Prémunition du Blé contre *Ophiobolus graminis* Sacc. Par des souches peu agressives du même parasite. *Ann. Phytopathol.*, **6:**395–406.

Trolldenier, G. (1981). Influence of soil moisture, soil acidity and nitrogen source on take-all of wheat. *Phytopathol. Zeit.*, **102:**163–177.

Vandenbergh, P. A., C. F. Gonzalez, A. M. Wright, and B. S. Kunka (1983). Iron-chelating compounds produced by soil pseudomonads: correlation with fungal growth inhibition. *Appl. Environ. Microbiol.* **46:**128–132.

Vojinović, Ž. D. (1970). Jedna metoda za odredivanje gustine populacije antagonista prema *Ophiobolus graminis* Sacc. u. zemljištu. *Zašt. Bilj.*, **110:**365–372.

Vojinović, Ž. D. (1972a). Biological antagonism as the cause of decline of *Ophiobolus graminis* Sacc. in prolonged wheat monoculture. *J. Sci. Agric. Res.*, **25:**31–41.

Vojinović, Ž. D. (1972b). Antagonists from soil and rhizosphere to phytopathogens. Final Technical Report, Institute of Soil Science, Beograd, Yugoslavia, 130 pp.

Vojinović, Ž. D. (1973). The influence of micro-organisms following *Ophiobolus graminis* Sacc. on its further pathogenicity. *Org. Eur. Med. Prot. Plantes Bull.*, **9:**91–101.

Vraný, J., V. Vančura, and M. Staněk (1981). Control of microorganisms in the rhizosphere of wheat by inoculation of seeds with *Pseudomonas putida* and by foliar application of urea. *Folia Microbiol.*, **26:**45–51.

Weller, D. M. (1983). Colonization of wheat roots by a fluorescent pseudomonad suppressive to take-all. *Phytopathology*, **73:**1548–1553.

Weller, D. M. (1984). Distribution of a take-all suppressive strain of *Pseudomonas fluorescens* on seminal roots of winter wheat. *Appl. Environ. Microbiol.*, **48**:897–899.

Weller, D. M. (1985). Application of fluorescent pseudomonads to control root diseases. In Ecology and Management of Soilborne Plant Pathogens (C. A. Parker, A. D. Rovira, K. J. Moore, P. T. W. Wong, and J. F. Kollmorgen, eds.). The American Phytopathological Society, St. Paul, pp. 137–140.

Weller, D. M. and R. J. Cook (1981). Pseudomonads from take-all conducive and suppressive soils. *Phytopathology*, **71**:264 (Abstract).

Weller, D. M. and R. J. Cook (1983). Suppression of take-all of wheat by seed treatments with fluorescent pseudomonads. *Phytopathology*, **73**:463–469.

Weller, D. M. and M. C. Graham (1984). Application of fluorescent pseudomonads to improve the growth of wheat. *Phytopathology*, **74**:806 (Abstract).

Weller, D. M., W. J. Howie, and R. J. Cook (1985a). Relationship of in vitro inhibition of *Gaeumannomyces graminis* var. *tritici* and in vivo suppression of take-all by fluorescent pseudomonads. *Phytopathology*, **75**:1301 (Abstract).

Weller, D. M., B.-X. Zhang, and R. J. Cook (1985b). Application of a rapid screening test for selection of bacteria suppressive to take-all of wheat. *Plant Dis.*, **69**:710–713.

Wildermuth, G. B. and A. D. Rovira (1977). Hyphal density as a measure of suppression of *Gaeumannomyces graminis* var. *tritici* on wheat roots. *Soil Biol. Biochem.*, **9**:203–205.

Wildermuth, G. B., A. D. Rovira, and J. H. Warcup (1979). Mechanism and site of suppression of *Gaeumannomyces graminis* var. *tritici* in soil. In *Soil-Borne Plant Pathogens* (B. Schippers and W. Gams, eds.). Academic, London, pp. 157–164.

Wilkinson, H. T., D. M. Weller, and J. R. Alldredge (1982). Enhanced biological control of wheat take-all when inhibitory *Pseudomonas* strains are introduced on inoculum or seed as opposed to directly into soil. *Phytopathology*, **72**:948–949 (Abstract).

Wilkinson, H. T., R. J. Cook, and J. R. Alldredge (1985a). Relation of inoculum size and concentration to infection of wheat roots by *Gaeumannomyces graminis* var. *tritici. Phytopathology*, **75**:98–103.

Wilkinson, H. T., J. R. Alldredge, and R. J. Cook (1985b). Estimated distances for infection of wheat roots by *Gaeumannomyces graminis* var. *tritici* in soils suppressive and conducive to take-all. *Phytopathology*, **75**:557–559.

Wong, P. T. W. and R. Baker (1984). Suppression of wheat take-all and Ophiobolus patch by fluorescent pseudomonads from a Fusarium-suppressive soil. *Soil Biol. Biochem.*, **16**:397–403.

Wong, P. T. W. and R. J. Southwell (1980). Field control of take-all of wheat by avirulent fungi. *Ann. Appl. Biol.*, **94**:41–49.

Xu, G.-W., and D. C. Gross (1986a). Selection of fluorescent pseudomonads antagonistic to *Erwinia carotovora* and suppressive of potato seed peice decay. *Phytopathology*, **76**:414–422.

Xu, G.-W, and D. C. Gross (1986b). Field evaluations of the interactions among fluorescent pseudomonads, *Erwinia cartovora*, and potato yields. *Phytopathology*, **76**:423–430.

Yarham, D. J. (1981). Practical aspects of epidemiology and control. In *Biology and Control of Take-All* (M. J. C. Asher and P. J. Shipton, eds.). Academic, New York, pp. 353–384.

Yarham, D. J. and J. M. Hirst (1975). Diseases in reduced cultivation and direct-drilling systems. *Eur. Metiterr. Plant Prot. Organ. Bull.*, **5**:287–296.

Zogg, H. and A. Fossati (1966). Studien über die biologische Bodenentseuchung. VI. Untersuchungen über die biologische Aktivität der Mikrofloren aus verschiedenen Bodenprofilen und Bodentiefen gegenüber *Ophiobolus graminis* Sacc. *Phytopath. Z.*, **57**:47–58.

Zogg, H. and W. Jäggi (1974). Studies on the biological soil disinfection. VII. Contribution to the take-all decline (*Gaeumannomyces graminis*) imitated by means of laboratory trials and some of its possible mechanisms. *Phytopath. Z.*, **81**:160–169.

4

SOIL SOLARIZATION

JAACOV KATAN

Department of Plant Pathology and Microbiology, The Hebrew University of Jerusalem, Faculty of Agriculture, Rehovot 76100 Israel

1. INTRODUCTION

This technique for the control of soilborne plant pests (pathogens, weeds, and arthropods) refers to a new approach for soil disinfestation by means of solar energy (Katan et al., 1976). At present, covering (tarping or mulching) the soil with transparent polyethylene, when appropriate climatic conditions prevail, is the only means for capturing solar energy to heat soil under field conditions. However, other techniques for heating the soil (or plant material) by solar energy, practised for decades and centuries ago, for example, by directly exposing them to the sun during the summer season (Avidov, 1956; Grooshevoy, 1939; Raghaven, 1964), can also be regarded as solarization for pest control. As early as 1939, Grooshevoy, a Russian plant pathologist, succeeded in establishing solar heating of the soil as a means of pathogen control. Unfortunately, plastic technologies were not available at that time. The use of polyethylene (or any suitable plastic material) as a *preplanting* soil treatment allows better controlled and more effective solarization. Moreover, only polyethylene soil cover enables the solarization of *wet* soil making a tremendous improvement because (1) when wet, resting structures are usually much more sensitive to heat (and many other killing agents), (2) heating is much better in wet soil (Mahrer et al., 1984), and (3) appropriate moisture promotes biological activity in the soil that may lead to beneficial processes that enhance pest control.

The basic process involved in soil solarization is the heating of the soil to relatively mild levels, usually 36–50°C in the upper 30 cm. Though heat is the major killing agent, the continuously accumulating evidence for the involvement of biological processes may explain the surprisingly good control sometimes achieved by solarization, even when the temperatures attained were not sufficiently high to justify such control (Davis and Sorensen, 1984).

In 1976 Katan et al. (1976), reporting successful results using "solar heating by polyethylene mulching" in Israel (based on observations by extension workers and farmers in the Jordan Valley), suggested the principles of this new approach to pathogen control. In 1977 American plant pathologists from Davis, California, used this approach to successfully control *Verticillium dahliae* (Pullman and DeVay 1977), thus denoting its possible wide applicability. Indeed, within less than a decade, over 100 articles, reviews, and abstracts on solarization, dealing with both the fundamental and practical aspects, reported the successful control of many pathogens, weeds, and mites in Israel, the United States, Japan, Greece, Jordan, Morocco, Italy, Australia, Egypt, Iraq, Portugal, Pakistan, the United Kingdom, South Africa, France, Spain, India, Mexico, and Peru (Table 2; Katan, 1981; Stapleton and DeVay, 1985b; other references in this article and unpublished data). This indicates not only the effectiveness of solarization, in spite of such limitations as its dependence on climate, but also the vital need for new nonchemical approaches to pest control. Soil solarization is commercially used in Israel, the United States (Ashworth et al., 1983), and Japan (Horiuchi, 1984).

It is only natural that in an era when the use of plastic in agriculture (e.g., plastic houses, soil mulching, insect control) was dramatically increasing and expanding, many investigators were independently inspired to proceed one step further towards the direct use of plastic for solarization to control pests. In 1960 Waggoner et al. (1960) published a comprehensive review on plastic mulching and also referred to possible pest control aspects. In 1971 Adams (1971) described studies on soil heating by polyethylene, during plant growth, to control *Thielaviopsis basicola*. According to Horiuchi (1984), studies on solarization, especially in greenhouses, developed in Japan independently of the work carried out in Israel.

To solarize is defined in Webster's *New International Dictionary* (1951) as "to expose to sunlight; to affect in some way by the action of the sun's rays." Of the many terms used since 1976, for example, solar heating of the soil, polyethylene or plastic mulching (tarping), solar pasteurization, and solar disinfestation, we prefer soil solarization, which was introduced by the American plant pathologists (Pullman et al., 1981b) because it is both widely accepted and concise.

Because a number of reviews describing soil solarization have already been published (Borges, 1982; Horiuchi, 1984; Katan, 1980, 1981, 1985; Pullman et al., 1984; Stapleton and DeVay, 1985b), in this chapter, after summarizing the principles and data on pest control, recent findings and strategies for further development are emphasized.

2. SOIL SOLARIZATION AS A SOIL DISINFESTATION METHOD

Soil solarization is a soil disinfestation practice. In principle, the goal is the eradication of soilborne inocula before planting. Solarization may, therefore, have to various degrees both the advantages and the disadvantages of traditional soil disinfestation. To ascertain effective killing of inocula at all niches and to the desired soil depth, soil disinfestation is carried out by drastic means. The possible short- and long-term negative side effects exerted on the chemical, physical, and biological properties of the soil is a major issue when evaluating any soil disinfestation method. Especially relevant, is the possible damage to beneficial soil microorganisms (e.g., mycorrhizae) and to antagonists, thus creating "a biological vacuum" (Domsch, 1964) which facilitates rapid reinfestation.

All agricultural practices disturb the biological equilibria in the plant ecosystem. Even in the most primitive agricultural system, plants of a single species are more crowded, homogeneous, and frequently grown on the same land as compared to a natural plant community. Thus the build-up of pest populations resulting from frequent cropping began in the early days of agriculture. Certain modern agricultural practices, such as the monoculture of highly profitable crops, aggravate this build-up, while others, for examples effective pest control practices, delay it. Of the latter, soil disinfestation is extreme; when adequately applied, it very effectively eradicates pests but, at the same time, may disturb the biological equilibria in the soil. The main goal of soil disinfestation is to control major soilborne plant pathogens. However, the elimination of accompanying detrimental agents, which have been accumulating in the soil throughout its long history of cropping, is an additional goal that might be simultaneously achieved (Katan, 1984). Minor pathogens (also referred to as weak parasites or deleterious soil microorganisms) and toxic substances, produced directly by soil microorganisms or originating from decomposed plant residues, are examples of such agents. The inhibitory or stimulatory effects of soil disinfestation on the detrimental and beneficial biotic and abiotic agents in the soil, are difficult to analyze and evaluate and are often overlooked. All these interactions are finally expressed as either increased or decreased plant growth concomitantly with pest control.

In the early days of modern plant pathology, two basic approaches to soil disinfestation were developed: heating (by steam) and fumigation (first by CS_2 and later by other chemicals). Soil solarization by polyethylene is a much younger approach to soil disinfestation and must be evaluated as such. Many of the concepts dealt with here are also relevant, with certain adaptations, to other methods of soil disinfestation.

3. THE PRINCIPLES OF SOIL SOLARIZATION

The major principles of soil solarization are:

1. Solarization heats the soil through repeated daily cycles. The maximal temperatures attained decrease with increasing soil depths.

2. The soil is tarped (mulched) with thin transparent polyethylene sheets or other plastic material (e.g., vinyl chloride). Any easily handled material which enables effective and economic heating of the soil under field conditions might

be used. A closed glasshouse (or plastic house), if climatic conditions allow and the soil is wet, is another version of solarization.

3. The best time for soil tarping, when climatic conditions are the most favorable, can be experimentally determined by tarping the soil and measuring the temperatures. Meteorological data from previous years and predictive models (see discussion given below) further facilitate this task.

4. At increasing soil depths, maximal temperatures are reached later in the day and maintained for longer periods of time. Typical temperatures of solarized soils from Israel are shown in Fig. 1. Higher soil temperatures were achieved in other climatic regions.

5. Extending the solarization period enables control at deeper soil layers (Table 1).

6. Adequate soil moisture during solarization, for reasons specified earlier, can be obtained by irrigation shortly before tarping (single irrigation) (Grinstein et al., 1979a,b). Additional irrigations during solarization (by a drip system) are usually not necessary, except for very light soils, and may lower soil temperatures, unless carried out during the night. (Naot, 1984). Good results were

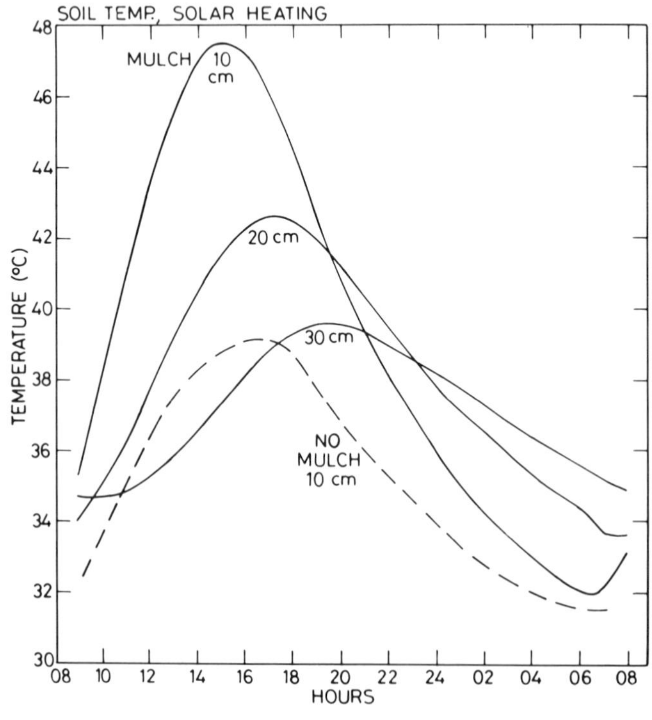

Fig. 1. The daily course of soil heating by polyethylene at three soil depths, as compared to nonsolarized (no mulch) soil at a depth of 10 cm. Typical results obtained during July–August in Rehovot, Israel.

TABLE 1. Length of Solarization Period Required for Killing 90–100% of *Verticillium dahliae* Sclerotia at Various Soil Depths[a]

Soil Depth[b] (cm)	*Number of Days for 90–100% Killing*
10	3–6
30	14–20
40	20–30
50	30–42
60	35–60
70[b]	35–60

[a]Natural sclerotia were buried under polyethylene or in nonsolarized plots. Samples were removed after various periods and the percent killed (as compared to the control) was determined. Data compiled from experiments carried out at various locations in Israel in July–August, 1979–1984.

[b]Data from the United States showed effective control of this pathogen also at lower soil depths.

achieved by the simpler single irrigation technique (Grinstein et al., 1979a,b; Jacobsohn et al., 1980; Katan et al., 1980, 1983). However, solarization controlled *V. dahliae* in a pistachio grove with and without sprinkler preirrigation before mulching (Ashworth and Gaona, 1982).

7. The temperatures attained in solarized soils are much lower than in steam-disinfested ones. Therefore, negative side effects associated with the latter technique are less likely to occur with solarization—though they should not be excluded.

4. PEST CONTROL AND YIELD INCREASE

Many experiments have been carried out and reported since 1976 to evaluate the potential of soil solarization in pathogen (and other pest) population reduction, disease control, and yield increase. Most of the studies were conducted under field conditions; therefore, their relevance and applicability to realistic conditions are high. This fortunate situation (not typical of many studies on nonconventional methods for disease control) was dictated since solarization can be carried out only under field conditions. The number of soilborne pests (pathogens, weeds, and mites) effectively controlled by solarization is very large and outnumbers that of uncontrolled pests (Table 2 and Fig. 2).

The ideal approach in studying pathogen and disease control is to record both pathogen populations (usually at various periods during solarization and at various soil depths) and disease level in solarized and nonsolarized plots. The detailed information thus obtained can be easily and reliably applied to other situations.

In some studies only pathogen populations were recorded. Inoculum density is a major but not a sole factor in determinating disease level and is much easier

TABLE 2. Effect of Soil Solarization on Soilborne Pests—Selected Examples

Pests Controlled	*References*
FUNGI	
Bipolaris sorokiniana	Smith et al., 1984
Didymella lycopersici	Besri, 1983
Fusarium oxysporum f.sp. *fragariae*	Horiuchi, 1984; Kodama and Fukai, 1982
F. oxysporum f.sp. *melonis*	Katan et al., 1981
F. oxysporum f.sp. *vasinfectum*	Katan et al., 1983
Orobanche spp.	Horowitz et al., 1983; Jacobsohn et al., 1980
Phytophthora cinnamomi	Pinkas et al., 1984
Pyrenochaeta lycopersici	Goisque et al., 1984; Katan 1981; Malathrakis et al., 1983; Tamietti and Garibaldi, 1981; Tjamos, 1984
Pyrenochaeta terrestris	Katan et al., 1980; Rabinowitch et al., 1981
Pythium ultimum	Pullman et al., 1981a,b
Sclerotinia minor	Porter and Merriman, 1983, 1985
Sclerotinia sclerotiorum	Porter and Merriman, 1985
Sclerotium cepivorum	El Yamani et al., 1983; Porter and Merriman, 1983
Sclerotium oryzae	Usmani and Ghaffar, 1982
Sclerotium rolfsii	Grinstein et al., 1979; Mihail and Alcorn, 1984
Verticillium dahliae	Ashworth and Gaona, 1982; Katan et al., 1976; Pullman et al., 1981a,b
NEMATODES	
Ditylenchus dipsaci	Siti et al., 1982
Globodera rostochiensis	Lamondia and Brodie, 1984
Helicotylenchus diaonicus	Stapleton and DeVay, 1983
Heterodera trifolii	Hadar et al., 1983
Meloidogyne hapla	Stapleton and DeVay, 1983
Meloidogyne javanica	Porter and Merriman, 1983; Cenis, 1984
Pratylenchus thornei	Grinstein et al., 1979b
MITE	
Rhizoglyphus robini	Gerson et al., 1981
WEEDS[a]	
Amaranthus spp., *Amsinckia douglasiana*, *Avena fatua*, *Chaenopodium* spp., *Cynodon dactylon*, *Digitaria sanguinalis*, *Echinochloa crus-galli*, *Lamium amplexicaule*, *Portulaca oleracea*, *Senecio vulgaris*, *Solanum halpense*, *Xanthium pensylvanicum*	Egley, 1983; Horowitz et al., 1983; Katan, 1981; Pullman et al., 1984; Rubin and Benjamin, 1984; Standifer et al., 1984; Stapleton et al., 1985; Stapleton and DeVay, 1985b

TABLE 2. (*continued*)

	References
Pests Partially Controlled, Not Controlled, or Controlled with Inconsistent Results[b]	
PATHOGENS	
F. oxysporum f.sp. *pini*	Old, 1981
Macrophomina phaseolina	Mihail and Alcorn, 1984; Old, 1981; Sheikh and Ghaffar, 1984
Meloidogyne	Katan, 1981; Overman, 1981
Monosporascus eutypoides	Reuveni et al., 1983
Paratylenchus neoamblycephalis	Stapleton and DeVay, 1983
Plasmodiophora brassicae	Horiuchi et al., 1982; Myers et al., 1983; Porter and Merriman, 1985; White and Buczaki, 1979
WEEDS	
Conyza canadensis, Cyperus rotundus, Malva niceaensis, Melilotus alba	Egley, 1983; Horowitz et al., 1983; Katan, 1981; Pullman et al., 1984; Rubin and Benjamin, 1984

[a]Details on many additional weeds that can be controlled by solarization are given in the references cited above.

[b]In certain studies, some of these pathogens were effectively controlled by solarization.

Fig. 2. Effect of soil solarization on the control of Egyptian broomrape in a carrot field 100 days after sowing. The nonsolarized plots were entirely destroyed (from Jacobsohn et al., 1980).

to assess. The value of this parameter for predicting *disease* control by solarization depends on our ability to relate inoculum density and potential to disease incidence. Another factor, though not related to the fate of inoculum during solarization, that should be considered, is the possibility of reinfestation, which will affect disease incidence in the future. The possibility of reinfestation of previously solarized soil depends, among other things, on agricultural practices, duration of the growth season, and whether the solarized soil becomes conducive or suppressive. Data based solely on inoculum density may, therefore, not actually predict disease control by solarization.

Special care should be taken when interpreting results from studies on pathogen population carried out in noninfested soils inoculated with either natural or artificial inocula. The heat sensitivity, behavior in the soil, and density of the artificial inoculum may differ from those of natural inoculum. Following population levels of a *natural* inoculum added to the soil (providing it was done using the right resting structures) usually involves much less labor than following natural pathogen populations and might provide more options and information.

Field microplots facilitate the work, but involve two limitations: (1) if they are too small (less than 2 × 3 m), solar heating efficiency is reduced, especially at the edges of the mulch (Mahrer and Katan, 1981); (2) reinfestation is difficult to evaluate.

With certain pathogens, solarization effectively reduced the populations in the soil profile at considerable soil depths. For example, *V. dahliae* populations were controlled at depths of 70–120 cm (Table 1, Ashworth and Gaona, 1982; Pullman et al., 1981b) *Phytophthora cinnamomi* to a 70-cm depth (Pinkas et al., 1984) and the nematodes *Paratrichodorus porosus*, *Pratylenchus hamatus*, and *Pratylenchus vulnus* to the depths of 46, 46, and 91 cm, respectively (Stapleton and DeVay, 1983).

Studies on solarization for controlling pathogens in simulation systems are very rare, because the daily course of heating by solarization at each soil depth is difficult to simulate in soil under controlled conditions. The determination of heat mortality curves, essential for studying the mechanisms of pathogen killing in the solarized soils, cannot be used for simulating heating by solarization because it involves heating at constant temperatures (usually in baths).

Studies on pathogen and disease control, conducted under different climatic and edaphic conditions, may yield various degrees of control of the same pathogen. However, certain pathogens (e.g., *V. dahliae* and *Pyrenochaeta lycopersici*, Table 2), were repeatedly reported as sensitive to solarization. Such pathogens might be used as a "reference" when testing solarization in a climatic region for the first time. Reproducible results, a prerequisite for developing any control measure against soilborne pathogens, are especially important for a method that is climate-dependent. The lengthy and tedious repetition of experiments under varying conditions and in different years, the main procedure for assessing reproducibility, should not be avoided or shortened.

As expected, disease and pest control have often been accompanied by a yield increase (Table 1 in Katan 1981, and others), which sometimes was infinitive, since yield in the nontreated control was entirely destroyed by the disease (Jacobsohn et al., 1980; Sitti et al., 1982). Increased yield by solarization may depend on (1) damage caused by the disease, (2) level of soil infestation, (3) effectiveness of disease control, (4) "compensation" by neighboring plants, as shown with Fusarium wilt in cotton (Katan et al., 1983), and (5) increased growth response phenomenon, beyond the control of the target pest (see discussion given below). Solarization may also improve the quality of products (Davis and Sorensen, 1984; Grinstein et al., 1979; Rabinowitch et al., 1985) and storage quality of onions (unpublished data).

As expected, certain pathogens and weeds cannot be controlled by solarization (Table 2). This may be due to high heat resistance, rapid recolonization, high sensitivity of the crop, and other factors.

4.1. Weed Control

Weed control, one of the very visible results of solarization, can even serve as an indicator of its success. Some solarization studies deal solely with weed control (Table 2). Annual weeds are usually more sensitive than perennials. The possible mechanisms of weed control by solarization are (1) thermal killing of seeds, (2) thermal killing of seeds induced to germinate, (3) breaking seed dormancy and consequent killing of the germinating seed, and (4) biological control through weakening or other mechanisms. Volatiles that accumulate under the mulch (Horowitz et al., 1983; Rubin and Benjamin, 1984) may play a role in direct killing or may affect dormancy or germination. Weed seeds or vegetative propagules, able to emerge from deeper layers, may escape the higher temperatures (Rubin and Benjamin, 1984). Weeds differ in their response to solarization, as expressed by the length of time required for obtaining an effective control (Horowitz et al., 1983) or by other parameters (Standifer et al., 1984). Purple nutsedge, usually not controlled by solarization, may even increase (Egley, 1983).

Solarization for weed control has certain advantages over herbicides in that (1) there is no residual effect, (2) certain weeds are killed at deep layers, and (3) there is a long-term effect usually only achieved upon applying highly persistent herbicides.

5. LONG-TERM EFFECT

The long-term effect of solarization on disease control and/or yield increase, extending for a second or even the third crop, was observed in various regions, with a variety of pathogens and crops, for example, Fusarium wilts of cotton and melon; Verticillium wilts of cotton, peanuts, and potatoes; corky root of tomatoes; and pink root of onions (Davis and Sorensen, 1984; Garibaldi and Tamietti, 1984; Goisque et al., 1984; Katan et al., 1981, 1983; Osman and Sahab, 1984; Pullman et al., 1981b; Rabinowitch et al., 1985). The incidence of Fusarium wilt of cotton, even in the third year after solarization, was significantly lower than in the nontreated plots, even though the solarized plots were exposed to heavy reinfestation from surrounding soil (Katan et al., 1983). Such a pronounced long-term effects, with a variety of pathogens, was not reported with other disinfestation methods. This means that either this phenomenon is unique to solarization, because of its very nature, or it was not studied with the other methods. Long-term disease control requires pronounced reduction in inoculum density, together with a shift in the biological equilibrium in favor of the antagonists, which retard soil reinfestation. Both phenomena are described herein. The long-term effect possibly indicates that as a rule solarization does not create a biological vacuum. It also has economic implications, that is, reducing the cost per crop accordingly. Combining solarization with other methods, or using it at earlier stages of inoculum build-up may enable a better and longer-lasting disease control that may require less frequent application.

Another approach to the long-term study of solarization is its continuous application in the same plot. Continuous application of solarization for 3–4 yr did not produce negative side effects with onions (Rabinowitch et al., 1985) or strawberries (unpublished data). On the contrary, its performance was somewhat improved in certain cases. However, as a rule, to avoid building up unexpected undesirable biotic or abiotic agents in the soil, continuous application of *any* control method in the same soil is not recommended. Alternations or combinations with other methods may not only improve control but also reduce the risks of such hazards.

6. MECHANISMS INVOLVED IN DISEASE CONTROL AND YIELD INCREASE

Soil solarization, in its present form, involves hydrothermal processes (Stapleton and DeVay, 1985), simultaneously causing many changes in the biotic and abiotic components of the soil, during and after solarization, which may finally lead to a change in disease, plant growth and yield, or both.

Solarized soils undergo significant changes in their temperature and moisture regimes, the inorganic and organic composition of their solid, liquid, and gaseous phases and their physical structure, all of which in turn affect the biotic components. The ecological model described by Park (1963) for soilborne pathogens and used for the analysis of pesticide-nontarget disease interactions (Katan and Eshel, 1973) is adapted to analyze the mode of action of solarization.

Disease severity and plant growth, at a given situation, are determined by the six possible interactions [each being either negative (detrimental) or positive (beneficial)] between the three living components coexisting in the soil (Fig. 3): the host plant (its resistance to the pathogen and capacity to produce yield), the pathogen [inoculum density (ID), inoculum potential (IP) and its virulence], and the soil microorganisms (their population density and activities toward pathogens and plants). Each of these living components interacts with the abi-

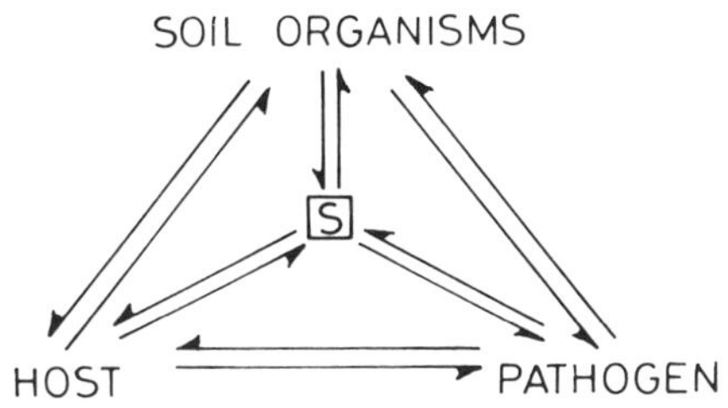

Fig. 3. The interactions between the three biotic components involved in a soilborne disease; S represents an external factor introduced into the system.

otic environment. Solarization, as well as any introduced agent, may interact with any of the biotic and abiotic components, creating a new temporary or lasting balance, which leads to an increase, decrease, or no change in disease or plant growth. Changes in the major component involved (e.g., pathogen population in heavily infested soil) do not necessarily denote the final results. Reduction in ID of various pathogens by many control measures (including solarization) does usually correlate with a corresponding reduction in disease incidence. However, in some cases, a reduction in pathogen ID or growth caused by an external factor was accompanied by increased disease incidence owing to stronger negative effects on host resistance or antagonistic activity (Cook and Baker, 1983; Katan and Eshel, 1973; Leach, 1947). Interpreting the relationship between changes in one component and plant yield is even more complicated, for reasons previously mentioned.

Assessing ID changes is usually the major tool for analyzing mechanisms of disease control by disinfestation. Unfortunately, detailed microbial analysis of the disinfested soil is much less used in such studies. The possibility that the third component, the plant host, is also affected should not be excluded, since microorganisms, which affect plant metabolism, may be stimulated or suppressed. In addition, significant changes in the mineral content of the soil (e.g., Ca^{2+}), which occur in disinfested soils, affect plant resistance.

Only three major aspects of the mechanisms of disease control and yield increase by solarization are dealt with below.

6.1. The Physical Effect: Thermal Inactivation of Pathogens

Heat, at temperatures exceeding the maximal for growth, has an inhibitory or lethal effect on organisms. Exposing the pathogen to elevated temperatures is an important but not exclusive mechanism of pathogen and pest control by solarization. Separate analyses of the physical and biological effects of solarization on pathogens, though much simpler, do not represent a realistic situation since in nature pathogen propagules do not exist in axenic conditions. Of the enormous amount of literature dealing with heat sensitivity of organisms, only a small fraction, that which deals with heating at mild temperatures (35–50°C) for long periods of time, might be relevant to soil solarization. When analyzing the physical effect of solarization on ID and IP of the pathogen and on disease control the following should be taken into consideration:

1. The thermal death rate of a population of an organism depends on both the temperature level and exposure time, which are inversely related. Various methods were developed for evaluating the thermal effect on organisms and for the quantitative definition of the time–temperature relationships and the inherent heat sensitivity of organisms. Certain thermal death curves have an exponential nature which indicates that a constant fraction of the population is killed per unit of time. In some cases, a linear relationship was revealed by plotting logarithms of time required to kill 90% of populations of various pathogenic fungi against temperature (Pullman et al., 1981a) (Fig. 4). A more detailed description of the various methods employed has been published (Dunn et al., 1985; Katan, 1981; Pullman et al., 1981a).

2. Propagules, which survive sublethal heating, may be partially damaged or weakened. However, measuring the reduced viability of a propagule and deter-

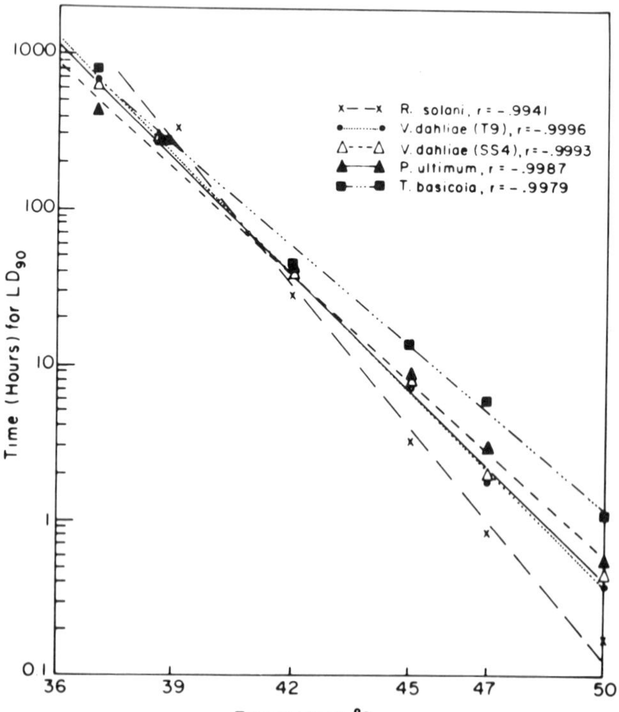

Fig. 4. Time and temperature exposures required to kill 90% of propagules of various fungal pathogens (from Pullman et al., 1981a).

mining its consequence by denoting its reduced survivability and pathogenicity is both an important and a difficult task. Potential parameters for assessing reduced viability are rate and speed of germination and enzymatic activity. Counts of colony-forming units on plates cannot provide such information, since in axenic cultures many weakened propagules may also form colonies. Some of the heated propagules may fully recover, while others may further deteriorate until death. Thus ID estimation measured shortly after heating may not represent the ID or IP at a later stage.

3. The course and pattern of heating during soil solarization vastly differ from those usually established for heat mortality curves under controlled conditions since with soil solarization propagules are subjected to varying temperatures in daily cycles in a split-alternate heating mode, which contrasts with constant heating at one temperature, the standard procedure under controlled conditions. For example, during 30 days of solarization, propagules at a depth of 10 cm are exposed to a total of approximately $30 \times A$ hr at 46°C + $30 \times B$ hr at 45°C + $30 \times C$ hr at 44°C, and so on. This obviously complicates the analysis of the net effect of temperature. A suitable elaborate computation is necessary to account for the cumulative effect of the fluctuations in temperature during the day.

4. The extent of the damage inflicted on the propagule depends on its inherent heat sensitivity and on environmental conditions, for example, moisture level, the protective effect of the soil, inoculum density, quality and age, nutritional conditions, and the presence of toxic substances.

5. When pathogen-infested soil is solarized or heated, three processes may simultaneously occur: propagule viability is reduced, propagule vulnerability to potential antagonists is increased (Henis and Papavizas, 1983; Lifshitz et al., 1983), and the activity of the antagonist is affected. Because heat mortality curves refer only to the first phenomenon, and for many other reasons specified above, the reliability of such curves as a tool for predicting the effectiveness of solarization should be evaluated in each and every case.

6.2. Biological Control in the Solarized Soils

Microbial activities occur in all agricultural soils and therefore a natural biological control, in various manners and to varying degrees, apparently operates in those soils, including the pathogen-infested ones. Since many agricultural practices, including many pest-control measures, affect soil microbial activities, a consequent stimulation or suppression of the natural biological control should be expected—a fact frequently overlooked. Soil disinfestation in general drastically reduces soil microbial activities and, therefore, often causes a biological vacuum. However, there were cases, especially when mild fumigants or doses were used, in which antagonists were stimulated and contributed to pathogen control; for example, the resistance of *Armillaria mellea* to *Trichoderma* was weakened by CS_2 (Garrett, 1956).

In the first study on solarization the possible involvement of biological processes that would explain the unexpected, pronounced and lasting control was raised and discussed (Katan et al., 1976). Later studies, carried out by various workers and under different conditions, basically supported that hypothesis.

Biological control, if generated by soil solarization (or any other soil disinfestation), may operate at one or both of the following stages (Katan, 1981):

1. Affecting, during or after solarization, ID, IP, or saprophytic activity of the pathogen *existing* in the infested soil.
2. Affecting inoculum introduced to the soil *after* solarization, from deeper soil layers, adjacent infested soil, or external sources.

Most studies on mechanisms of conventional biological control, soil disinfestation, or disease control of soilborne pathogens emphasize stage 1. Though eradicating the primary inoculum is a prerequisite for disease control, studying the fate and behavior of inoculum introduced to previously treated soil is crucial for the evaluation of disease control *throughout* the growing season, since under field conditions reinfestation can be restricted but not totally avoided. Studies on soil suppressiveness are especially relevant to the mechanisms of stage 2.

6.2.1. Biological Control of the Existing Pathogens' Inocula by Solarization

a. Microbial suppression or killing of those propagules previously weakened by sublethal heating. Sublethal heating increased the leakage of water-soluble organic compounds from *S. rolfsii* sclerotia, increased their microbial colonization and the frequency of surface cracks, and resulted in decreased disease incidence (Lifshitz et al., 1983). Sublethal heating affected the survival of *Armillaria mellea* (Munneke et al., 1976), since less heating was required for indirect killing of the pathogen. *Trichoderma* was the dominant colonizer of the heated infected roots. The weakening effect is, apparently, a widespread phenomenon, since whenever an eradication measure is employed, a zone of a lower and marginal dosage should exist between the treated and nontreated area. A better picture of this phenomenon might be obtained by assessing not only the number but also the growth capacity of the propagules, as well as ID and IP analyzed at various periods after the termination of solarization (or disinfestation). For example, the population of the nematode *Helicotylenchus dignonicus* was significantly reduced 3 months after solarization, but not immediately after treatment (Stapleton and DeVay, 1983). There are several major questions regarding the weakening effect. How is it determined? What is a "partially alive" propagule? Is there a "no-return point" of sublethal damage, after which recovery is impossible? Is sublethal damage cumulative? Which combinations of sublethal agents are synergistic, which are additive, and what would be the best sequence, if such combinations are used? These questions are also relevant for developing integrated control systems.

b. Diminished fungistasis. A partial or complete nullification of fungistasis in the absence of the host is regarded as harmful to resistant resting structures, which, in such a situation, are rendered as germinating propagules, vulnerable to microbial action. Biological control, especially by organic amendments, aims at achieving such a situation. Solarization reduced fungistasis toward *S. rolfsii* in various soils (Greenberger et al., 1985) and toward *Fusarium* and *Penicillium frequentatus* in preheated soils (Dobbs and Hinson, 1953; Katan et al., 1976). Ashworth and Gaona (1982) suggested a hypothesis of suicidal germination of *V. dahliae* microsclerotia by a substance (s) that accumulates under the mulch. Solarization may reduce fungistasis through suppression of competitors, release of mineral and organic nutrients in the soil, or nullification of possible inhibitory substances. Though the extent of this phenomenon of reduced fungistasis in solarized soils is unknown, controlling it may improve solarization under marginal conditions.

c. Constant soil moisture provided by the plastic tarp. This ascertains better conditions for the activity of the natural antagonists in the soil and hence an improved biological control. The partial anaerobic conditions prevailing in moist mulched soils (Horiuchi, 1984; Rubin and Benjamin, 1984; Stapleton and DeVay, 1984) may also have an impact on biological control.

d. Volatiles. The mulch cover seals the soil and causes an accumulation of volatiles such as CO_2, ethylene, and possibly other substances (Greenberger et al., 1984; Horowitz et al., 1983; Rubin and Benjamin, 1984). Volatile compounds toxic to *S. rolfsii* and *Fusarium*, accumulated in heated soils and microorganisms, producing antifungal toxic volatiles, were isolated from solarized and heated soils and sclerotia (Greenberger et al., 1984). Decreased nematode populations were observed in moist, plastic-covered but shaded (nonheated) soil; hence both the high soil moisture and the accumulation of volatiles may contribute to nematode control (Stapleton and DeVay, 1983). This aspect offers many options for improving solarization, such as using mulch that is less permeable to volatiles or amending with suitable organic matter before mulching.

e. A shift in the microbial composition, in favor of antagonists, is detrimental to pathogens. In certain cases, populations of the potential antagonist *Trichoderma* increased in the solarized soil (Elad et al., 1980). Pathogenic isolates of *F. oxysporum* f. sp. *vasinfectum* in infected root pieces buried in solarized soil were replaced by saprophytic ones during solarization (Katan et al., 1983). The inoculum potential of *Rhizoctonia solani* and its buildup was suppressed by solarization (Elad et al., 1980). Extensive studies on microbial changes in the soil, during and after solarization, by Stapleton and DeVay (1982, 1984) show that while populations of many microorganisms were reduced immediately following solarization, actinomycetes, thermophilic and thermotolerant fungi, and *Bacillus* spp. were less affected or even increased. Fluorescent pseudomonads (known as beneficial root colonizers) quickly colonized solarized soils and plant roots. The percentage of gram-positive bacteria exhibiting *in vitro* antibiosis increased 20-fold in solarized soil. Antagonists (e.g., lytic and antibiotic-producing microorganisms) were frequently stimulated in solarized soils (Greenberger et al., 1985). Solarization reduced the populations of many fungi; however, tolerant fungi such as *Talaromyces* survived the treatment (Kodama et al., 1980). Speculation was raised regarding the reasons for the frequent stimulation of beneficial microorganisms in heated soils. Temperatures in solarized soils are not too high, allowing many saprophytes, some of which are potential antagonists, to survive (Baker and Cook, 1974; Stapleton and DeVay, 1982). It appears that soil solarization also causes changes in soil biota and substrate in favor of organisms with greater saprophytic ability, usually saprophytes, rather than phytopathogens, which have more specialized growth requirements (Stapleton and DeVay, 1985b).

6.2.2. The Fate of Pathogen Introduced in Solarized Soils

Studies have frequently shown that solarized soils become hostile and less receptive to pathogen reinfestation. This can be regarded as an induced suppressiveness (Greenberger et al., 1985), similar to the natural one (Cook and Baker, 1983). The incidence of *Fusarium*, *Sclerotium rolfsii*, and *Verticillium* diseases was lower in solarized and subsequently inoculated soils than in nontreated inoculated ones, while populations of lytic and antibiotic microorganisms were higher. The suppression of chlamydospore formation was frequently observed (Greenberger et al., 1985). A similar phenomenon of induced suppressiveness in solarized soils was found with Fusarium wilt of carnation (Hardy and Sivasithamparam, 1985) and *Phytophthora cinnamomi* (Pinkas et al., 1984). Kodama et al. (1980) observed that the rate of development of Fusarium yellows in strawberries planted in mildly heated soil was much less pronounced than in fumigated soil. They suggest that many thermal-tolerant microorganisms survive in soils solarized in closed vinyl houses, and that the newly established microflora prevent reinfestation. Baker and his associates advocated using lower temperatures, which are less detrimental to many saprophytes, for heat disinfestation of soils, thus reducing the chance of reinfestation (Baker, 1962; Olson and Baker, 1968; and others). Baker (1983) described cases in which mild soil heating preserved or even increased its suppressiveness. Bollen (1974) described four sources of reinfesting microflora in steamed soil, discussing the feasibility of controlling recolonization in such soils to enable a rapid rebuild-up of a stable antagonism. The widespread solarization-induced suppressiveness, somewhat similar to the natural one but less specific, should not be regarded as a universal phenomenon. The possibility that in certain soils a conduciveness might be induced should not be excluded. In Japan solarization was not effective against certain soil bacteria; in fact the treatment sometimes even increased disease incidence (Horiuchi, 1984). Populations of the nematode *Paratylenchus hamatus* drastically reduced immediately after solarization, and increased greatly 7 months later (Stapleton and DeVay, 1983). This, however, was an exception to the rule. The capacity of soils to develop or nullify suppressiveness upon solarization or, for that matter, any disinfestation method, must be determined.

6.3. Increased Growth Response (IGR) Beyond Pest Control

The phenomenon of plant growth enhancement in disinfested soils in the absence of known pests, discovered at the end of the last century, has since been repeatedly reported with all disinfestation methods, including solarization (Chen and Katan, 1980; Katan, 1981; Stapleton and DeVay, 1982, 1983, 1985; Stapleton et al., 1985; and others). The term "absence of known pests" is somewhat vague and difficult to evaluate. Some cases [e.g., replanting problems (Sewell, 1984), soil sickness, and yield decline] caused by complexes of biotic and abiotic agents, and treated by disinfestation, are on the border line between typically defined soilborne diseases and those cases of IGR in noninfested soils. Some IGR cases might be found to result from the control of hitherto unknown pests which could not be identified by the procedures available at that time. To draw rigid lines between these categories is difficult, if necessary at all. The mechanisms for explaining IGR are either chemical (release of mineral nutrients or growth factors and nullification of toxic products) or biological (elimination of minor pathogens and stimulation of beneficial microorganisms) or both.

Similar to other disinfestation methods, solarization increased the concentrations of many minerals—though not always consistently (Chen and Katan, 1980; Stapleton et al., 1985). Higher concentrations of soluble organic substances were found in extracts of solarized soils and the growth of tomato seedlings in these extracts was enhanced (Chen and Katan, 1980).

The increase in plant growth in the studies described above ranged from a few to several hundred percent—depending on the soil, the plant, and the parameter used. The effect on yield components were not always analyzed. The correlation between various parameters of plant growth and yield is very important in this context, but it is beyond the scope of this review. The yield of onions in a solarized field with negligible levels of known pathogens was increased by solarization by 29% (Rabinowitch et al., 1985).

Increased growth response has very important economic implications which should be taken into account when considering the use of solarization. The major difficulty is that we cannot predict whether and to what extent a soil will respond with an IGR. Developing predictive methods for this purpose and developing measures to further increase IGR will improve the economic benefit of solarization.

Increased growth rate definition much depends on that of a healthy plant. It is now widely accepted that plant health is a much broader concept than merely disease control. As Browning (1983) states: "Plant health is far more than the opposite of plant disease as used in plant pathology. Plant health describes the relative freedom of the green plant and its ecosystem from biotic and abiotic stresses that limit its producing to the maximum of its genetic potential over time." Is the more developed plant in a solarized (or disinfested) soil a healthier one? Interpreting Garrett's (1956) conclusion that "the normal crop is one that suffers an appreciable and regular degree of root damage by parasites," leads us to a positive answer.

7. COMBINING WITH OTHER METHODS—INTEGRATED CONTROL

There is an increasing tendency in plant protection sciences to integrate various methods of control. The expected benefits are an improved, more lasting, wider spectrum of control; a reduced dosage (or shorter period for solarization) without decreased effectiveness; and the possibility of using solarization in marginal cases (e.g., cooler regions). Such a combination may result in either a synergistic or an additive effect. We should search for the synergistic one. The weakening phenomenon described above is such a mechanism leading to synergism. Soil solarization might be, theoretically, combined with many biological, chemical (especially at reduced dosages), and cultural methods of control. However, combining with biological control agents is especially promising and tempting. Combining the antagonist *T. harzianum* with heating or solarization in *Rhizoctonia solani*-infested soils improved disease and pathogen control and delayed inoculum build-up (Chet et al., 1982; Elad et al., 1980). Combining solarization with vapam resulted in a synergistic effect in controlling delimited shell spots of peanut pods (Frank et al., 1986). Combining solarization with a suitable crop sequence resulted in an improved and longer-lasting control of Fusarium wilt of cotton (Katan et al., 1983). Combining with fumigants (possibly at reduced dosages) also provides a potential combination. An effective control of the corky root disease of tomatoes was obtained by combining solarization with a reduced dosage of methyl bromide (Tjamos, 1984).

8. SOLARIZATION IN COOLER REGIONS AND SEASONS

Various attempts have been made to adapt solarization in climatically marginal regions or periods of the year. Combining with other methods of control, as discussed in the previous section, is a promising approach for this purpose.

An interesting approach developed independently in Italy, Japan, and other countries, is mulching the soil with polyethylene or vinyl *inside* closed glasshouses or plastic houses for intensifying soil heating. In Northern Italy soil solarization in the open field was not effective, but when it was carried out inside glasshouses it controlled corky root of tomatoes in two successive crops (Tamietti and Garibaldi, 1981; Garibaldi and Tamietti, 1984). Mulching inside plastic houses was less effective. This approach was extensively developed in Japan, especially for the control of Fusarium in strawberries (Horiuchi, 1984; Horiuchi et al., 1982; Kodama and Fukui, 1982; Kodama et al., 1980). In certain cases solarization was combined with organic amendments or the fertilizer calcium cyanamide.

A third approach would be to look for those pathogens that can be controlled by solarization even under cooler conditions, either because of their heat sensitivity or for other reasons. Davis and Sorensen (1984) found that solarization effectively controlled Verticillium wilt disease in potatoes in Idaho under marginal conditions of temperatures, and disease control lasted for two successive years.

The fourth approach would be the use, when available, of novel plastic materials that are more efficient in heating.

9. SOLARIZATION OF PERENNIAL CROPS

Using solarization in existing plantings of trees, as a post-plant treatment for controlling soilborne pathogens, is a departure from the standard preplanting use of solarization for annual crops. This approach was pioneered by Ashworth and his associates and was tested for controlling Verticillium wilt in established pistachio nut groves (Ashworth and Gaona, 1982). The prerequisites for achieving an effective control by solarization in an existing orchard (providing that the pathogen is a priori sensitive to solarization) (Ashworth and Gaona, 1982; Katan, 1981) are (1) solarization does not cause root damage owing to elevated temperatures, (2) the shaded area does not consist of a large proportion of the treated area and does not considerably reduce heating efficiency, (3) the pathogen is killed at the desired soil depth, and (4) when pathogen population is effectively reduced upon solarization, the possibility of salvaging trees infected before solarization still depends on whether symptom expression in these trees in the following years requires reinfection from soil inoculum. We can assume that solarization in a young orchard (where the shaded area is still small and the infection is not yet in an advanced stage) will be more effective than in an older one. Since various species of trees differ in their heat sensitivity and in their interaction with pathogens, drawing general rules for solarization in trees is difficult. Solarization controlled *V. dahliae* in a pistachio plantation to the depth of 120 cm. Complete mulching reduced the disease by 75% and was better than a noncontinuous mulch (Ashworth and Gaona, 1982). The temperature did not account for the control of the pathogen at the deeper soil layers. Though solarization was not successful in all cases, 6000–8000 acres of trees in California have been treated on a commercial basis and special equipment was developed (Ashworth et al., 1983). A remarkable control of Verticillium in olive trees in Greece, on a small scale, was reported by Tjamos (1983). Post-plant solarization of peach seedlings improved plant growth (Stapleton and DeVay, 1985a).

It appears that the use of solarization in existing orchards has great prospects, but its realization requires continuous and tedious work. Solarization also has been used as a preplanting control means for orchards. The frequently observed phenomena, long-term effect and induced suppressiveness, make solarization a good potential candidate for preplanting treatment, though some doubts were raised regarding this possibility (Ashworth et al., 1983). Only long-term study can settle these misgivings.

10. UNUSUAL USES OF SOIL SOLARIZATION

1. Besri (1983) developed a very unique and efficient use of solarization. He solarized tomato supports, which are the source of inoculum for *Didymella lycopersici* in tomatoes. Solarization proved to be simple, inexpensive, and effective

in controlling the disease. Adapting this approach to other cases, for example, eradication of infected plant material left in the field, is worth considering.

2. Solarization of plant-growth media and potting mixes, some of which are difficult to fumigate, is also useful. Solarization of tuff medium used for growing plants in greenhouses in Israel resulted in an increased yield of cucumbers (unpublished data).

3. The use of old polyethylene (i.e., previously used) provides an extremely inexpensive method for solarization. Surprisingly, Avissar et al. (1986b) found that used polyethylene heats even more effectively than new material. This was attributed to changes in the photometric properties of the aged mulch (Avissar et al., 1986a,b). Effectiveness of used polyethylene was shown in an another study (Al Raddad, 1979).

4. There is a variety of potential uses of solarization for various purposes which have not yet been fully explored—nurseries, home gardens, landscape improvement, improving soil structure, and others.

11. PESTICIDE BEHAVIOR IN SOLARIZED AND DISINFESTED SOILS

The literature on studies dealing with the fate and biodegradation of pesticides in agricultural soils is voluminous. Only very few studies were carried out in disinfested soils. Such soils differ in many aspects from soils without disinfestation, especially regarding microbial activity, and therefore, the fate of a pesticide here might be different. Indeed, the degradation of the fungicide benomyl and the herbicide terbutryn incorporated in a previously solarized (or fumigated) soil was much slower than in a nontreated one (Ahronson et al., 1982; Avidov et al., 1985; Yarden et al., 1985). Thus soil disinfestation might be used for extending the effectiveness of pesticides. Also, the dosage of pesticides (especially herbicides) applied to a previously disinfested soil must be adjusted (mostly reduced) to avoid phytotoxicity. When a pesticide is applied to soil before mulching, it is consequently exposed to the intensive physical, chemical, and biological processes taking place during solarization which can affect its fate in various ways (Ahronson et al., 1982; Rubin and Benjamin, 1983). Since certain pesticides were decomposed more rapidly in such a situation, an interesting possibility was raised, that is, to use solarization as a tool for the decontamination of toxic residues from soils (Ahronson et al., 1982).

12. SOIL TEMPERATURE MODELS

The principles of soil heating have been described (see more details in Katan, 1981; Waggoner et al., 1960). The major factors involved in soil heating are climatic (e.g., solar irradiation intensity, air temperature, air humidity, and wind velocity) and soil properties. The basic equation that describes the energy balance of bare, nonmulched soil is

$$(1 - \alpha) R_s + \epsilon R_L - H - E - S - \epsilon\sigma T_s^4 = 0$$

where α = the albedo (depending on soil characteristics); R_s = incoming solar radiation; ϵ = soil emissivity; R_L = incoming long-wave radiation; H = sensible heat flux (indicating heat transfer to or from the atmosphere); E = latent heat flux (indicating evaporation); S = soil heat flux (indicating heat transfer through the soil); σ = Stephan–Boltzman's constant; and T_s = soil surface temperature. Mahrer (1979) developed a one-dimensional numerical model for predicting soil temperatures in mulched and bare soils at each depth and at each hour of the day. The agreement between observed and predicted soil temperatures was very good with a high accuracy (magnitude of error being at the range of 0.6–1.2°C). Such a model takes into consideration polyethylene properties too, for example, transmissivity values for short-(solar) and long-wave atmospheric infrared radiation. Calculations show that in wet, mulched soils, increased temperatures are due primarily to the reduction of heat loss by evaporation and heat convection during the daytime and partially to the greenhouse effect, that is, preventing part of the long-wave radiation from leaving the ground (owing to the formation of small water droplets). A further improved two-dimensional numerical model for predicting the spatial temperature regime in mulched soil was also developed (Mahrer and Katan, 1981). The results showed that soil heating at the edges of the mulch is lower than in the center (a phenomenon also reported earlier by Jacobsohn et al., 1980). It was also found that a narrow mulch provides less efficient heating than a wider one. Again, the predicted results agree well with the observed ones (Fig. 5). Recently, a numerical one-dimensional model for studying and predicting temperature and moisture regimes during solarization and for studying the interactions between soil moisture and temperature was developed (Mahrer et al., 1984). This model accurately predicted soil and moisture regimes in the two tested soils and showed that soil temperature maxima increased with increasing soil moisture content. These models were also applied to study the efficiency of aged polyethylene in soil heating (Avissar et al., 1986a,b). In conclusion, such models, if accurate, can replace many tedious temperature measurements and enable us to choose the most appropriate time of the year for solarization and evaluating the importance of various factors involved in soil heating. They provide an extremely valuable tool for improving solarization.

13. TECHNOLOGY AND COST BENEFIT CONSIDERATION

Mulching might be carried out either manually or by machine and can be either continuous to cover the whole field or in strips. Though the latter technique is

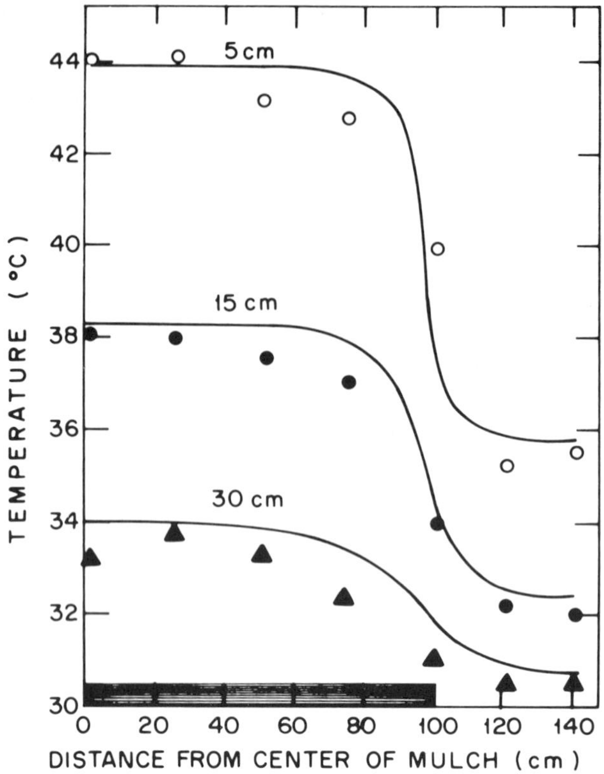

Fig. 5. Observed and predicted average daily maximum soil temperatures at soil depths of 5, 15, and 30 cm, as a function of the distance from the center of a 200-cm wide mulch. Observed values are indicated by open and closed circles or triangles. The polyethylene mulch is represented by a thick band (from Mahrer and Katan, 1981).

easier to apply by machine, it is less effective since soil heating near the mulch edge is reduced because of the border effect described above. Also, the chances of reinfestation are higher; therefore, a long-term effect is possible mainly with continuous mulching. The polyethylene sheets can be connected by welding (Hetzroni et al., 1984) or with glue (Ashworth et al., 1983; Pullman et al., 1984). Machines for removing the plastic material after use have also been developed (Ashworth et al., 1983).

The cost of solarization was estimated as \$200–\$350 per acre, which falls within a medium price range for soil disinfestation (Pullman et al., 1984). The economic feasibility of solarization depends also on income increase, which is affected by yield increase and crop price. Additional factors to consider are the possibility of a long-term effect and benefits such as reduced usage of herbicides, prevention of outbreaks of new diseases, assurance for stable yields, and safety and reduced hazards to the environment. Methods for cost reduction were reviewed by Katan (1981).

14. ADVANTAGES AND LIMITATIONS

The advantages, difficulties, and limitations of solarization have been discussed in many publications. The major advantages are that it is nonchemical and a simple method with apparently no major undesirable disturbances in soil biological balance. Difficulties stem from restriction to certain climatic areas and seasons and land occupation for several weeks. As with *any* control method, solarization should be adapted only to those situations where its effectiveness is obvious and economically feasible. Possible negative side effects should be detected, investigated, and identified. Long-term experiments with continuous solarization are helpful for such a task. Should any undesirable phenomenon be detected in a certain soil, solarization should be either avoided or used with a suitable combination.

Krikun et al. (1982) expressed some reservations regarding solarization. First, they claimed that "many diseases were not controlled" by solarization. Table 2 and the numerous publications referred to in this chapter clearly show that most of the diseases studied were controlled by solarization. Moreover, the concept concerning a method that controls many diseases should not be accepted. We have to consider any new method for control, even if its effectiveness is limited to only one disease of a single crop. Developing a super effective control measure is not feasible and might even be too drastic. Second, the authors stated that solarization is several times more expensive than methyl bromide treatment, which is rather doubtful. As stated many times, solarization (or any disinfestation method) should not be expected to be a magic tool capable of curing every disease without any side effect (or judged as such), but should be regarded as an additional option for control, which in the right circumstances is both effective and economical. In an era of increasing tendency to ban pesticides and great environmental concerns, any new option, even if of limited use, should be welcomed and thoroughly investigated. Since solarization is a pest management method, it suffers from lack of backing by a commercial company which could promote it in large areas. However, it is most encouraging that, in spite of this limitation, solarization has been successfully applied in thousands of acres in Israel and the United States in less than a decade and that so many scientists from many disciplines are concentrating on its study, improvement, and implementation.

15. STRATEGIES FOR THE FUTURE AND GENERAL CONCLUSIONS

The effectiveness of solarization under a variety of conditions, demonstrated by many investigators in five continents, is now well established. Studies on the physical, chemical, and biological changes created by solarization and the mechanisms involved, are carried out in parallel. Many questions raised in the first publications were challenged, some of them were reasonably answered, and naturally, new ones arose. These studies, especially the long-term ones, must be continued and even repeated to confirm the previous results and to open the doors for new ones. Solarization has the features of an integrated pest management method. Its study requires coordinated interdisciplinary research. In this chapter, ideas for future lines of research were suggested and certainly many others exist. In addition, there is a need to put emphasis on certain directions or to pave the way for new ones. Following are four examples.

1. The final stage and the most difficult one in the realization of a new method of control on a large scale is its development and implementation. With pesticides, this is done by well-organized chemical companies. With pest-management methods, this can be achieved only through coordinated efforts by the extension service and other authorities, the growers and the research personnel who have to analyze the causes for possible failures and find ways to overcome difficulties that may arise. This is not the traditional duty of research workers, but it is a most rewarding and stimulating task. Without this, even the best pest-management method may fade away before reaching maturity.

2. Progress in plastic and other technologies related to solarization can revolutionize this technique. We need plastic materials with better heating efficiency, durability, and precise timing of degradation; thinner and therefore less expensive plastic materials with lower permeability to certain volatiles; plastic materials that can be sprayed; and methods for plastic recycling. So far, the plastic industry has concentrated on developing plastics for heating the plants and soil in the cool seasons. We hope that it will also respond to the needs of solarization, that is, novel plastics for heating in hot seasons.

3. Developing means for the use of solarization in climatically marginal cases. This should be attempted even if the present methods for control in those regions are completely satisfactory. We need additional methods both because alternation of methods is most recommended to avoid possible buildup of known and unknown harmful agents, and also for preparing alternative methods for replacement in case of banning.

4. The soil is the arena where the pathogens, their hostile antagonists, and the controlling agent meet and interact. It is very difficult, however to measure the impact of disinfestation on the antagonists, which consist of different groups possessing a variety of antagonistic mechanisms. We have to give higher priority to the investigation of the effect of solarization on the biological components of the soil. Without this, the resultant picture might be distorted. This suggestion, as well as most of the principles discussed here, is relevant and can be adopted to fumigation and other methods for the control of soilborne diseases.

Progress in plant pathology has always been the result of a combination of breakthroughs linked by continuous systematic construction, of concepts and theories cemented with experimental data. Since overall methods of control do not exist, further success in developing solarization and other control methods will also depend on our ability to adequately compromise with what can be achieved while wisely utilizing knowledge obtained.

ACKNOWLEDGMENTS

I wish to express my deep appreciation and thanks to A. Greenberger, A. Grinstein, H. Alon, and many other colleagues and students for cooperating in a joint effort to develop the solarization method. Part of our studies were supported by grants provided by the Israel Ministry of Agriculture, The United States-Israel Binational Agricultural Research and Development Fund, and the United States-Israel Binational Science Foundation, Jerusalem, Israel.

REFERENCES

Adams, P. B. (1971). Effect of soil temperature and soil amendments on Thielaviopsis root rot of sesame. *Phytopathology*, **61:**93–97.

Ahronson, N., B. Rubin, J. Katan, and A. Benjamin (1983). Effects of methyl bromide or solar heating treatment on the persistence of pesticides in the soil. In *Pesticide Chemistry*, Vol. 3 (J. Miyamoto and P. C. Kearny, eds.) pp. 189–194.

Al-Raddad, A. M. M. (1979). Soil disinfestation by plastic tarping. M.Sc. thesis, University of Jordan. 95 pp.

Ashworth, L. J., Jr. and S. A. Gaona (1982). Evaluation of clear polyethylene mulch for controlling Verticillium wilt in established pistachio nut groves. *Phytopathology*, **72:**243–246.

Ashworth, L. J., Jr., D. P. Morgan, S. A. Gaona, and A. H. McCain (1983). Control of Verticillium wilt of pistachio. *Plasticulture*, **58:**33–44.

Avidov, E., N. Aharonson, J. Katan, B. Rubin, and O. Yarden (1985). Persistence of terbutryn and atrazine in soil as affected by soil disinfestation and fungicides. *Weed Sci.*, **33:**457–461.

Avidov, Z. (1956). Bionomics of the tobacco white fly (*Bemisia tabaci*) in Israel. *Ktavim*, **7:**25–41.

Avissar, R., Y. Mahrer, L. Margulies, and J. Katan (1986a). Field ageing of transparent polyethylene mulches: I. Photometric properties. *Soil Soc. Am. J.* **50:**202–205.

Avissar, R., O. Naot, Y. Mahrer, and J. Katan (1986b). Field ageing of transparent polyethylene mulches: II. Influence on the effectiveness of soil heating. *Soil Sci. Am. J.* **50:**205–209.

Baker, K. F. (1962) Principles of heat treatment of soil and planting material. *J. Aust. Inst. Agric. Sci.*, **28:**118–126.

Baker, K. F. (1983). The future of biological and cultural control of plant disease. In *Challenging Problems in Plant Health* (J. Kommedahl and P. H. Williams, eds.). APS, St. Paul, pp. 422–430.

Baker, K. F. and R. J. Cook (1974). *Biological Control of Plant Pathogens.* Freeman, San Francisco, 433 pp.

Besri, M. (1983). Lutte contre le chancre a *Didymella lycopersici* de la tomate par chauffage solaire (Solarisation) des tuteurs. *Phtopathol. Zeitschrift*, **108:**333–340.

Bollen, G. T. (1974). Fungal recolonization of heat-treated glasshouse soils. *Agro-Ecosystems*, **1:**139–155.

Borges, L. V. (1982). Solarizacon do solo novo metodo de pasteurizacao do solo. *Rev. Cienc. Agrar.*, **5:**1–15.

Browning, A. (1983). Goal for plant health in the age of plants: a national plant health system. In *Challenging Problems in Plant Health* (T. Kommedahl and P. H. Willams, eds.). APS, St. Paul, pp. 45-57.

Cenis, J. L. (1984). Control of the nematode *Meloidogyne javanica* by soil solarization. *Proc. Sixth Cong. Mediter. Phytopathol. Union*, Cairo, Egypt, pp. 132.

Chen, Y. and J. Katan (1980). Effect of solar heating of soils by transparent polyethylene mulching on their chemical properties. *Soil Sci.*, **130:**271-277.

Chet, I., Y. Elad, A. Kalfon, Y. Hadar, and J. Katan (1982). Integrated control of soilborne and bulborne pathogens in iris. *Phytoparasitica*, **10:**229-236.

Cook, R. J. and K. F. Baker (1983). *The Nature and Practice of Biological Control of Plant Pathogens.* APS, St. Paul, 539 pp.

Davis, J. R. and L. H. Sorenson (1984). Effect of soil solarization with moderate air temperatures on potato production. *Am. Potato J.*, **61:**520 (Abstract).

Dobbs, S. G. and W. H. Hinson (1953). A widespread fungistasis in soils. *Nature*, **172:**197-99.

Domsch, K. H. (1964). Soil fungicides. *Annu. Rev. Phytopathol.*, **2:**293-320.

Dunn, P. H., S. C. Barro, and M. Poth (1985). Soil moisture affects survival of micro-organisms in heated chaparral soil. *Soil Biol. Biochem.*, **17:**143-148.

Egley, G. H. (1983). Weed seed and seedling reduction by soil solarization with transparent polyethylene sheets. *Weed Sci.*, **31:**404-409.

Elad, Y., J. Katan, and I. Chet (1980). Physical, biological and chemical control integrated for soilborne diseases in potatoes. *Phytopathology*, **70:**418-422.

El Yamani, T., M. F. Abdel Rahim, K. Y. Michael, A. Grinstein, and J. Katan (1983). Soil solarization—a potential method for controlling white rot and other soilborne diseases in onions. *Proceedings of the 2nd International Workshop on Allium White Rot*, Beltsville, Maryland, 126 pp.

Frank, Z. R., Y. Ben Yephet, and J. Katan (1986). Synergistic effect of metham and solarization in controlling delimited shell spots of peanut pods. *Crop Prot.* (in press).

Garibaldi, A. and G. Tamietti (1984). Attempts to use soil solarization in closed glasshouses in Northern Italy for controlling corky root of tomato. *Acta Hortic.*, **152:**237-243.

Garrett, S. D. (1956). *Biology of Root-Infecting Fungi.* Cambridge University Press, Cambridge, 293 pp.

Gerson, U., S. Yathom, and J. Katan (1981). A demonstration of bulb mite control by solar heating of the soil. *Phytoparasitica*, **9:**153-155.

Goisque, M. J., H. Lormet, C. Martin, J. Lagier, P. Davet, Y. Conteadier, and J. Lormet (1984). La disinfection solaire du sol. *Rev. Hort.*, **247:**1-4.

Greenberger, A., A. Yogev, and J. Katan (1984). Biological control in solarized soils. *Proc. Sixth Cong. Mediter. Phytopathol. Union.* Cairo, Egypt, pp. 112-114.

Greenberger, A., A. Yogev, and J. Katan (1985). Induced suppressiveness in solarized soils. *Phytopathology*, **75:**1291 (Abstract).

Grinstein, A., J. Katan, A. Abdul-Razik, O. Zeidan, and Y. Elad (1979a). Control of *Sclerotium rolfsii* and weeds in peanuts by solar heating of soil. *Plant Dis. Rep.*, **63:**1056-1059.

Grinstein, A., D. Orion, A. Greenberger, and J. Katan (1979b). Solar heating of the soil for the control of *Verticullium dahliae* and *Pratylenchus thornei* in potatoes. In *Soilborne Plant Pathogens* (B. Schippers and W. Gams, eds.). Academic, New York, pp. 431-438.

Grosshevoy, S. E. (1939). Disinfection of seed-bed soil in cold frames by solar energy. The A. I. Mikoyan Pan-Soviet Sci. Res. Inst. Tob. and Indian Tob. Ind. (VITIM), Kransnadar, Publ., 137, pp. 51-56. *Rev. Appl. Mycol.*, **18:**635-636.

Hadar, E., S. Sofer, S. Brosh, M. Mordechai, E. Cohn, and J. Katan (1983). Control of clover cyst nematode on carnation. *Hadesseh*, **63:**1698-1700 (Hebrew with English summary).

Hardy, G. E., St., K. Sivasithamparam (1985). Soil solarization effects on Fusarium wilt of carnation and Verticillium wilt of eggplant. In *Biology and Management of Soilborne Plant Pathogens* (C. A. Parker, A. D. Rovira, K. J. Moore, P. T. W. Wong, and J. F. Kollmorgen, eds.). APS, St. Paul, pp. 279-281.

Henis, Y. and G. C. Papavizas (1983). Factors affecting germinability and susceptibility to attack sclerotia of *Sclerotium solfsii* by *Trichoderma harzianum* in field soil. *Phytopathology*, **73:**1469-1474.

Hetzroni, A., A. Grinstein, Y. Alper, and H. Frankel (1984). A continuous plastic film covering and welding machine for soil solarization. *Acta Hortic.*, **152:**259-265.

Horiuchi, S. (1984). Soil solarization for suppressing soilborne diseases in Japan. *FFTC Book Series* No. 26, Soilborne Crop Diseases in Asia. Food and Fertilizer Technology Center, Taiwan, ROC, pp. 11-23.

Horiuchi, S., M. Hori, S. Takashi, and K. Shimuzi (1982). Factors responsible for the development of clubroot suppressing effect in soil solarization. *Bull. Chugoku Natn. Agric. Exp. Sta.*, Series E, No. 20, pp. 25-47.

Horowitz, M., Y. Regev, and G. Herzlinger (1983). Solarization for weed control. *Weed Sci.*, **31:**170-179.

Jacobsohn, R., A. Greenberger, J. Katan, M. Levi, and H. Alon (1980). Control of Egyptian broomrape (*Orobanche aegyptiaca*) and other weeds by means of solar heating of the soil by polyethylene mulching. *Weed Sci.*, **28:**312-16.

Katan, J. (1980). Solar pasteurization of soils for disease control: Status and prospects. *Plant Dis.*, **64:**450-1454.

Katan, J. (1981). Solar heating (solarization) of soil for control of soilborne pests. *Ann. Rev. Phytopathol.*, **19:**211-236.

Katan, J. (1984). The role of soil disinfestation in achieving high production in horticultural crops. *Br. Crop. Prot. Conf.*, pp. 1189-1196.

Katan, J. (1985). Solar disinfestation of soils. In *Biology and Management of Soilborne Plant Pathogens* (C. A. Parker, A. D. Rovira, K. J. Moore, P. T. W. Wong, and J. F. Kollmorgen, eds.). APS, St. Paul, pp. 274-278.

Katan, J. and Y. Eshel (1973). Interactions between herbicides and plant pathogens. *Resid. Rev.*, **45:**145-177.

Katan, J., G. Fishler, and A. Grinstein (1983). Short- and long-term effects of soil solarization and crop sequence on Fusarium wilt and yield of cotton in Israel. *Phytopathology*, **73:**1215-1219.

Katan, J., A. Greenberger, H. Alon, and A. Grinstein (1976). Solar heating by polyethylene mulching for the control of diseases caused by soilborne pathogens. *Phytopathology*, **76:**683-688.

Katan, J., A. Grinstein, G. Fishler, A. Z. Frank, H. D. Rabinowitch, A. Greenberger, H. Alon, and U. Zig (1981). Long-term effects of solar heating of the soil. *Phytoparasitica*, **9:**236 (Abstract).

Katan, J., I. Rotem, Y. Finkel, and J. Daniel (1980). Solar heating of the soil for the control of pink root and other soilborne diseases in onions. *Phytoparasitica*, **8:**39-50.

Kodama, T. and T. Fukai (1982). Solar heating in closed plastic house for control of soilborne diseases. V. Application for control of Fusarium wilt of strawberry. *Ann. Phytopath. Soc. Jpn.*, **48:**570-577 (Japanese with English summary).

Kodama, T., T. Fukai, and Y. Matsumoto (1980). Solar heating sterilization in the closed vinyl houses against soilborne diseases. III. Influence of the treatment on the population level of soil microflora and the behavior of strawberry yellows pathogen, *Fusarium oxysporum* f. sp. *fragariae*. *Bull. Nara Pref. Agric. Exp. Sta.*, **11:**41-52 (Japanese with English summary).

Krikun, J., D. Orion, A. Nachmias, and R. Reuveni (1982). The role of soilborne pathogens under conditions of intensive agriculture. *Phytoparasitica*, **10:**247-258.

LaMondia, J. A. and B. B. Brodie (1984). Control of *Globodera rostochiensis* by solar heat. *Plant Dis.*, **68:**474-476.

Leach, L. D. (1947). Growth rates of host and pathogen as factors determining the severity of pre-emergence damping-off. *J. Agric. Res.*, **75:**161–179.

Lifshitz, R., M. Tabachnik, J. Katan, and I. Chet (1983). The effect of sublethal heating on sclerotia of *Sclerotium rolfsii*. *Can. J. Microbiol.*, **29:**1607–1610.

Mahrer, Y. (1979). Prediction of soil temperature of a soil mulched with transparent polyethylene. *J. Appl. Meteorol.*, **18:**1263–1267.

Mahrer, Y. and J. Katan (1981). Spatial soil temperatures regime under transparent polyethylene mulch: numerical and experimental studies. *Soil Sci.*, **131:**82–87.

Mahrer, H., O. Naot, E. Rawitz, and J. Katan (1984). Temperature and moisture regimes in soils mulched with transparent polyethylene. *Soil Sci. Soc. Am. J.*, **48:**362–367.

Malathrakis, N. E., G. E. Kapetanakis, and D. C. Linardakis (1983). Brown root rot of tomato and its control in Crete. *Ann. Appl. Biol.*, **102:**251–256.

Mihail, J. D. and S. M. Alcorn (1984). Effects of soil solarization on *Macrophomina phaseolina* and *Sclerotium rolfsii*. *Plant Dis.*, **68:**156–159.

Munnecke, D. E., W. Wilbur, and E. F. Darley (1976). Effect of heating or drying on *Armillaria mellae* or *Trichoderma viride* and the relation to survival of *A. mellea* in soil. *Phytopathology*, **66:**1363–1368.

Myers, D. F., R. N. Campbell, and A. S. Greathead (1983). Thermal inactivation of *Plasmodiophora brassicae* Woron, and its attempted control by solarization in the Salinas Valley of California. *Crop Prot.*, **2:**325–333.

Naot, O. (1984). The effect of transparent polyethylene mulches on soil moisture and temperature regimes in solar heating: numerical and experimental studies. MSc thesis, The Hebrew University of Jerusalem, 67 pp.

Old, K. M. (1981). Solar heating of soil for the control of nursery pathogens of *Pinus radiata*. *Aust. For. Res.*, **11:**141–147.

Olson, C. M. and F. K. Baker (1968). Selective heat treatment of soil and its effect on the inhibition of *Rhizoctonia solani* by *Bacillus subtilis*. *Phytopathology*, **58:**79–87.

Osman, A. R. and A. F. Saheb (1984). Control of *Rhizoctonia solani* by soil solarization. *Acta Hortic.*, **152:**245–251.

Overman, A. J. (1981). Off-season land management and soil fumigation for tomato on sandy soil. *J. Nematol.*, **13:**455 (Abstract).

Park, D. (1963). The ecology of soilborne fungal disease. *Annu. Rev. Phytopathol.*, **1:**241–258.

Pinkas, Y., A. Kariv, and J. Katan (1984). Soil solarization for the control of *Phytophthora cinnamomi*: thermal and biological effects. *Phytopathology*, **74:**796 (Abstract).

Porter, I. J. and L. R. Merriman (1983). Effects of solarization of soil on nematode and fungal pathogens at two sites in Victoria. *Soil Biol. Biochem.*, **15:**39–44.

Porter, I. J. and P. R. Merriman (1985). Evaluation of soil solarization for control of root diseases of row crops in Victoria. *Pl. Pathol.*, **34:**108–118.

Pullman, G. S. and J. E. DeVay (1977). Control of *Verticillium dahliae* by plastic tarping. *Proc. Am. Phytopathol. Soc.*, **4:**210 (Abstract).

Pullman, G. S., J. E. DeVay, C. L. Elmore, and W. H. Hart (1984). Soil solarization—a nonchemical method for controlling diseases and pests. Cooperative Extension, Division of Agriculture and Natural Resources, Leaflet 21377, University of California, Davis, CA.

Pullman, G. S., J. E. DeVay, and R. H. Garber (1981a). Soil solarization and thermal death: A logarithmic relationship between time and temperature for four soilborne plant pathogens. *Phytopathology*, **71:**959–964.

Pullman, G. S., J. E. DeVay, R. H. Garber, and A. R. Weinhold (1981b). Soil solarization: Effects on Verticillium wilt of cotton and soilborne populations of *Verticillium dahliae*, *Pythium* spp., *Rhizoctonia solani* and *Thielaviopsis basicola*. *Phytopathology*, **71:**954–959.

Rabinowitch, H. D., J. Katan, B. Ben David, I. Rotem, and U. Zig (1985). Soil solarization in onion: Effects in successive years. *Hassadeh*, **65:**1792–1794 (Hebrew).

Rabinowitch, H. D., J. Katan, and I. Rotem (1981). The response of onion to solar heating, agricultural practices and pink root disease. *Sci. Hort.*, **15:**331–340.

Raghaven, D. (1964). *Agriculture in Ancient India.* ICAR Publications, New Delhi, 164 pp.

Reuveni, R., J. Krikum, and U. Shani (1983). The role of *Monosporascus eutypoides* in a collapse of melon plants in an arid area of Israel. *Phytopathology*, **73:**1223–1226.

Rubin, B. and A. D. Benjamin (1983). Solar heating of the soil: Effect on weed control and on soil-incorporated herbicides. *Weed Sci.*, **31:**819–825.

Rubin, B. and A. Benjamin (1984). Solar heating of the soil: Involvement of environmental factors in the weed control process. *Weed Sci.*, **32:**138–142.

Sewell, G. W. F. (1984). Replant diseases: Aetiology and importance. *Proc. Br. Crop. Prot. Conf.*, pp. 1175–1182.

Sheikh, A. H. and A. Ghaffar (1984). Reduction in viability of sclerotia of *Macrophomina phaseolina* with polyethylene mulching of soil. *Soil Biol. Biochem.*, **16:**77–79.

Siti, E., E. Cohen, J. Katan, and M. Mordechai (1982). Control of *Ditylenchus dipsaci* in garlic by bulb and soil treatments. *Phytoparasitica*, **10:**93–100.

Smith, E. W., F. C. Wehner, and J. M. Kotze (1984). Effect of soil solarization and fungicide soil drenches on crater disease of wheat. *Plant Dis.*, **68:**582–584.

Standifer, L., P. H. Wilson, and R. Porche-Sorbet (1984). Effects of solarization on soil weed seed populations. *Weed Sci.*, **32:**596–573.

Stapelton, J. J. and J. E. DeVay (1982). Effect of soil solarization on populations of selected soilborne microorganisms and growth of deciduous fruit tree seedlings. *Phytopathology*, **72:**323–326.

Stapleton, J. J. and J. E. DeVay (1983). Response of phytoparasitic and free-living nematodes to soil solarization and 1,3-dichloropropene in California. *Phytopathology*, **73:**1429–1436.

Stapleton, J. J. and J. E. DeVay (1984). Thermal components of soil solarization as related to changes in soil and root microflora and increased plant growth response. *Phytopathology*, **74:**255–259.

Stapleton, J. J. and J. E. DeVay (1985a). Soil solarization as a post-plant treatment to increase growth of nursery trees. *Phytopathology*, **75** (Abstract).

Stapleton, J. J. and J. E. DeVay (1985b). Soil solarization: a nonchemical approach for management of plant pathogens and pests. *Crop Protection* (in press).

Stapleton, J. J., J. Quick, and J. E. DeVay (1985). Soil solarization: effect on soil properties, crop fertilization and plant growth. *Soil Biol. Biochem.*, **17:**369–373.

Tamietti, G. and A. Garibaldi (1981). Control of corky root in tomato by solar heating of the soil in greenhouse in Riviera Ligure. *La Difesa delle Piente*, **3:**143–150 (in Italian).

Tjamos, E. C. (1983). Prospects for controlling wilt of olive trees by soil solarization. *Hellenic Congress on Plant Diseases and Pests*, Athens, p. 15.

Tjamos, E. C. (1984). Control of *Pyrenochaeta lycopersici* by combined soil solarization and low dose of methyl bromide in Greece. *Acta Hortic.*, **152:**253–258.

Usmani, S. M. H. and A. Ghaffar (1982). Polyethylene mulching of soil to reduce viability of sclerotia of *Sclerotium oryzae. Soil Biol. Biochem.*, **14:**203–206.

Waggoner, P. E., P. M. Miller, and H. C. DeRoo (1960). Plastic mulching principles and benefits. *Connecticut Agricultural Experiment Station Bulletin*, **634,** 44 pp.

White, J. G. and S. T. Buczacki (1979). Observations on suppression of clubroot by artificial or natural heating of soil. *Trans. Br. Mycol. Soc.*, **73:**271–275.

Yarden, U., J. Katan, N. Aharonson, and B. Ben Yephet (1985). Delayed and enhanced degradation of benomyl and carbendazim in disinfested and fungicide-treated soils. *Phytopathology*, **75:**763–767.

5

FUNGAL ANTAGONISM IN AERIAL AGROBIOCENOSES

BERNADETTE DUBOS
Station de Pathologie Végétale, Institut National de la Recherche Agronomique, Centre de Bordeaux, France

The term biocenosis, or "group of living organisms occupying the same space and exhibiting complex relationships of interdependence with regard to each other and to the environment" is modified in its range of significance when aerial agronomic types of biocenoses are considered.

The quality and quantity of saprophytic and parasitic microflora found on plant aerial organs is closely dependent on interrelationships existing between the plant and its environment. This is generally true, but becomes of even greater importance where plants cultivated in an agronomic system are concerned. Such microbial populations are less abundant and diverse than in the soil. They cohabit or confront each other in a limited space on the surface of plant organs, and their chances of survival are reduced, as a result of severe competition, to a greater extent.

The abiotic environmental factors that affect microflora are numerous and their fluctuations are brutal (climate, cultivation methods, and crop protection treatments in particular). The equilibrium in this aerial agrobiocenosis is thus very fragile and the phytopathologist may regard its components in a simplified manner as forming two groups—the phytoparasites that destroy the harvest, and those that oppose the phytoparasites, which are termed antagonists. This concept of antagonism needs to be clearly defined. Only a few years ago it was customary to confine its significance to the phenomenon of antibiosis, as opposed to symbiosis, and paradoxically to exclude hyperparasitism for example. As indicated by Olivier (1983), the concept of antagonism is taking on a wider significance as our knowledge increases. The current approach goes beyond this limited definition to include both the direct and indirect actions exerted by microorganisms, which result in a reduction of disease expression. Man's exploitation of antagonism consists of careful modification of the biocenotic equilibria to the benefit of the cultivated plant.

The object of this chapter is not to provide an exhaustive review of the prospects offered by fungal antagonism for controlling plant parasites. There are already excellent references on this subject: Baker and Cook (1974), Dubos and Bulit (1981), Corke and Risbeth (1981), Ponchet (1982), Blakeman and Fokkema (1982) and Olivier (1983). Instead, the intention is to show how the possible utilization of this phenomenon originally attracted the attention of phytopathologists and to illustrate with a few specific examples of the principal pathways available how antagonism may be domesticated for use in plant protection.

1. DEMONSTRATION OF MICROBIAL ANTAGONISM

The utilization of microbial antagonism is in keeping with the concept of a more intelligent management of natural resources; however, effective management necessitates a thorough understanding of both qualitative and quantitative aspects of the situation that one can expect to dispose of.

Until 1961, when Ruinen drew the attention of scientists to the lack of knowledge about microorganisms that colonize the aerial surfaces of plants, this particular sector had been neglected. The subsequent increase in research was considerable, however, and in 1976 Dickinson published a review of the fungal inhabitants of the phylloplane together with a summary of information concerning other plant parts. The ecological conditions controlling such fungal populations were also described.

It is relevant to the rest of our subject to employ Dickinson's classification of the nonpathogenic fungal flora found on leaves. Such fungi may be classified into three groups:

1. "Phylloplane inhabitants which are able to grow and sporulate under fluctuations in climatic and nutritional conditions: *Cladosporium*, sooty molds, species of Chaetothyriaceae."
2. "Phylloplane invaders which may grow extensively only when conditions are particularly favorable: *Alternaria, Epicoccum, Stemphylium*."
3. "Primary saprophytes which are unable to grow extensively until the onset of senescence: *Ascochytula, Leptosphaeria, Pleospora, Phoma*."

After acquiring the skills of a taxonomist and an ecologist, and learning to think in "ecosystem" terms, the phytopathologist was in a better position for understanding the variations observed in these aerial agrobiocenoses.

The role of environmental factors in controlling the biological equilibrium existing between the saprophytic and parasitic microflora is reported in numerous publications. The reviews by Hislop (1976), Fokkema (1976), Blakeman and Fokkema (1982), and Olivier (1983) should be mentioned in this respect.

The iatrogenic effect of pesticides is sometimes evoked to explain the development of new diseases (Griffiths, 1981). Well-known examples include coffee berry disease caused by *Colletotrichum coffeanum* (Griffiths, 1971; Firman and Waller, 1977) and *Eutypa* decline in apricot (Carter, 1971), in which the saprophytic microflora, antagonists of these particular parasites, are partially destroyed by cupric treatments. The buffering role played by the saprophytic microflora and the danger of destroying this population, especially by the irrational use of fungicides such as benomyl, was described in 1975 by Fokkema et al.

Benomyl was used by the author, under experimental conditions in the field, to vary the saprophytic colonization of rye leaves. The results are summarized in the following Table 1.

When leaves treated in this way were inoculated with the parasite *Cochiobolus (Helminthosporium) sativum* just after flowering, disease incidence was less important in the control plots (a diminution in the number of necroses of about 60%). This reduction in number of necroses may be correlated with a higher density of saprophytic microflora antagonistic to the parasite.

It is well known that pollen (Fokkema, 1971) and insect honeydew (Ripagen et al., 1979) encourage the development of certain pathogens, not only because they serve as sources of nutrition, but also, and this is especially true of pollen, because they exude substances that stimulate infection. As in the case illustrated above, the authors demonstrated the beneficial role, in relation to the plant, of the saprophytic microflora, which, here, acts as a competitor, starving the parasite and, thus, reducing disease incidence.

These few intentionally simplified examples demonstrate, in the terms of the definition described at the beginning of this chapter, the antagonistic effect of the saprophytic microflora with regard to the parasites in aerial agrobiocenoses.

2. DOMESTICATION OF ANTAGONISM IN PLANT PROTECTION

It is now a well-established fact that the normal innocuous microflora found on the aerial organs serves as a stabilizing element which it is always dangerous to ill-treat. It is not alone sufficient, however, to ensure the high degree of plant

TABLE 1. Variation in the Saprophytic Microflora of Rye Leaves Following an Application of Benomyl

Year of Experiment	*Evaluation of Microflora (Propagules/cm²)*	
	Control	*Benomyl*
1972	10,000	1,200
1973	3,000	400

protection required in modern agriculture, which must, in turn, be highly productive and competitive. It would therefore seem logical to try to modify it in a direction unfavorable to the parasites. However, work already carried out has rarely, if ever, been concerned with fungal organisms (or at least not the fungi involved in aerial biocenoses).

The remaining alternative is to take a particular antagonist and introduce it into a given biocenosis in massive quantities. The initial problem is, thus, the choice of a suitable antagonist.

2.1. Choice of Antagonist

Various methods are available for detecting a suitable antagonist. Baker and Cook (1974) recommend seeking the antagonists of a given parasite in cases of nondevelopment or decline of the corresponding disease, the ideal situation for detection of hypovirulent strains. In the case of hyperparasitic antagonists, the pathogen serves as a trap and the antagonists should, therefore, be sought in those biotopes in which the disease is at an advanced stage of development. More often the scientist resorts to genuine screening of microbes belonging to the same ecosystem as the parasite or, alternatively, of microbes already known to exhibit antagonistic properties with respect to other pathogens (as in the now classic case of *Trichoderma*). Such screening generally relies on evaluation *in vitro* of the antagonistic potential of strains as a function of well-defined criteria representative of ability for competition, antibiosis, or mycoparasitism. It must be stressed, and this is a general rule, that results obtained *in vitro* cannot be considered as an extrapolation of phenomena likely to occur in nature. A microbe that is antagonistic to a given parasite *in vitro* is not necessarily effective against the corresponding disease. The inverse is also true and it is in this respect that the phenomenon of plant immunization induced by a nonpathogenic microorganism or by a slightly aggressive form of the pathogen should be considered. This latter form of microbial antagonism in aerial agrobiocenoses is quite new and promising, not only for utilization in biological control but also in understanding host–parasite interrelationships. It is the subject of a special chapter in this work by Kuc.

2.2. General Scheme for the Practical Utilization of an Antagonist

The various stages involved in the practical application of a method of biological control based on microbial antagonism are identified in their major points, irrespective of the type of microbe employed. They are represented schematically in Table 2. The problems encountered at each particular stage differ, however, according to the interacting pathogen/antagonist pair involved. In fact, the different means of detecting antagonists already provide some idea of the probable diversity and complexity of their modes of action.

The three specific examples described are representative, with the exception of immunization, of the types of antagonism known to occur in aerial agrobiocenoses at the present time.

TABLE 2. General Outline for the Practical Utilization of Microbial Antagonism in a Method of Biological Control

Detection of the antagonist	
	Observation of natural phenomena
	Systematic screening *in vitro* according to clearly-defined criteria
Investigation of modes of action	
	In vitro
	In vivo as a function of the different development phases of the parasite
Evaluation of potential	
	As agent of control
	For multiplication on an industrial scale
Decision	
Large-scale multiplication	
Production of method of utilization	
	Compatibility with cultivation methods
	Application: Alone
	With fungicides

3. HYPOVIRULENCE

The variations affecting fungal organisms may sometimes be expressed in a diminution or loss of pathogenic ability. Several examples are known in which a highly attenuated form of a particular parasite is able to compete with and dominate virulent forms of the same parasite. This is a genuine phenomenon of antagonism between different strains of the same microorganism and has been named hypovirulence. The term hypoaggressivity would be more accurate (as this is a quantitative phenomenon) but common usage has tended to prefer hypovirulence. An excellent example is *Endothia parasitica*, the agent of chestnut canker.

3.1. Demonstration of Hypovirulence

Originally from the Far East, *E. parasitica* was introduced into the United States at the close of the last century, then to Italy during World War I and to France in 1957. Almost all the chestnut forests in these countries have been severely damaged by canker.

Hypovirulence was discovered as a result of an observation made by Biraghi (1950) about coppiced chestnuts that had been cut back several times following attacks by *Endothia*. On the same stump, branches could be found with vigorously developing cankers and others in which injury was restricted to the superficial bark tissues.

This spontaneous recovery and natural regression of the disease was investigated in Italian chestnut forests by Grente and Sauret (1969). The explanation was found in the appearance of *Endothia* strains that had lost some of their virulence. Such strains were isolated from cankers undergoing cicatrization. The authors found that the proportion of samples producing the hypovirulent rather than the virulent strain was positively correlated with this canker cicatrization character. In addition, the relative proportion of cankers from which isolation of a virulent strain was possible diminished as the state of health improved. It was difficult to isolate virulent strains from fully recovered chestnuts, which was also true for the hypovirulent strains, but to a lesser degree.

It would seem, therefore, that before disappearing themselves, the hypovirulent strains might have eliminated the virulent ones. Mixed inoculations were made to confirm this hypothesis and the following results were obtained:

1. Simultaneous inoculations showed that the virulent strain was often unable to develop in the presence of the other strain.
2. Consecutive inoculations showed that the virulent strain, present in the bark, was eliminated by the hypovirulent one.

Thus it was possible, by means of mixed inoculations, to reproduce the phenomenon of spontaneous recovery observed in Italian chestnut woods. This principle is the basis of the biological control method developed.

3.2. Characteristics and Mode of Action by Hypovirulent Strains

The hypovirulent (or so-called type B) strains exhibit certain features different from the normal (or type N) strains. In particular, they do not produce pycnidia within the normal period of time, nor the orange pigment associated with this. The convergency between hypovirulence and type B morphology is, in fact, fortuitous, as illustrated by studies of the results obtained with a biological control, which demonstrated the appearance in chestnut woods of hypovirulent strains with normal morphology.

Grente showed *in vitro* and *in vivo* that the hypovirulence of B strains could be transmitted to N strains by a process of vegetative anastomosis between hyphae which results in mixing of the cytoplasm. Three sectors with different morphologies could thus be detected following confrontation of the strains *in vitro*:

1. The N thallus with pycnidia.
2. The B thallus with colorless mycelium.
3. An "intermediate sector" of rather variable morphology, corresponding to the N thallus, which had been transformed as a result of acquisition of the hypovirulence character.

Because the hypovirulence character could be acquired, the author described such hypovirulence as "contagious."

It is important to note that the cytoplasmic mixtures resulting from vegetative anastomosis may degenerate at different rates. This brings us to the idea of "compatibility" in the transmission of hypovirulence and may be compared with the "vegetative compatibility groups" detected by Anagnostakis (1977) in N strains of *E. parasitica*. If one of the strains in an anastomosis iso group is hypovirulent, it will transmit this character to the other strain in each and every confrontation. In contrast with hetero-group combinations, Grente showed that the chances of transmission are inversely proportional to the frequency and rapidity of degeneration. This implies that for each N strain a compatible B strain must be found. If the chances of hetero-group anastomoses are increased, however, by multiplying the number of confrontations, cases of transmission may occur. As a result of this process, which the author describes as "forced anastomosis," new compatible strains may be produced for transmission of hypovirulence to the virulent strains one wishes to eliminate.

Although the American scientists (Anagnostakis et al.) know that the vegetative compatibility groups are controlled genetically by a nuclear system, the exact nature of hypovirulence is, in contrast, not yet understood. Grente clearly demonstrated its extranuclear character. Day et al. (1977) demonstrated the presence of RNA-bicatenaries and, in certain cases, "virus-like particles" (Dodds, 1978) that do not exist in the virulent strains. These findings have been confirmed by Lapierre and L'Hostis (1983) but do not prove that such cytoplasmic heredity factors exert a depressive effect on pathogenic ability, as no direct transmission has been successful.

Grente (1983) looked at different aspects of the mode of action of hypovirulent strains:

1. It is now clearly evident that the contagious character of hypovirulence is transmitted by anastomosis from the vegetative hypha of an "infected" thallus to that of a healthy thallus.
2. In the case of the hypovirulent strains, the plant is able to block progression of the fungus by means of its natural defence system. This is apparent in the differentiation and subsequent suberization of subero-phellodermic barriers. Such reactional processes are reported by the author to be due to a reduced production of toxins of the "Diaporthine" complex.
3. The propagation of hypovirulence in nature originates from an emissive source constituted by a canker that has become hypovirulent.

Thus the practical mode of action of hypovirulence is "essentially the transmission of 'something' over distance which results in the modification of a population of strains of *Endothia* to the benefit of the hypovirulents."

3.3. Development of a Method of Biological Control and its Evaluation

The aim of employing a biological control method, such as this, is to accelerate the rate of substitution, observed under natural conditions in Italy, of virulent strains by hypovirulent strains.

The first stage is to develop a suitable culture technique. The process involving "pelleting" of mycelium obtained from fermentors has now been abandoned in favor of manufacture in "pasta" form. To produce this inoculum the mycelium is cultured on 1% malt medium in Petri dishes and ground up. The paste is mixed with a simple corn-flour ("Maizena") medium. The resulting paste is processed in aluminum tubes with a glazed inner lining, in the same way as certain pharmaceutical products. Such tubes may be conserved for 4–6 months at ambient temperature (Grente, 1981).

Fig. 1. Treatment of canker (Courtesy of J. Grente).

Control consists of inoculating a portion of the actively developing cankers with the compatible hypovirulent strain. The inoculum is, in practice, introduced into a hole (5 mm in diameter) punched in the bark and protected with a band of adhesive tape immediately after inoculation (Fig. 1). Highly positive results are obtained and the chances of recovery are proportional to the distance separating the cicatrizing cankers and to the number of such cankers. The treated cankers are apparent as unmistakable sources of dissemination of hypovirulence.

This method of control is currently being applied to French chestnuts and 18,000 ha are receiving treatment at the present time.

It should be stressed that, although hypovirulence exists naturally in France and Italy and is effective in natural control of the disease, this is not the case in the United States. The reason seems to be the existence of a very large number of anastomoses groups. Whereas in France only one group per region is generally found and only nine groups are known in all, in the United States up to 10 different groups have been found in the same canker and 77 compatibility groups in all have been recorded. The practical application in such countries of a biological control method involving hypovirulence will therefore have to be considered in different terms, although the problem could possibly be resolved by using a mixture of strains belonging to different anastomotic groups (Anagnostakis, 1982; Van Alfen, 1982).

4. HYPERPARASITISM

A distinction should first be made between the terms hyperparasitism and mycoparasitism.

Hyperparasitism *sensu stricto* refers to the parasitism exerted by certain fungi on other fungi which are themselves parasites of higher plants. This is encountered in the Mastigomycotina, Ascomyticotina, and Deuteromycotina (mainly the Hyphomycetes).

Mycoparasitism has a much wider significance, covering the various parasitic phenomena exhibited by fungi with regard to other fungi, whether parasitic or saprophytic in nature.

The phenomenon of hyperparasitism is very widespread in nature. A review of the literature existing on this subject has been published by Kranz (1981). It is, nevertheless, true that the phenomenon has been little exploited, with the exception of the utilization of *Ampelomyces quisqualis* to control cucumber powdery mildew (*Sphaerotheca fuliginea*, Jarvis and Slingsby, 1977) and *Erysiphe cichoracearum* (Sundheim, 1983) under glasshouse conditions.

Research carried out on the second example has been widely published by the authors and also mentioned in other articles (Blakeman, 1981; Cook and Baker, 1983; Dubos and Olivier, 1983). For this reason we prefer to describe work done with *Hansfordia pulvinata*, a hyperparasite of *Cladosporium fulvum* in tomato.

4.1. Demonstration of the Hyperparasite

In 1962 Matta reported *Hansfordia pulvinata* as a hyperparasite of *Cercospora unamunoi* in sweet pepper. It was first observed on *Cladosporium fulvum* in France in 1978 (Peresse and Le Picard, 1980) and in Texas in 1981 (Taber et al., 1981), and then in Florida on *Cercosporidium personatum* of groundnut (Shokes and Taber, 1983).

The presence of the hyperparasite is apparent as an abnormally light coloration of the spots produced by the olive-brown mycelial mat of *Cladosporium* on the underside of tomato leaves. This coloration is due to development of the white mycelium of *Hansfordia*, which covers the thallus of the parasite and sporulates abundantly, producing a light grey powdery appearance. A pure culture of the fungus may easily be obtained on a conventional nutritive agar medium.

4.2. Characteristics and Mode of Action of *H. pulvinata*

The biological characteristics of *Hansfordia* and its mode of action with regard to *C. fulvum* have been described by Peresse and Le Picard (1980, 1983). The spores are small (3.5–5 μm), light and poorly wettable; hence, they are easily carried and disseminated by air currents. Their germinative ability is high and long lasting. They are able to germinate in sterile, deionized water (an indication of ample nutritive reserves), in an atmosphere of saturated humidity and in the absence of free water.

Mycelial growth can occur within a wide range of temperatures, similar to those for *Cladosporium* and perfectly compatible with the temperatures usually maintained in glasshouses. According to Baker and Cook (1974), its characteristics are those of a "good antagonist." Studies have been carried out *in vitro* and *in natura* to provide details of the interrelationships existing between the two fungi.

A confrontation of the protagonists *in vitro* demonstrates a retardation of mycelial growth in *C. fulvum*, indicating the presence in the culture medium of an inhibitory metabolite. Tirilly and Trique (1983) demonstrated the emission of a fungitoxic sesquiterpene, 13-desoxyphomenone, which has also been detected *in vivo* (Tirilly and Trique, 1983). Simultaneous spraying of tomato leaves with desoxyphomenone and spores of *C. fulvum* has demonstrated the twofold biological activity (fungistatic and phytotoxic) of this metabolite.

The results shown in Table 3 suggest participation of the plant in this hyperparasitic interrelationship, although the mechanisms of action have not yet been determined.

The establishment of a parasitic relationship is apparent in the stimulated growth and rapid cover of the mycelial thallus of *Cladosporium* by *Hansfordia*. The hyperparasite is able to become established on the tomato leaf from the time of emergence of *C. fulvum* onward (Fig. 2).

TABLE 3. Biological Activity of 13-Desoxyphomenone on Tomato Leaves

Dose ($\mu g/cm^{-2}$)	*Reduction in Number of Leaf Spots due to Cladosporium (%)*	*Phytotoxicity*
0	0	0
0.05	36 ± 12	Not visible
0.10	82 ± 8	Not visible
0.15	93 ± 5	Not visible
0.20	95 ± 5	Not visible
0.25	100	Small necroses on veins
0.50	100	Small necroses on veins
0.75	100	Vein and foliar necroses

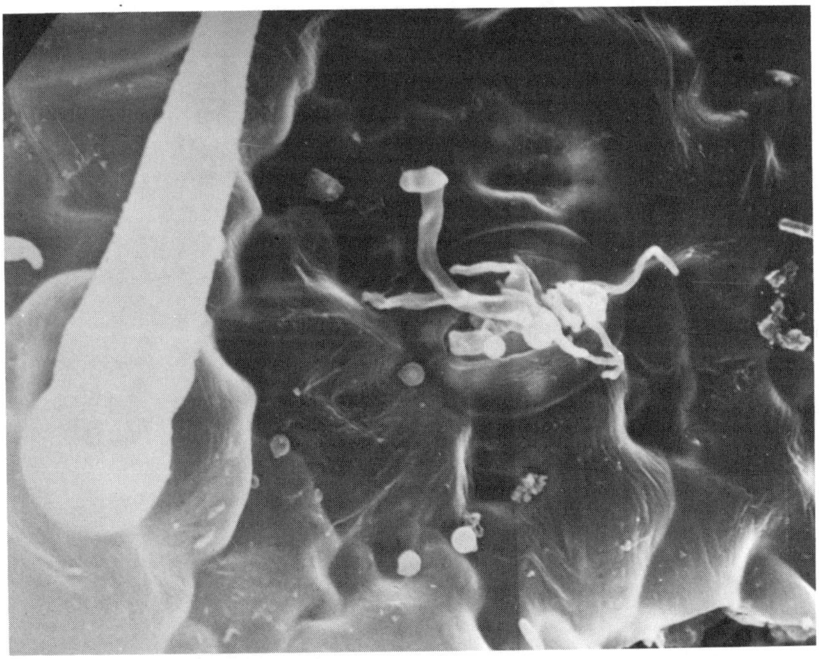

Fig. 2. The hyperparasite becomes established after the initiation of growth by *C. fulvum* (Courtesy of Y. Tirilly).

The proximity of a hypha of *Cladosporium* induces the formation of telemorphic branching in *Hansfordia*, but only over short distances. The authors suggest the existence of volatile substances that are able to bring about these phenomena and that exist only at the level of the aerial hyphae. These short branches curl around the hypha of the parasite in a sort of knot and the tip expands to form an appressorium (Fig. 3). The hyphae encircled in this way exhibit modification in their cytoplasm and the mycelium is ultimately killed (Fig. 4).

The *Cladosporium* conidia are massed together into packets by the hyphae of *Hansfordia* which send out filaments and penetrate the conidia, which seem to collapse and be emptied of their contents (Fig. 5).

The mechanisms operating at the hyperparasite–host interface remain unknown. A trophic relationship seems to be set up which ensures the acquisition of nutritive substances or growth factors as the development of *Hansfordia* is strongly stimulated in the presence of *Cladosporium*. In addition, Tirilly and Trique (1983) have shown that superinfection with *Hansfordia* leads to an augmentation in β-1-3 glucanase activity in the tomato leaf over the first few days, an indication of lytic potential with regard to the parasite's cell walls.

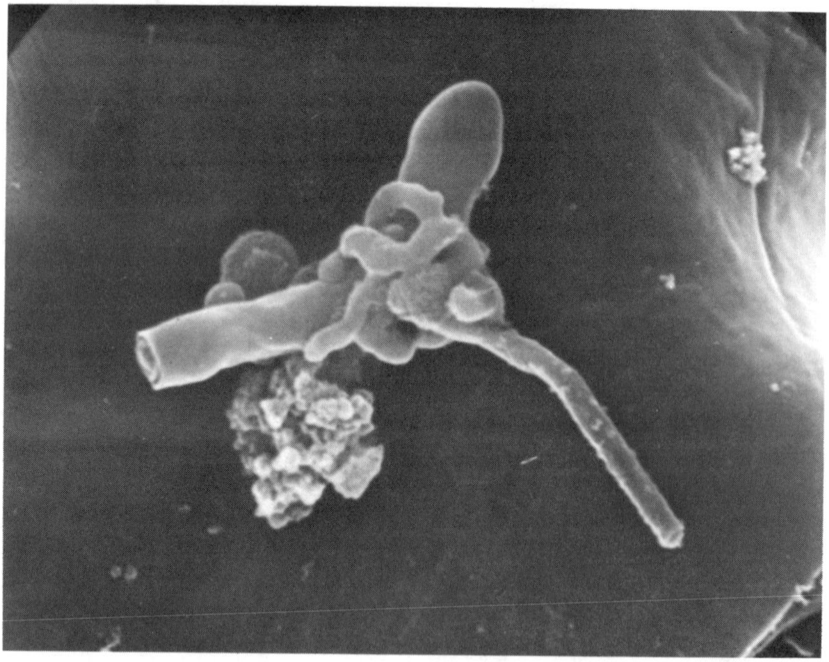

Fig. 3. Coiling of hyphae of *H. pulvinata* around a hyphal segment of *C. fulvum* (Courtesy of Y. Tirilly).

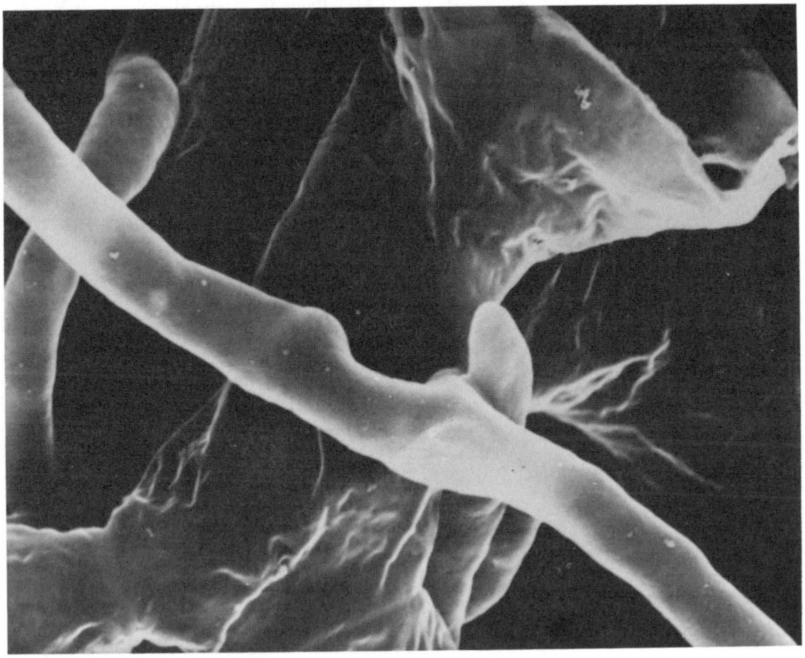

Fig. 4. Mycelium of *C. fulvum* killed by *H. pulvinata* (Courtesy of Y. Tirilly).

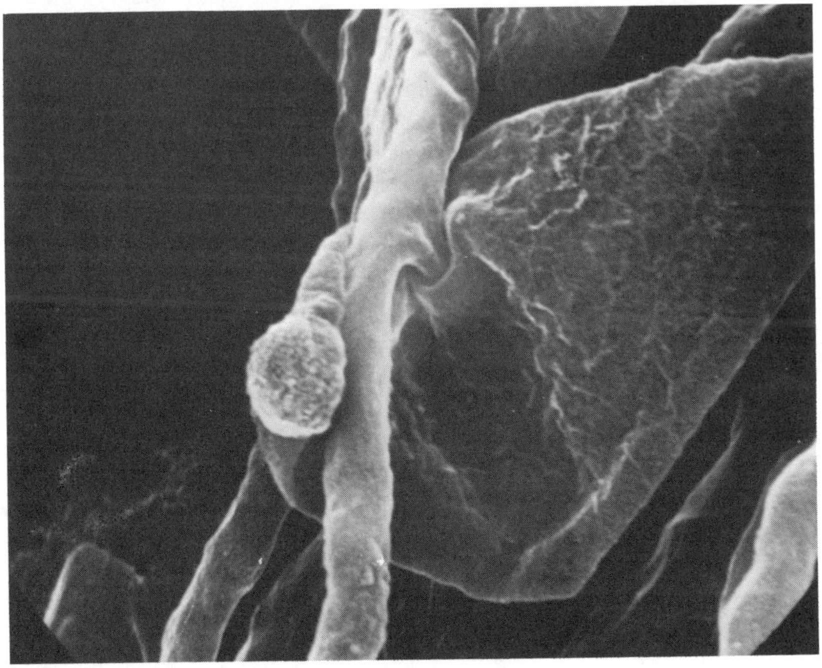

Fig. 5. Conidia of *C. fulvum* collapsed and emptied of its contents (Courtesy of Y. Tirilly).

The participation of the plant host in hyperparasitic relationships, as already evoked in connection with desoxyphomenone, has been well illustrated in enzymatic studies carried out by Tirilly and Trique (1983). The results suggest that the pectinolytic enzymes of *Hansfordia* participate in the production of yellowing and accelerated senescence of the tomato leaves. In this way the hyperparasite is able to guarantee a proportion of its nutritional requirements through its own pectinolytic ability. The authors conclude that "*H. pulvinata* may be considered not only as a parasite of *C. fulvum* but also, to a certain extent, of the plant host. This foliar parasitism is only exhibited in the zones already attacked by *Cladosporium* however. Punctual exacerbation of the initial symptoms of *Cladosporium* ultimately results in overall attenuation of the disease."

4.3. Evaluation of the Hyperparasitic Potential of *H. pulvinata* and Development of a Method for the Biological Control of *C. fulvum*

The *in vitro* culture of *Hansfordia* is relatively easy on ordinary nutritive media and should not pose any problems on an industrial scale.

Two types of biopreparation were used by the authors, one derived from conidia and the other from ground-up mycelium. When sprayed on the foliar epidermis in the absence of the parasite, the conidia germinate. The germinative filaments exhibit limited development and emit a few spores. However, the *H. pulvinata* does not prevent infection of the plants with *C. fulvum* nor does it affect the incubation phase. Only when the hyphae emerge does it come into action, producing two types of propagules (conidia or mycelium) which ensure its implantation on the thallus of the parasite and have a negative effect on development and the production of spores that would otherwise be the source of secondary infections.

As a result of its presence in a tomato production, glasshouse *Hansfordia* is able to reduce the pathogenic inoculum carried in the air, a source of new infectious foci. It should, therefore, in theory, slow down the development of an epidemic. An important technical problem has been encountered by researchers however. Treatments in the glasshouse using pressurized sprayers (required to attain the *Cladosporium* on the underside of the tomato leaves) also result in the aerial dissemination of the spores of the parasite. Under favorable climatic conditions numerous secondary infections become apparent over the next 15 days even though the initial leaf spots have been satisfactorily covered over by *H. pulvinata*.

Tirilly (1985) tried to solve this problem by inhibiting the germination of spores of *C. fulvum* disseminated during treatment by a concurrent application of fungicide. Fosetyl Al is suitable for this purpose and is compatible with *Hansfordia*.

This provides a rather original means of using the hyperparasite in an integrated control program in which the thalli of the parasite produced as a result of primary infection would be covered over (to about 80%) by the hyperparasite and the development of secondary infections inhibited by the Fosetyl Al.

5. UTILIZATION OF SAPROPHYTIC FUNGI

There are numerous examples in the literature relating to the utilization of certain components of the innocuous microflora of aerial organs to control the pathogens of these organs. The fungal species most often cited belong to those listed by Dickinson, the most representative being various *Alternaria* spp. (Newhook, 1957; Spurr, 1977) and *Cladosporium* spp. (Bhatt and Vaughan, 1962; Newhook, 1957; Corke and Hunter, 1979; Royse and Ries, 1978). Despite the very interesting results obtained, no attempts have been made to develop these species. This is probably due to the efficacity of synthetic organic fungicides with regard to target parasites and because the examples were only considered by the authors to be of academic interest. It is hoped that these studies will be taken up again in view of the new information available and the progress achieved in the field of biological control.

The most advanced studies involving the use of saprophytic fungi involve, paradoxically, antagonists that are either foreign to or only occur occasionally in the aerial biocenosis. The species *Trichoderma* has received the most study, and the two examples described concern its antagonism to *Botrytis cinerea*, which causes grey rot in grapes, and *Chondrostereum purpureum*, the agent of silver leaf in fruit trees.

5.1. General Information on *Trichoderma*

The genus *Trichoderma* consists of a number of saprophytic fungi commonly found in the soil, on dead wood and on plant debris, and which may easily be recognized in culture by their usually green spores.

Although recognition of a *Trichoderma* is relatively easy, identification of the species is much less simple. Rifai (1969) arranged the various *Trichoderma* into nine "species groups," within which there is still a certain degree of variability. The perfect forms are generally assigned to the Hypocreales or, sometimes, to the Eurotiales, Clavicipitales and Spheriales. A given species of *Hypocrea* may have several different *Trichoderma* as anamorphs; inversely, a same "species group" of *Trichoderma* may be attached to several different species of *Hypocrea*. Thus the form *T. harzianum* corresponds to six different sexual forms, while certain of these sexual forms may also have other anamorphs, such as *T. viride*.

This complexity in the fine determination of *Trichoderma* easily explains the confusion between various different "species groups" and in particular between *harzianum* and *viride*, which seem to be of most frequent occurrence under the temperate climatic conditions of Western Europe.

Studies with the electron scanning microscope (Roquebert, 1982) and polyacrylamide gel electrophoresis of structural proteins (Dubos and Mouches, 1982) would, however, imply that the groups distinguished by Rifai do, in fact, correspond to real morphological and biochemical differences. These two techniques identify the *Trichoderma* employed in the studies described in this chapter as belonging to the *harzianum* species group.

The antagonistic properties of *Trichoderma* have been known for a long time; the first publication to mention them is dated 1887 (Vuillemin). Examination of the phenomenon of antagonism and its possible application as a means of controlling parasites of cultivated plants was only started between the two world wars. The examples studied dealt essentially with soil parasites, although as early as 1951 Wood reported the efficacity of *T. viride* against *B. cinerea* in lettuce.

The first positive results obtained by Grosclaude in 1970 in controlling silverleaf may be considered as the starting point of research into the utilization of *Trichoderma* in the biological control of plant aerial parasites.

5.2. Modes of Action of *Trichoderma* and Choice of Antagonistic Strain

As with most fungi the properties and abilities of the *Trichoderma* genus are exhibited to various degrees in different strains. It is therefore essential to detect those strains that appear to possess the properties that determine "antagonistic ability."

Antagonistic ability is the result of several different components:

1. A competitive action, due to concurrent consumption of the same energy sources in a given medium.
2. Antibiotic activity due to the secretion of antibiotic substances or metabolites that inhibit parasite activity. A suitable reference may be found in work by Dennis and Webster (1971a,b).
3. A predatory effect or direct parasitic action against the pathogen (Dennis and Webster, 1971c). This phenomenon seems to involve recognition of the mycelial wall by lectins (Chet and Elad, 1983). Most authors also agree that degradation of the parasite's mycelial walls may be attributed to the presence of β1-3-glucanases and extracellular chitinases (Davet, 1983).

Other mechanisms are also sometimes evoked, such as indirect effects resulting in modifications of the environment that are adverse to the parasite.

Analysis of the silver leaf and grey rot examples demonstrates the existence of specificity, however, in the antagonistic ability of *Trichoderma*, which means that choice of strain depends on the target organism. Thus, a thorough understanding of the modes of action of the antagonist with regard to the parasite is required, implying that certain qualities will be of greater importance than others.

In the case of silver leaf in fruit trees, development of the antagonist in the pruning wound and underlying wood leads to a nonspecific reaction of the tree in the form of vascular gummosis, which inhibits further development of the parasite (Grosclaude, 1983). Competitive ability seems, therefore, to be the most suitable parameter to use in screening antagonistic strains. The temperature factor should also be taken into account as Grosclaude (1970) has shown that the activity of a strain poorly adapted to cold conditions proves inadequate if the winter is harsh.

Of the various forms of confrontation occurring between *Trichoderma* and *Botrytis*, those demonstrated in vines involve both competition and mycoparasitism (Dubos et al., 1982).

Competition is especially active during the colonization of floral debris, and supply of the antagonist at the end of flowering is of prime importance. It should be stressed that one of the characteristics of *B. cinerea* is to exhibit powerful saprophytic properties as well as parasitic ones. The fungus initially colonizes various sorts of plant debris where it develops vigorously, spreading from this nutritive base to contaminate healthy organs. The antagonist is a pioneer colonizer of the senescing floral caps on the developing grapes, which thus prevents the saprophytic colonization of the parasite. Isolations from floral caps sampled, from plots either treated with *Trichoderma* or left untreated showed (Table 4) that 60.5% of the floral caps in control plots were carriers of *Botrytis*, compared with 11.5% in treated plots. In addition *Trichoderma* was present in 94.2% of the samples from these latter plots. Examination of the floral parts with the scanning microscope shows that the conidia of the antagonist germinates and colonizes the tissues. Under favorable climatic conditions, the fruiting cushions characteristic of *T. harzianum* may be observed on the vine.

In addition the examination of *Botrytis* population dynamics in the vineyard (Table 5) shows that the parasite becomes apparent during the closure of the bunch phase in control plots, whereas its isolation from the treated plots is only possible from the ripening stage onward.

The application of *Trichoderma*, therefore, retards the installation of *Botrytis* in the vineyard. This retardation is real in epidemiological terms, as the development of the first disease foci is similarly delayed. The effect of *Trichoderma* on conservation of the parasite as sclerotia on the ripened wood may also be attributed to the phenomenon of trophic competition.

The phenomena of mycoparasitism are revealed when the percentages of infected bunches and the percentages of rot on the bunches are compared for the treated and control plots. The percentages of infected bunches do not differ statistically between the two variables (treated and control plots), in contrast to the rot percentages. This indicates that from the time of ripening onward (when the inoculum is highly abundant) the *Trichoderma* does not reduce the number of

TABLE 4. Percentage Colonization of the Floral Caps by Different Fungi

Fungi Isolated	*Treated Floral Caps*	*Untreated Floral Caps*
Trichoderma	86.2	5
Trichoderma + *B. cinerea*	8.0	0.5
B. cinerea	3.5	60
Divers (common microflora)	2.3	34

TABLE 5. Evolution of *B. cinerea* Populations in Control and Treated Plots[a]

	Sampling Dates						
	6/17	*6/25*	*7/3*	*8/13*	*9/1*	*9/6*	*9/17*
Control							
Stalk	0	0	25	30	89	97	96
Grape	0	0	4	36	68	95	97
Treated							
Stalk	0	0	0	0	0	28	100
Grape	0	0	0	0	12	32	100

[a]The figures represent the percentages of grapes and stalks contaminated by *B. cinerea*. Dates of treatments: before flowering, 6/14; fruit set, 6/22; closure of the bunch, 7/8; ripening, 8/13; three weeks before harvesting, 8/30.

foci of infection but prevents their development. The antagonist is observed to fructify on the margin between the necrotic and healthy area on the rotting grapes, thus preventing extension of the parasite to healthy grapes in the bunch. In this case *Trichoderma* behaves as a hyperparasite, and under the microscope the usual evidence of predation (some coiling and numerous penetrations) may be observed on the borders of the necroses.

Destruction of the sclerotia and the conidiophores derived from their germination is dependent on the mycoparasitic properties of the antagonist.

It has been shown, however, (Dubos et al., 1983) that the expression of competitive phenomena and hyperparasitism *in vitro* and under natural conditions between *Trichoderma* and *Botrytis* is strictly controlled by temperature. The range of temperatures in which this antagonism may be expressed is very narrow and is considered to be between 18 and 25°C. Few strains exhibit such properties below and above these temperatures. The thermal activity spectrum of *Botrytis*, however, ranges from 5 to 28°C, so that an antagonistic strain of *Trichoderma* should cover these thermal requirements at least in part.

The choice of antagonistic strain as a function of the target organism and its environment is thus extremely important and fraught with consequences for the potential user—not every *Trichoderma* preparation can be used against every disease.

Hence in the two examples studied, competitive ability and, to a lesser degree, hyperparasitic properties should be privileged. Their expression should be feasible at low temperatures in the case of silver leaf, and over the widest possible range for grey rot.

These requirements occur only in part in wild strains. They should be included in a program for the genetic improvement of antagonistic ability in *Trichoderma* from now on.

Antibiotic phenomena are easy to demonstrate *in vitro*, but difficult to detect under natural conditions. It is quite probable, in view of the way the antagonist is introduced, that they only have a minor role.

5.3. Production of a *Trichoderma* Biopreparation

Adequate quantities of a biopreparation of *Trichoderma* are required for experiments in the field.

5.3.1. *Laboratory Preparation*

The fungus is cultured in Roux flasks or horizontally-placed 1-L milk bottles containing about 50 ml of oat agar medium. After incubation for 3 wk under laboratory conditions (20 to 22°C) the culture is ground up in deionized water. The conidial concentration is adjusted as a function of the experimental objectives.

5.3.2. *Industrial Preparation*

By means of a new technique for culture on solid media (Boirie et al., 1983), a dehydrated product containing 10^{10} viable spores/gram may be obtained. This powder is put into suspension in water prior to utilization.

Culture in liquid media is also feasible. The biomass thus obtained, which consists mainly of chlamydospores, is dried and finely ground. The suspension of revivable elements must be prepared in water that does not have any inhibitory effects on germination. In particular, water with a high mineral content should be avoided (Grosclaude, 1983).

5.4. Conditions for Field Application

5.4.1. *Example of Silver Leaf* (Grosclaude, 1983)

The biological protection of pruning wounds consists of applying a protective treatment as soon as possible after pruning and before infection by the pathogen. A small quantity of *Trichoderma* spore suspension may be applied to the wounds as they are made, by means of special pruning secateurs. Spore application with an ordinary sprayer would produce the same results, but would require much larger quantities of spores.

Two types of application equipment are currently available:

1. A pneumatic sprayer/pruning tool.
2. A hand-operated secateur/sprayer supplied with spore suspension from a low-volume recipient under pressure.

5.4.2. *Example of Grey Rot* (Dubos, 1982)

Experiments were carried out in vines according to the standards defined by the Commission des Essais Biologiques (CEB, Biological Trials Commission) for testing chemical fungicides (1970), involving a block method with six replicates and assessment of rot by the Desaymard method in 100 bunches per unit plot.

Treatments were carried out according to the "standard method" (CEB, 1970) with application at (*A*) end of flowering/beginning of fruit-set, (*B*) closure of the bunch, (*C*) beginning of ripening, and (*D*) 3 wk before harvest.

The biopreparation was applied with a pneumatic sprayer at a rate of 450 1/ha with the nozzle directed toward the bunches.

It is important to ensure that the other chemicals used to control vine diseases are compatible with *Trichoderma*. The degree of tolerance or susceptibility of *Trichoderma* to various pesticides was measured by Davet (1983). Problems of incompatibility may be relatively easily overcome by employing resistant clones obtained by mutation. The antagonistic ability of such clones does not appear to be reduced.

When the concentrations of 10^6, 10^7, 10^8 and 10^9 spores/ml were compared in "standard method" treatments, 10^8 and 10^9 exhibited comparable efficacity, whereas a rapid decline was observed with 10^7 and 10^6. This may be explained by the work of Henis (1984), in which a concentration of 10^8 spores/ml proved to be optimal to rapid germination and increased germination tube vigor (Table 6).

A concentration of 10^8 spores/ml should be employed in vineyard treatments.

The different treatments (involving various combinations of the four "standard method" applications and an earlier one approximately 1 wk before flowering) all provide an improvement in comparison to the control (Table 7).

Only the combination of all four "standard method" treatments gave a significantly different result from the others.

The primordial importance of early treatments, especially those around flowering time, should be noted. It has already been reported that *Trichoderma* colonizes the nutritive base constituted by the senescing floral caps and thus limits the initial installation of *Botrytis* as a saprophyte in the bunch.

It is possible from this observation to envisage a mixed control strategy in which chemical treatments are included from the time of ripening onward: either by (1) maintaining the "standard method" treatments; or (2) reducing the number of treatments and only retaining those that, according to our knowledge of the parasite's epidemiology and the antagonist's modes of action, seem of greatest importance, namely treatment *A* at the end of flowering and treatment *C* at the time of ripening.

TABLE 6. Determination of Optimal Spore Concentration of the Biopreparation

Treatments (Standard Method)	*Overall Rot (%)*	*Efficacity (%)*
Trichoderma		
10^6 spores/ml	22.8	27.3
10^7 spores/ml	14.8	53.7
10^8 spores/ml	9.5	70.8
10^9 spores/ml	9.9	69.4
Untreated control	32.5	—

TABLE 7. Determination of Most Suitable Time of Application

Treatments	*Timing of Application*[a] *a*	*A*	*B*	*C*	*D*	*Overall Rot (%)*	*Efficacity (%)*
Trichoderma		X	X	X	X	9.5	70.8
10^8 spores/ml	X	X		X		12.7	51.2
	X	X				18.0	46.4
	X		X			16.5	47.4
		X		X		14.9	54.1
		X	X			15.4	52.7
	X					19.6	39.5
				X		29.2	10.2
Untreated control						32.5	

[a]*a*, Before flowering; *A*, fruit set; *B*, closure of the bunch; *c*, ripening; *D*, 3 wk before harvesting.

5.5. Evaluation of this Control Method

5.5.1. *Silver Leaf*

Experiments in industrial orchards, under the conditions of natural infection, are hazardous and always long-term. For this reason it is better to assess the efficacity of *Trichoderma* in nursery trials, where the disease symptoms make their appearance a few months after inoculation with the parasite.

The results presented in Table 8 concern the protection of wounds produced in the cutting back of young plum trees (*Prunus mariana*, clone G.F. 8.1) (Grosclaude, 1984, unpublished).

The *Trichoderma* was applied immediately after cutting back and certain of the variables were inoculated with 50 μl of a basidiospore suspension of *C. purpureum* containing 10^4 spores/ml. The wounds in some of the variables were protected with a coating of neutral bituminous grout (Bituplastic) after inoculation.

Trichoderma applied to fresh wounds 2 days before inoculation was totally effective against *C. purpureum*. These results (1984) are in agreement with those obtained each year since 1970 by Grosclaude.

5.5.2. *Grey Rot*

Table 9 illustrates the results obtained in vineyard trials using the experimental design defined by the CEB (Biological Trials Commission) in 1970.

In Bordeaux vineyard trials, the active ingredient in the form of a fresh preparation of *Trichoderma* or a semi-industrial preparation produced percentage efficacities of 60–70%, with the exception of 1983. Similar results were obtained by Vercesi (1981) and Gullino and Garibaldi (1983) in Italian vineyards.

TABLE 8. Demonstration of the Efficacity of *T. harzianum* Against Silver Leaf Disease

			Assessment (5/21/84)	
Variables	*Treatment— Day D (1/9/84)*	*Treatment— Day D + 2 (1/11/84)*	*Number of Silver-Leaved Plants (Out of 18)*	*Disease Intensity (Out of 90)*[a]
11	Pruning	Bituplastic	0	0
12	Pruning then *Trichoderma*	Bituplastic	0	0
14	Pruning then *Trichoderma*	*Curpureum* then bituplastic	0	0
17	Pruning	*Curpureum* then bituplastic	16	26.5
20[b]	Pruning	*Curpureum*	2	1.5

[a]0, Healthy plant; 5, plant killed by silver leaf. Maximum, 5 × 18 = 90.

[b]The few silver-leaved plants in control plot 20 (compared to 17) demonstrate the utility of an application of neutral protective coating to prevent the inoculum from drying out during dry periods in this type of trial.

TABLE 9. Summary of 8 Yr of Field Experiments

Year	*Rot in Treated Plots (%)*	*Rot in Control Plots (%)*	*Efficacity of Treatments (%)*
1976[a]	9.4	33.6	70
1977[a]	9.5	31.4	71.7
1978[a]	7.6	24.6	69.3
1979[a]	4.2	22.5	81.1
1980[a]	7	15.9	57
1981[a]	9.7	24	59.6
1982[b]	4	27	84
1983[b]	25.2	34	26

[a]Fresh preparation obtained in the laboratory.

[b]Formulation (Phior) produced by the ORSAN Company, 80190 NESLE.

The protection provided is excellent but lower than that obtained chemically with specific fungicides, and especially with highly active ones such as procymidone, vinchlozoline, or iprodione. It is also more irregular and was ineffective in 1983. This poses the problem of method reliability, which is widespread, owing to the varied conditions associated with vineyards in general.

Studies carried out on the effect of temperature on the biological behavior of *Trichoderma* under laboratory conditions (Pallas, 1981) and in the field (Dubos et al., 1983) help to explain the results obtained in 1983. The particularly hot and dry climatic conditions rendered the biological treatments inoperative, at least insofar as the strains usually employed were concerned.

Table 10 shows that strains belonging to the biological group described as "thermosensitive," which includes the strain present in "Phior" and B.11, gave mediocre results. In contrast B.30, which belongs to the "tolerant" group, and 303, which is "thermophilic," gave acceptable results.

The relatively buffered climatic conditions usually associated with the Bordeaux region are well suited to the exploitation of the antagonistic ability of *Trichoderma* in biological control. These results demonstrate the great importance of choice of strain, however, and above all the need to further improve such antagonistic ability so as to make this method of control reliable and usable in other vineyards.

The data in Table 11 are taken from studies by Gullino and Garibaldi (1983) on such a program used in vineyards in Northern Italy. These initial results are very encouraging as they permit improvement of the efficiency of biological con-

TABLE 10. Results Obtained in 1983 with Strains of *Trichoderma* Exhibiting Different Biological Reactions to Temperature

Biological Group	*Strain*	*Rot in Treated Plots*[a] *(%)*	*Efficacity of Treatments (%)*
Thermosensitive	Phior	25.2	26
	B.11	25.8	24
Tolerant	B.30	15	56
Thermophilic	303	10.9	68

[a]Rot percentage in control plots = 34.

TABLE 11. Results Obtained with a Mixed Control Program

Treatments	*Timing of Application* A	B	C	D	*Overall Rot*[a] *(%)*	*Efficacity (%)*
T. viride (3R)	+	+	+	+	14	59.4
Vinchlozoline (750 g a.i./ha)	+	+	+	+	8.3	75.9
T. viride (3R)	+	+	+		14.8	57.1
Vinchlozoline (750 g a.i./ha)	+	+	+		12.5	63.8
T. viride (3R)	+	+				
Vinchlozoline (750 g a.i./ha)			+		10.2	70.4
Control					34.5	

[a]For 100 bunches examined per unit plot (block design with four replicates).

trol and at the same time inhibit the possible development of *Botrytis* strains that are resistant to cyclic imides.

The use of *Trichoderma* to control silver leaf in fruit trees and grey rot in vine are examples of biological control which could lead rapidly to practical application for a number of reasons:

1. *Trichoderma* is suited to industrial multiplication involving culture on solid media or fermentation in liquid media.
2. There are no simple and effective treatments available for the protection of wounds resulting from pruning.
3. Comparable results to those obtained in vines have been obtained for grey rot in other plants, particularly strawberry and apple, by Tronsmo (Tronsmo and Dennis, 1977; Tronsmo, 1983). In addition chemical control has now attained certain limits due to the development in *Botrytis* of resistance first to the benzimidazoles and second to the cyclic imides.

Without being overoptimistic, it is feasible to imagine that after improving adaptability to the particular ecosystem in which they are to be used, or even of the actual antagonistic properties, the strains obtained should become independent of environmental constraints, acquire a certain polyvalency and, thus, become more reliable.

This characteristic of polyvalency is not to be neglected, as other possibilities for using *Trichoderma* against diseases caused by aerial parasites have already been confirmed. For example, research on Excoriosis in vine (Dubos et al., 1978; Bugaret et al., 1983; Jailloux et al., 1983) at an advanced stage; and *Cytospora*, which produces canker in fruit trees (Schulz, 1979), lopping wounds in forest trees (Pottle et al., 1977) and wound parasites that cause decline in vine (a problem of ever-increasing economic importance, which has attained the danger point in some areas) is also a candidate for *Trichoderma* application. These considerations should provide motivation for the industrialists as they open up new potential marketing prospects.

6. CONCLUDING REMARKS

Greater attention is currently being paid to the occurrence of microflora in aerial agrobiocenoses. Pesticide utilization and modification of cultural methods are now more rational and are increasingly accompanied by studies of subsequent effects on the resident microflora. Thus components of the aerial fungal microflora are relatively well known. However, various interrelationships existing between these microorganisms are poorly understood and are, therefore, seldom exploited. In addition the relationships existing at the plant/microorganism interface are frequently ignored. It is probable that mechanisms other than those cited in this chapter remain to be discovered; furthermore, other models might be found within the principal systems already identified.

The first of the three examples discussed is already in practice, and the other two are at the onset of industrial development. Developing the techniques for practical application of these methods in agriculture was often filled with obstacles. The results obtained so far should stimulate not only the study of new models but also the refinement of knowledge already acquired, giving these methods wider application. Nevertheless, several factors must be considered in the practical application of these methods.

1. Selection of an antagonistic strain is of primary importance and has to be considered at different levels of complexity. An antagonist should not be characterized in terms of species, because in nature populations are heterogeneous—certain strains possess marked antagonistic properties, whereas others possess little or no activity in this respect (e.g., not all strains of *T. harzianum* are antagonistic to *B. cinerea*). Furthermore, in the case of hypovirulence, we have observed incompatibility between hypovirulent and virulent strains of *Endothia*.

2. Antagonistic properties may also vary according to the environmental conditions that affect the development of the parasite and, consequently, the interactions between the two protagonists. Therefore, we must understand the modes of action of the antagonist with regard to the parasite. Such knowledge allows the use of antagonists to their fullest potential, especially in pinpointing the best conditions for their application, and permits recognition of the qualities of the antagonist, which might be useful as a base for genetic improvement.

3. Knowledge of the epidemiology of the parasite is also a prerequisite to practical application of biological control. Progress in this area, partially due to the development of mathematical models, allows integration of parasite/antagonist interaction into control strategies which are then better adapted to the environment and the economic constraints of practical agriculture.

These novel approaches to plant protection do not possess the same conceptual uniformity as methods of classical control, such as those based on the use of fungicides. Unlike the techniques used in chemical control, which frequently rely on the sole use of pesticides, these methods are based on different concepts. They are heterogeneous and diverse; however, this diversity is full of promise and has great potential. Although practical application of these methods is complicated, their benefits are many. They are less disruptive to the environment and, thereby, allow biological equilibrium with greater stability.

REFERENCES

Anagnostakis, S. L. (1977). Vegetative incompatibility in *Endothia parasitica*. *Exp. Mycol.*, **1**:306–316.

Anagnostakis, S. L. (1982). Biological control of chestnut blight. *Science*, **215**:466–471.

Baker, K. F. and R. J. Cook (1974). *Biological Control of Plant Pathogens*. Freeman, San Francisco, 433 pp.

Bhatt, D. D. and E. K. Vaughan (1962). Preliminary investigations on biological control of Gray-mold (*Botrytis cinerea*) of strawberries. *Plant Dis. Rep.*, 46, **5:**342–345.

Biraghi, A. (1950). La distribuzione del cancro del castagno in Italia. *Ital. For. Montana*, Ann. 5, Fasc. 1.

Blakeman, J. P. (1981). *Microbial Ecology of the Phylloplane. Academic Press*, London.

Blakeman, J. P. and N. J. Fokkema (1982). Potential for biological control of plant diseases on the phylloplane. *Annu. Rev. Phytopathol.*, **20:**167–192.

Boirie, M., B. J. Pons, and C. Renaud (1983). Production, modes d'action at application d'une nouvelle préparation à activité biopesticide à base de *Trichoderma. Colloq. l'INRA*, **18:**255–262.

Bugaret, Y., B. Dubos, and J. Bulit (1983). L'utilisation du *Trichoderma harzianum* Rifaï dans la pratique viticole pour lutter contre l'Excoriose (*Phomopsis viticola* Sacc). Les antagonismes microbiens. *Colloq. l'INRA*, **18:**297–302.

Carter, M. V. and T. V. Price (1974). Biological control of *Eutypa armeniacae*. II. Studies of the interaction between *E. armeniacae* and *Fusarium lateritium*, and their relative sensitivities to benzimidazole chemicals. *Aust. J. Agric. Res.*, **25:**105–119.

Chet, I. and Y. Elad (1983). Mechanism of mycoparasitism. *Colloq. l'INRA*, **18:**35–40.

Commission des Essais Biologiques (1970). Méthode d'essai d'efficacité pratique de fongicides destinés à combattre la Pourriture grise des raisins (*Botrytis cinerea* Pers.), 37.

Cook, R. J. and K. F. Baker (1983). The nature and practice of biological control of plant pathogens. *The American Phytopathological Society*, St. Paul, MN, 539 pp.

Corke, A. T. K. and T. Hunter (1979). Biocontrol of *Nectria galligena* infection of pruning wounds on apple shoots. *J. Hortic. Sci.*, **5:**47–55.

Corke, A. T. K. and J. Risbeth (1981). Use of microorganisms to control plant diseases. In *Microbial Control of Pests and Plant Diseases* (Burges, ed.). Academic Press, New York, pp. 717–736.

Davet, P. (1983). Les *Trichoderma*. Exemple de champignons antagonistes d'agents pathogènes. *Faune et Flore auxiliaires en agriculture*. ACTA, Paris, 193–205.

Day, P. R., J. A. Dodds, J. E. Elliston, R. A. Jayne, and S. L. Anagnostakis (1977). Double-stranded RNA in *Endothia parasitica. Phytopathology*, **67:**1393–1396.

Dennis, C. and J. Webster (1971a). Antagonistic properties of species-groups of *Trichoderma*. I. Production of non-volatile antibiotics. *Trans. Br. Mycol. Soc.*, **57:**25–39.

Dennis, C. and J. Webster (1971b). Antagonistic properties of species-groups of *Trichoderma*. II. Production of volatile antibiotics. *Trans. Br. Mycol. Soc.*, **57:**41–48.

Dennis, C. and J. Webster (1971c). Antagonistic properties of species-groups of *Trichoderma*. III. Hyphal interactions. *Trans. Br. Mycol. Soc.*, **57:**363–369.

Dickinson, C. H. and T. F. Preece (1976). *Microbiology of Serial Plant Surfaces*. Academic Press, New York, 669 pp.

Dodds, J. A. (1978). Doubled-stranded RNA and Virus-like particles in *Endothia parasitica. American Chestnut Symposium Proc.*, West Virginia University, Morgantown, January 4–5, 1978, pp. 108–110.

Dubos, B. (1982). Unpublished

Dubos, B. and J. Bulit (1981). Filamentous fungi as biocontrol agents on aerial plant surfaces. In *Microbial Ecology of the Phylloplane* (Blackman, ed.). Academic Press, New York, pp. 353–367.

Dubos, B. and Mouches (1982). Unpublished

Dubos, B. and J. M. Olivier (1983). Les antagonismes microbiens: modes d'action et application à la lutte biologique contre les maladies des plantes. *Colloq. l'INRA*, Paris, 360 pp.

Dubos, B., J. Bulit, Y. Bugaret, and D. Verdu (1978). Possibilités d'utilisation du *Trichoderma viride* Pers. comme moyen biologique de lutte contre la Pourriture grise (*Botrytis cinerea* Pers.) et l'Excoriose (*Phomopsis viticola* Sacc.) de la Vigne. *Acad. Agric. France*, 1159–1168.

Dubos, B., F. Jailloux, and J. Bulit (1982a). L'antagonisme microbien dans la lutte contre la Pourriture grise de la Vigne. *Bull. O.E.P.P.*, **12:**171–175.

Dubos, B., F. Jailloux, and J. Bulit (1982b). Protection du vignoble contre la Pourriture grise: les propriétés antagonistes du *Trichoderma* à l'égard du *Botrytis cinerea. Colloq. l'INRA*, **11:**205–219.

Dubos, B., J. Roudet, and J. Bulit (1983). Influence de la température sur les aptitudes antagonistes des *Trichoderma* à l'égard du *Botrytis cinerea* Pers. de la Vigne. *Colloq. l'INRA*, **18:**81–89.

Firman, I. and J. M. Waller (1977). Coffee berry disease and other *Colletotrichum* diseases of coffee. *Phytopath. Pap., CMI.*, **20,** 53 p.

Fokkema, N. J. (1971). The effect of pollen in the phyllosphere of rye on colonization by saprophytic fungi and on infection by *Helminthosporium sativum* and other leaf pathogens. *Neth. J. Plant Pathol.*, **77,** Suppl. 1, 60 p.

Fokkema, N. J. (1976). Antagonism between fungal saprophytes and pathogens on aerial plant surfaces. In *Microbiology of Aerial Plant Surfaces.* (Dickinson and Preece, eds.). Academic, New York, pp. 487–506.

Fokkema, N. J., J. A. J. Van de Laar, A. L. Nelis-Blomberg, and B. Schippers (1975). The buffering capacity of the natural mycoflora of rye leaves to infection by *Cochliobolus sativus* and its susceptibility to benomyl. *Neth. J. Plant Pathol.*, **81:**176–186.

Grente, J. (1981). Les variants hypovirulents de l'*Endothia parasitica* et la lutte biologique contre le chancre du châtaignier. Thèse d'Etat. University of Brest, 194 pp.

Grente, J. (1983). Problèmes concernant le déterminisme et le mode d'action de l'hypovirulence chez *Endothia parasitica. Colloq. l'INRA*, **18:**157.

Grente, J. and S. Sauret (1969). L'hypovirulence exclusive, phénomène original en pathologie végétale. *CR Acad. Sc. Paris*, **268:**2347–2350.

Griffiths, E. (1971). Negative effects of fungicides in coffee. *Trop. Sci.*, **14:**79–89.

Griffiths, E. (1981). Iatrogenic plant diseases. *Annu. Rev. Phytopathol.*, **19:**69–82.

Grosclaude, C. (1970). Premiers essais de protection biologique des blessures de taille vis-à-vis de *Stereum purpureum* Pers. *Ann. Phytopathol.*, **2**(3):507–516.

Grosclaude, C. (1983a). Utilisation du *Trichoderma* dans la protection des blessures chez les arbres. *Faune et Flore Auxiliáires en Agriculture*. ACTA, Paris, pp. 217–218.

Grosclaude, C. (1983b). Activité du *Trichoderma harzianum* vis-à-vis du *Stereum purpureum. Colloq. l'INRA*, **18:**115–118.

Gullino, M. L. and A. Garibaldi (1983). Situation actuelle et perspectives d'avenir de la lutte biologique et intégrée contre la Pourriture grise de la Vigne en Italie. *Colloq. l'INRA*, **18:**91–97.

Henis, Y. (1984). Premier séminaire international sur les *Trichoderma* et *Gliocladium*, Beltsville (USA). Communication non publiée.

Hislop, E. C. (1976). Some effects of fungicides and other agrochemicals on the microbiology of the aerial surfaces of plants. In *Microbiology of Aerial Plant Surfaces* (Dickinson and Preece, eds.). Academic, New York, pp. 41–74.

Jailloux, F., Y. Bugaret, and G. Froidefond (1983). Recherches sur le mode d'action du *Trichoderma harzianum* Rifai à l'égard de l'Excoriose de la vigne (*Phomopsis viticola* Sacc.). *Colloq. l'INRA*, **18:**99–108.

Jarvis, W. R. and K. Slingby (1977). The control of powdery mildew of greenhouse cucumber by watersprays and *Ampelomyces quisqualis. Pl. Dis. Rep.*, **61:**728–730.

Kranz, J. (1981). Hyperparasitism of biotrophic fungi. In *Microbial Ecology of the Phylloplane* (Blackman, ed.). Academic, New York, pp. 327–352.

Lapierre, H. and B. L'Hostis (1983). Lutte biologique contre les champignons par l'utilisation de certains de leurs virus ou d'autres matériels génétiques à hérédité cytoplasmique. *Colloq. l'INRA*, **18:**155–156.

Matta, A. (1962). Segnalzione della *Cercospora unamunoi* (Unam.) E. Castel. e di un suo parassita su peperone in serra. *Ann. Fac. Sc. Agr. Univ. Stud. Torino*, **I:**307–312.

Newhook, F. J. (1957). The relationship of saprophytic antagonism to control of *Botrytis cinerea* Pers. on tomatoes. *New Zealand J. Sci. Technol.*, **38**(5):473–481.

Olivier, J. M. (1983). Les organismes antagonistes d'agents phytopathogènes. *Faune et Flore Auxiliaires en Agriculture*. ACTA, Paris, pp. 145–164.

Pallas, G. (1981). Caractérisation des souches de *Trichoderma*. Mem. D.E.A., Université de Bordeaux II.

Peresse, M. and D. Le Picard (1980). *Hansfordia pulvinata*, mycoparasite destructeur du *Cladosporium fulvum. Mycopathologia*, **71:**23–30.

Peresse, M. and D. Le Picard (1983). Perspectives actuelles d'utilisation du *Hansfordia pulvinata*, hyperparasite du *Cladosporium fulvum*, contre la Cladosporiose de la Tomate. *Faune et Flore Auxiliaires en Agriculture*. ACTA, Paris, pp. 283–288.

Ponchet, J. (1982). Réalités et perspectives de la lutte biologique contre les maladies des plantes. *Agronomie*, **2**(4):305–314.

Pottle, W. H., A. L. Shigo, and R. O. Blanchard (1977). Control of wound hymenomycetes by *Trichoderma harzianum. Pl. Dis. Reptr.*, 61, 8, 687–690.

Rifai, M. A. (1969). A revision of the genus *Trichoderma. Mycological papers*, 116, 56 p.

Riphagen, I., N. J. Fokkema, W. J. Kastelein, and P. H. Vereijken (1979). Effect of aphid honeydew on saprophytic and pathogenic Fungi of wheat leaves under controlled environmental conditions. *Acta Bot. Neer.*, **28:**240–241 (Abstract).

Roquebert, (1982). Unpublished

Royse, D. J. and S. M. Ries (1978). The influence of fungi isolated from peach twigs on the pathogenecity of *Cytospora cincta. Phytopathology*, **68:**603–607.

Ruinen, J. (1961). The phyllosphere. I. An ecologically neglected milieu. *Plant Soil*, **15:**81–109.

Schulz, U. (1979). Untersuchungen zur biologischen Bekämpfung der "ValsaKranKheit." *Mitt. Biol. Bundes Land. Forstwirtsch*, p. 191.

Shokes, F. M. and R. A. Taber (1983). Occurrence of a hyperparasite of *Cercosporidium personatum* on peanut in Florida. *Phytopathology*, **73:**505.

Spurr, H. W. (1977). Protective applications of conidia of nonpathogenic *Alternaria sp.* isolates for control of tobacco brown-spot disease. *Phytopathology*, 67, 128–132.

Strizyk, S. (1982). Modélisation et évaluation du risque parasitaire. *Phytoma*, 335, 6–7.

Sundheim, L. (1983). L'hyperparasite *Ampelomyces quisqualis* dans la lutte contre l'Oïdium du Concombre. *Les Colloques de l'INRA*, 18, 145–154.

Taber, R. A., D. H. Smith, R. F. Pettit, and J. D. Johnson (1981). Potential for biological control of *Fulvia fulvum* by *Hansfordia* in Texas. *Phytopathology*, **71:**908.

Tirilly, Y. (1985). Fosetyl. Al and *Hansfordia pulvinata* as integrated control agents of *Fulvia fulva* on tomato leaves. Bordeaux Mixture Centenary Meeting (In Press).

Tirilly, Y. and B. Trique (1983). Production d'enzymes lytiques et d'une toxine par l'*Hansfordia pulvinata*, hyperparasite du *Cladosporium fulvum. Colloq. l'INRA*, **18:**51–58.

Tronsmo, A. (1983). *Trichoderma harzianum* used as a biocontrol agent against *Botrytis cinerea* apple. *Colloq. l'INRA*, **18:**109–113.

Tronsmo, A. and Dennis C. (1977). The use of *Trichoderma* species to control strawberry fruit rot. *Neth. J. Pl. Path.*, **83:**449–455.

Van Alfen, N. K. (1982). Biology and potential for disease control of hypovirulence of *Endothia parasitica. Annu. Rev. Phytopathol.*, **73:**132–135.

Vercesi, A. (1981). Efficacité pratique du *Trichoderma* à l'égard de la Pourriture grise de la Vigne (communication personnelle).

Vuillemin, P. (1887). Etudes biologiques sur les champignons. *Bull. Soc. Sci. Nancy*, Sér. 11, **8**(20):113.

Wood, R. K. S. (1951). The control of diseases of lettuce by the use of antagonistic organisms. Control of *Botrytis cinerea* Pers. *Ann. Appl. Biol.*, **38:**203-216.

6

TRICHODERMA— APPLICATION, MODE OF ACTION, AND POTENTIAL AS A BIOCONTROL AGENT OF SOILBORNE PLANT PATHOGENIC FUNGI

ILAN CHET

Department of Plant Pathology and Microbiology, The Hebrew University of Jerusalem, Faculty of Agriculture, Rehovot, Israel

In their pioneering work, Weindling and Emerson (1936) isolated antifungal substances from a culture of *Trichoderma lignorum*, later identified as a *Gliocladium* (Webster and Lomas, 1964), which controls damping-off disease in citrus caused by *Rhizoctonia solani.* Weindling and his co-workers (Weindling, 1934; Weindling and Fawcett, 1936; Weindling and Emerson, 1936) suggested, as early as 50 years ago, that *Trichoderma* may serve as a biocontrol agent against soilborne plant pathogenic fungi. However, only now, with the increasing interest in biological control, owing to ecological and economic concerns, and with the rapid development of biotechnology in agriculture, do we begin to see commercial products that utilize *Trichoderma* as the active biocontrol agent. Dennis and Webster (1971a–c) described the antagonistic properties of *Trichoderma* in terms of antibiotic production and hyphal interactions. Wells et al. (1972) were the first to report field control of *Sclerotium rolfsii* Sacc. by infesting soil with *T. harzianum* grown on an autoclaved mixture of ryegrass seeds and soil.

It is clear that the basic findings and description of the phenomenon, reported years ago, could not be further developed because of the lack of efficient wide-spectrum isolates, the know-how to scale up fungal growth, and appropriate and economic techniques for formulation and delivery.

During the last few years both plant pathologists and commercial companies have shown significant interest in the potential of *Trichoderma* spp. Baker and Cook have dedicated two excellent books (Baker and Cook, 1982; Cook and Baker, 1983) to biological control, in which they summarize some of the key publications about *Trichoderma* during the last 50 years.

In this chapter I have concentrated mainly on studies published within the last few years to show and evaluate research achievements related to the capability of *Trichoderma* to reduce disease incidence, caused by soilborne plant pathogenic fungi, both from the basic and applied points of view.

1. SELECTIVE MEDIA FOR TRICHODERMA

Quantitative estimation of *Trichoderma* spp. in soil is often difficult because of the relatively rapid growth of other soil fungi on conventional agar media.

A selective medium is a very useful tool for isolating *Trichoderma* from soils as well as for estimating its survival in soil. Elad et al. (1981b) developed a *Trichoderma*-selective agar medium (TSM) for the quantitative isolation of *Trichoderma* spp. from soil. Selectivity was obtained by using chloramphenicol as a bacterial inhibitor and pentachloronitrobenzene, *p*-dimethylaminobenzenediazo sodium sulfonate, and rosebengal as selective fungal inhibitors. The glucose concentration in the TSM, though low, still allowed relatively rapid growth and sporulation of *Trichoderma*, enabling convenient and rapid identification of *Trichoderma* colonies. The percentage of *Trichoderma* recovered from artificially inoculated soils was high.

When the medium was used with soils containing a high population of *Fusarium* spp., they were found to grow on it also. Modifications were, therefore, made to improve the *Trichoderma*-selective medium by supplementing it with either benomyl, for efficient isolation of *Fusarium* spp. from soil, or captan, for specific selective isolation of *Trichoderma* spp. even in the presence of *Fusarium* in the soil (Elad and Chet, 1983).

Two other effective selective media were developed by Davet (1979) and Papavizas and Lumsden (1982). The selectivity of Davet's medium is based on the presence of allyl alcohol (0.5 ml/l) and vinchlozoline (2.5 mg/l). Papavizas and Lumsden (1982) improved the semi-selective medium for *Trichoderma* developed by Papavizas (1981), which is ineffective with soils containing Mucorales, by adding alkylaryl polyether alcohol alone or in combination with sodium propionate. The basal medium contained (per liter): V-8 juice, 200 ml; water, 800 ml; agar, 20 g; and glucose, 1 g. The improved medium contained the following antimicrobial agents (μg/ml): neomycin sulfate, bacitracin, penicillin G, and chloroneb, 100; chlortetracycline hydrochloride, 25; nystatin, 20; and sodium proprionate, 500. Alkylaryl polyether alcohol was added at 2.0 ml/l. All three available media are effective for the detection and estimation of *Trichoderma* spp. population. Different conditions or soils may prove one relatively superior to the others.

2. TRICHODERMA ISOLATE IDENTIFICATION AND EFFICACY TESTS

The genus *Trichoderma* includes nine different species aggregates (Rifai, 1969). Kuter has recently presented a scheme enabling relatively rapid identification (*Trichoderma* Workshop, Beltsville, Maryland, 1984). However, many isolates found in the soil do not exactly fit any of the nine species, but are on the "border line" between two species. Moreover, conventional taxonomy does not always help identify and especially compare isolates. Since many *Trichoderma* isolates are tested for their biological activity, it is essential to differentiate among them.

Zamir and Chet (1985) recently found that different *Trichoderma* isolates contain characteristic isoenzymes which can serve as markers. Barak et al. (unpublished data) prepared antibodies against different *Trichoderma* isolates that agglutinate conidia and enable distinction between isolates even in the same species aggregate. It appears that two systems should be applied for the identification and characterization of isolates: taxonomic differentiation together with the application of physiological and biochemical markers.

When isolating *Trichoderma* from the soil, one of the crucial problems is the rapid evaluation of its efficacy. Hadar et al. (1979a) and Elad et al. (1980a) tested the antagonistic ability of *Trichoderma* against soilborne plant pathogens in the dual culture system described by Dennis and Webster (1971a), using a synthetic medium agar (Okon et al., 1973). This system was rather efficient and rapid when tested against *Rhizoctonia solani.* However, Sivan and Chet (unpublished data) found that when tested against *Fusarium*, no correlation could be drawn between results in plates and in the soil. Henis et al. (1983) found no clear correlation between the penetration of *S. rolfsii* sclerotia by *T. harzianum* and the antagonistic ability of the latter. However, dried sclerotia, placed in Petri plates containing natural soil, were consistently degraded by the antagonist (Henis and Papavizas, 1983). Studying this interaction, Henis et al. (1984) found a high negative correlation between plant stand in the greenhouse and sclerotial germination on soil plates, but not with antagonism on agar plates. This dual culture technique can be used for rapid evaluation of the antagonistic capacity of *Trichoderma* strains against *S. rolfsii.* It appears, however, that, as yet, we have no general, rapid, reliable test, and, considering the variability in pathogens, soils and temperatures, that the ultimate tests are still those using "test plants."

Baker and his co-workers (Henis et al. 1978; Liu and Baker, 1980; Chet and Baker, 1981) tried the radish system and found it easy and rapid—diseases, such as *Rhizoctonia solani* damping-off, could be detected within 7–9 days. Elad et al. (1980a) used bean seedlings and Sivan et al. (1984) utilized cucumbers as test plants for *Pythium aphanidermatum*.

3. GROWTH AND FORMULATION OF *TRICHODERMA* FOR APPLICATION AS A BIOCONTROL AGENT

Backman and Rodriguez-Kabana (1975), using molasses-enriched clay granules as both a food base for growing the same antagonist and a carrier to facilitate dispersal in the field, observed a significant decrease in *S. rolfsii* damage to peanuts. These granules, however, were ineffective when used by Kelley (1976) for the control of *Phytophthora cinnamomi* in pine seedlings.

Hadar et al. (1979a,b) tried different substances for growing *Trichoderma*. Of the various agricultural wastes tested, such as manure and sawdust, wheat bran proved the best medium for the growth and sporulation of *T. harzianum*. It was, therefore, chosen as a foodbase and carrier for the application of *T. harzianum* to the soil. The fungal preparation contained 2.9×10^9 conidia/g dry wt. Eighty percent of *T. harzianum* conidia in this preparation were viable even after storage for 6 months at either 4°C or room temperature (24–30°C)

It was later found that the increased pH of the preparation, which occurred during fungal growth, sometimes enhanced bacterial contamination.

In another study Sivan et al. (1984) found that a mixture of peat:wheat bran (1:1 v/v) was a much better medium and carrier since its pH was kept at about 5.5 during the entire growth period and the culture survived at 25°C for 1 yr.

Huang et al. (1981) grew *T. viride* on various autoclaved seeds at room temperature for several weeks. Dunn et al. (1983) reported that large batch production of *Talaromyces* and *Trichoderma hamatum* was performed by liquid fermentation in 20-l vessels. The system simulated industrial conditions by utilizing commercially available ingredients consisting of molasses plus either dried brewers yeast, cottonseed flour, or corn-steep liquor. *Trichoderma hamatum*, grown initially in a 1-l fermentor containing a molasses–yeast medium, yielded a 7- 8-g dry-weight mat composed of hyphae and chlamydospores (Lewis and Papavizas, 1984). Mixed with Pyrax (10% by weight), the formulation contained 10^7 propagules/g. Addition of NaCl or $CaCl_2$ to the various media did not stimulate conidial production. Survival of *Trichoderma* in the Pyrax formulations was monitored over time, with approximately 90% of the propagules remaining viable after 1 month. Ricard (1981) produced a mixture of *T. polysporum* and *T. harzianum* as elongated pellets which were introduced into trees to combat silverleaf disease. Later, he also found that these pellets could reduce the incidence of the Dutch elm disease (Ricard, 1983).

Recently, Eveleigh reported (*Trichoderma* Workshop, Beltsville, Maryland 1984) that he grew *T. reesei* in fermentors and the fungus produced 10^7 conidia/ml after several days. This finding may solve the problem of how to produce conidia of biocontrol agents in liquid fermentation.

4. APPLICATION AND EFFECTIVENESS OF TRICHODERMA IN CONTROLLING SOILBORNE PATHOGENS

4.1. Greenhouse Experiments

Hadar et al. (1979a,b) isolated *T. harzianum* and applied it in the form of wheat-bran culture to *R. solani*-infested soil in the greenhouse. This isolate effectively controlled damping-off of bean, tomato and eggplant seedlings, but not *S. rolfsii.* Chet et al. (1979) found a more efficient isolate, capable of decreasing the incidence of plant diseases caused by *S. rolfsii* and *R. solani* in naturally infested soils. The wheat-bran preparation of *T. harzianum* increased bean plant growth in noninfested soil and controlled *S. rolfsii* more efficiently than the conidial suspension of the same antagonist. An uninoculated wheat-bran preparation increased disease incidence. Incorporating the wheat-bran inoculum preparation of this isolate into pathogen-infested soil significantly reduced bean diseases caused by *S. rolfsii, R. solani* or both.

Henis et al. (1978) found that successive radish planting induced soil suppressiveness to *R. solani.* Liu and Baker (1980), working with a similar system, found that *T. harzianum* population increased with the increase in suppressiveness. The suppressiveness to *R. solani* of soil infested with large ($> 589\ \mu m$) propagules of this fungus was generated only in acidified or naturally acidic soils after replanting radish, alfalfa, or sugar beet at weekly intervals (Chet and Baker, 1981). Indeed, Colombian soil, naturally suppressive to *R. solani,* was found to be acidic, containing a naturally high population of *T. hamatum.* This antagonist rendered conducive soil suppressive to *R. solani* (Chet and Baker, 1981).

To reduce the amount of *Trichoderma* added to soil to control soilborne plant pathogenic fungi, Harman et al. (1980) developed a seed treatment. In laboratory studies, treating pea and radish seeds with *T. hamatum* conidia in a Methocel slurry, protected seeds and seedlings from *Pythium* spp. and *R. solani*, respectively, nearly as effectively as fungicide seed treatments. *Trichoderma* propagules increased approximately 100-fold in soils planted with *T. hamatum*-treated seeds; replanting these soils once or even twice with untreated seeds yielded lower disease incidence than soils originally planted with untreated seeds. *R. solani* and *Pythium* densities were lower in soils containing *T. hamatum* than in soils without this antagonist (Harman et al., 1980). In a further study, the same group found that, at soil temperatures between 17 and 34°C, the addition of chitin or *R. solani* cell walls to the coating of seeds previously treated with a conidial suspension, increased both the ability of the mycoparasite, *T. hamatum*, to protect the seeds against *Pythium* spp. or *R. solani* and the population density of *Trichoderma* in the soil. *T. hamatum* with chitin, but without *R. solani* cell walls, effectively reduced damping-off caused by *Pythium* spp., compared to a seed treatment containing only *T. hamatum.* The addition of a nonpathogenic isolate of *R. solani* had little effect either on seed protection or levels of *Trichoderma* in the soil, whereas the addition of cellulose tended to decrease both (Harman et al., 1980). Similarly, coating iris bulbs with *T. harzianum* was highly effective against *R. solani* and *S. rolfsii* (Chet et al., 1982).

Harman and his co-workers (Harman, 1982; Hadar et al., 1984) applied *T. hamatum* to New York State soils but failed to achieve biological control. Coated seeds introduced into these soils, which have a pH level close to 7 and low iron availability, were colonized by *Pseudomonas* spp., which produced siderophores. These, in turn, apparently inhibited the development and activity of *Trichoderma*.

Sivan and Chet (1982) found that a preparation of *Trichoderma hamatum* (Bon.) Bain, based on the wheat-bran/peat mixture, controlled *Pythium aphanidermatum* (Edson) Fitz. in pea, cucumber, tomato, pepper and gypsophila, and *R. solani* Kühn and *Sclerotium rolfsii* Sacc. in beans. Coating pea and cucumber seeds with a *T. hamatum* conidia suspension reduced *Pythium* disease incidence by 67–88% and increased the level of *Trichoderma* population in the soil 1.5–4.7 times.

A new isolate of *T. harzianum*, capable of controlling *P. aphanidermatum* better than *T. hamatum* and at higher temperatures, was also found. This led to the idea that a mixture of several isolates should be prepared in order to achieve activity at a wide temperature (or pH) range (Sivan et al., 1984).

Marois and Mitchell (1981b) found that chlamydospores of *Fusarium oxysporum* f. sp. *radicis lycopersici* added to fumigated soil were augmented by conidia of *Trichoderma harzianum, Aspergillus ochraceus* and *Penicillium funiculosum*. Adding conidia of three antagonists reduced *Fusarium* crown rot of tomato in the field too (Marois et al., 1981a).

Marois and Locke (*Trichoderma* Workshop, Beltsville, Maryland, 1984) reported that *T. viride* controlled *Fusarium* wilt of chrysanthemum in freshly steamed bench mixes.

Recently, Sivan and Chet (1985) found a new isolate of *T. harzianum* and applied it, either as conidial seed coating or a bran:peat (1:1) preparation, to naturally infested soil. The isolate significantly reduced disease incidence caused by *Fusarium oxysporum* f. sp. *melonis* in melons, *F. oxysporum* f. sp. *vasinfectum* in cotton and *F. culmorum* in wheat. It was also effective against *R. solani* and *S. rolfsii* attacking seedlings.

It is interesting to note that not much work was done with *Armillaria mellea*. However, Dubos et al. (1978) have found that *T. viride* reduced the initiation and growth of the underground rhizomorphs of this pathogen.

4.2. Field Experiments

Many of the promising results in biological control were achieved in steamed or naturally infested soil under a controlled environment in the greenhouse. However, during the last several years, more and more field experiments were carried out revealing the biological activity of *Trichoderma* under natural conditions. A wheat-bran preparation of *T. harzianum* was applied along with peanut seeds at a concentration of 15 g preparation/m^2. The results, similar to those reported by

Wells et al. (1972) and Backman and Rodriguez-Kabana (1975), who achieved biological control using ryegrass and molasses, respectively, as a foodbase for *T. harzianum*, showed significant control of *S. rolfsii* and *R. solani* (Chet et al., 1979).

In three field experiments the wheat-bran preparation of *T. harzianum* inoculum was mixed with the soil to a depth of 7–10 cm, with a rotary hoe. Dead seedlings with typical symptoms of damping-off caused by either *S. rolfsii* or *R. solani* were recorded throughout the growth period. In the first experiment, carried out with beans in soil naturally infested with *R. solani*, plants were removed eleven weeks after sowing. Disease indices for *R. solani*, in the untreated and *T. harzianum*-treated plots, were 1.12 and 0.87, respectively. The green pod yield of the *T. harzianum*-treated plots increased by 20%, compared with the control.

The second field experiment was carried out with cotton in soil naturally infested with *R. solani.* Twenty-nine days after planting, the proportions of diseased plants in the untreated and *T. harzianum*-treated plots were 44 and 23%, respectively.

In a third field experiment carried out with tomatoes in soil naturally infested with *S. rolfsii, T. harzianum* had significantly reduced the percentage of diseased plants (by 20%) 123 days after planting (Elad et al., 1980a).

Black-root rot of strawberries, caused by *Rhizoctonia solani*, is common in strawberry fields. *Trichoderma harzianum* preparations were used in two successive field experiments in commercial strawberry nurseries and fruiting fields. The severity of *R. solani* in daughter plants was reduced by 18–46% in the treated nursery plots. Infestation of nursery soil with the pathogen, as tested by planting beans in soil samples, was reduced by the *Trichoderma* treatment by up to 92%, as compared with the untreated control. A rapid decline of the disease was observed in soil from *T. harzianum*-treated plots successively planted with bean seedlings. More isolates of *Trichoderma* spp. antagonistic to *R. solani* were found in the infested field as compared with the noninfested one. *Trichoderma harzianum*-treated plants, transferred to the commercial field, gave a 21–37% increase in early strawberry yield. A combined treatment in the nursery and fruiting field resulted in a 20% yield increase as compared to control plots (Elad et al., 1981a). The common practice for controlling stem rot caused by *R. solani* in carnations in Israel is soil fumigation followed by incorporation of pentachloronitrobenzene (PCNB) at 78–80 kg/ha (Hadar et al., 1979a,b). Apparently all the commercial plots are thus fumigated and the *Trichoderma* preparation was tested for its capability to replace PCNB. A linear correlation was obtained between the rate of *T. harzianum* preparation applied to the soil and the degree of disease control. Disease incidence was reduced by 70% when the preparation was applied at 150 g (dry wt)/m^2. *Trichoderma harzianum* gave the best disease control when applied and established in the rooting mixture before transplanting in the field.

When *Trichoderma* was applied to the soil, the antagonist had no contact with the plants' roots. However, in this study, loading the biocontrol agent to the root zone before rooting in the nursery allowed the establishment and multiplication of *Trichoderma* in the root system, thus protecting the root collar. The total amount of preparation used was extremely low—13.5 g/m^2 as compared with 150 g/m^2 used for similar control in infested plots (Elad et al., 1981c). The efficacy of seed coating under a controlled environment (Harman et al., 1980) may lead to the use of low doses of *Trichoderma* for disease control under field conditions as well. Indeed, *T. hamatum* was applied as seed coating for controlling *R. solani* in cotton grown in naturally infested soil. Seeds were treated with either Pelgel (Nitragin, USA) solution (10% w/v), containing 5×10^8 conidia per milliliter as an adhesive (Harman et al., 1980), or, for comparison, ethylmercury chloride (350 g/ton seeds). Treatment with *T. hamatum* reduced damping-off caused by *R. solani* by 60% and bare patches by 39%, 23 days after planting, and increased plant stands by 14% 17 days later. Spraying 2.8 kg/ha terraclor (PCNB 75% w.p.) in seed furrows gave similar results, whereas seeds treated with ethylmercury chloride did not differ from the untreated control (Elad et al., 1982c).

The coating technique requires only 500 ml of conidial suspension per hectare, which is very low when compared with the amounts of *Trichoderma* preparation previously applied (Elad et al., 1980a,b; 1981a).

Ruppel et al. (1983) found that in Colorado fields infested with *Pythium* spp., stands of sugar beet, from *T. harzianum*-treated seeds, tended to be greater than those from nontreated seeds and equal to those from maneb-treated seeds. However, the differences were not significant. In Colorado, *Rhizoctonia solani*, the causal agent of root rot, was also present in the soil. A combined analysis of variance of treatments across 2 yr showed that a preplanting-incorporated in-row application of *T. harzianum* in a wheat-bran carrier or a maneb seed treatment slightly, but significantly, reduced the severity of *Rhizoctonia* root rot in sugar beet as compared with the nontreated control. Seed treatments with *T. harzianum* had little or no effect. In plots where the agent was applied to the soil, *Trichoderma* propagules increased 600-fold by harvest time; where the maneb seed treatment was used, they increased only five-fold.

Elad et al. (1983f) reported again the effectiveness of seed coating in the field. The effectiveness of chemical seed coating was compared with biological seed coating or broadcast application of *T. harzianum*. *Trichoderma* population level in field soil treated with the fungus was 20 times higher than in other soils. *Aspergillus niger*, an important pathogen of peanuts (Jackson and Bell, 1969), generally appears with other *Aspergillus* spp., including *A. flavus*, which produces aflatoxins. *Trichoderma harzianum* application to soil or seeds reduced *Aspergillus* disease better than the commercial chemical seed treatment (43–60 and 31%, respectively). Respective percentages of yield increases in the chemical, *Trichoderma* seed coating and broadcast application treatments were 1.3, 4.2, and 13.9 (Elad et al., 1983f). The difference in the effectiveness of seed coating between Colorado and Israel may be due to the higher temperatures in Israel, which may enhance the activity of *Trichoderma*. Moreover, the slightly higher pH of Colorado soils may reduce the activity of the biocontrol agent (Chet and Baker, 1980).

Fruit rot of tomato and cucumbers caused by *R. solani* is a serious disease in warm, humid regions. Under field conditions, *Trichoderma* preparation reduced tomato fruit rot incidence from 25% in the untreated control to 14.6% 2 wk before harvesting. Consecutive *Trichoderma* treatments, first to the soil and later broadcast application over plants, significantly reduced *R. solani* inoculum potential from a disease index of 3.6 in the control to 0.5 in the combined treatment (Strashnov et al., 1985a).

4.3. Integrated Control

Different fungicides and soil fumigants are widely used for controlling soilborne plant pathogens. Because of the concern regarding the toxicity of these compounds, there is a general trend to reduce the amounts applied to soil. Methyl bromide is a typical example of an efficient fumigant about which the public is concerned because its use causes the accumulation of bromide residues, especially in water (Hoffman and Malkomes, 1974). Moreover, the use of fumigants may enhance reinfestation of the treated soil by pathogens (Marois and Mitchell, 1981b).

One of the most attractive ways of reducing the amounts of fungicide is the integration of sublethal doses of chemicals with *Trichoderma* spp., which is resistant to relatively high doses of methyl bromide (Munnecke et al., 1973) and PCNB (Elad et al., 1981a–c). Davet et al. (1981) found that benomyl inhibited *Trichoderma* growth while thiram enhanced it. Papavizas et al. (1982) and Baker (*Trichoderma* Workshop, Beltsville, Maryland, 1984) found mutants resistant to benomyl and captan. These indicate the possibility of combining some appropriate chemicals with this biocontrol agent.

Integrating biological and chemical control seems a very promising way of controlling pathogens with minimal interference with the biological equilibrium (Baker and Cook, 1982; Henis and Chet, 1975; Papavizas, 1973). Curl et al. (1977) obtained only a slight additive benefit to biological control of *R. solani* by adding PCNB at doses of 1 and 10 mg/g soil together with *Trichoderma* in sterilized soil. Hadar et al. (1979a), however, applied small noneffective doses (1–2 μg/kg) of PCNB to soil along with a *Trichoderma* preparation (2 g/kg) and found that the incidence of eggplant disease caused by *R. solani* declined from 40 to 13%, while *T. harzianum* alone reduced disease incidence to only 26%. Pentachloronitrobenzene alone at this rate had no effect on disease incidence of eggplant seedlings. Similarly, Chet et al. (1979) reported a synergistic effect resulting from the interaction between *T. harzianum* and sublethal doses of PCNB when applied against *S. rolfsii* in peanuts. The control achieved in the combined treatment was greater than the sum of the different treatments (Elad et al., 1983d). This synergism is apparently due to partial suppression of soil microflora, enabling a more effective activity of the biocontrol agent.

Solar heating of the soil or the fumigant methyl bromide, when tested under field conditions, significantly reduced diseases caused by *R. solani* and *Verticillium dahliae* in potatoes. The biocontrol agent *T. harzianum* reduced *R. solani* disease. Combining solar heating or methyl bromide with *T. harzianum* improved their efficiency and also resulted in the control of *S. rolfsii.* Combining heat treatment and *T. harzianum*, both at sublethal doses and under greenhouse conditions, resulted in 90–100% control of *S. rolfsii* disease in beans (Elad et al., 1980b).

The recommended practice for controlling *R. solani* in carnation in Israel is still soil fumigation by methyl-bromide followed by application of PCNB. This fumigation controls *Fusarium* and nematodes, but is not efficient in controlling *Rhizoctonia.* Under field conditions, diseased plants appeared in the untreated control plots 2 wk after transplanting; disease incidence increased to 9% at 4 wk and 59% at 14 wk. Disease incidence was significantly lower in plots treated with either *T. harzianum* or PCNB. Yield was 107, 137, and 161 flowers/m^2 in the control, PCNB, and *T. harzianum* treatments, respectively (Elad et al., 1981c), thus indicating the potential of replacing PCNB by *Trichoderma* after soil fumigation.

Moreover, the common control treatment of *R. solani* in strawberry is soil fumigation of nurseries and fruiting fields with methyl bromide (Wilhelm and Poulus, 1980). However, because planting material is usually taken from last year's fields, a build-up of the pathogen population in fields and in the strawberry plants themselves can result, a phenomenon which can enhance adverse soil infestation. Successful application of *Trichoderma* in commercial nurseries and fields (Elad et al., 1981a) revealed the possibility of using this biocontrol agent for reducing such reinfestation. Indeed, *Trichoderma* spp. is the most common antagonist to appear after soil fumigation, since they have few competitors and can reproduce rapidly (Munnecke et al., 1981). Application of *T. harzianum* after soil fumigation with methyl bromide improved the control of both *S. rolfsii* Sacc. and *R. solani* in a peanut field. Although soil fumigation controlled the diseases caused by these pathogens, rapid reinfestation followed. *Trichoderma harzianum* prevented reinfestation of the fumigated soil (88% reduction) both in a controlled environment and under field conditions. The combined treatment, fumigation and application of *T. harzianum*, caused an almost total mortality of sclerotia in the soil. Transplanting tomato plants treated with *T. harzianum* into soil fumigated with methyl bromide reduced disease incidence caused by *S. rolfsii* and *R. solani* by 93% and increased yield by 160% (Elad et al., 1982b).

Recent results show that under greenhouse conditions, a combination of *T. harzianum* and a reduced dose of methyl bromide (200 kg/ha) totally decreased disease incidence of *R. solani* in bean seedlings as compared with untreated controls. Similar disease control was achieved with the recommended dose of the fumigant. Under field conditions, this combination gave a significant synergistic effect in carrot seedling damping-off caused by *R. solani* and had a similar effect on growth, yield and disease control as that of the recommended dose (Strashnov et al., 1985b).

The positive effect of integrated control was also observed with *Pythium.* Application of *T. harzianum* preparation combined with the fungicide Prothiocarb (as Previcur) to a rooting mixture infested with *P. aphanidermatum* resulted in a synergistic effect, reducing disease incidence in gypsophila cuttings better than either treatment alone (Sivan et al., 1984). These results were similar to those obtained with cucumbers grown in *Pythium*-infested soil. Integration of Dexon with the seed coating of *T. hamatum* reduced the percentage of diseased plants from 53 to 9, results superior to those of each treatment alone (Elad et al., 1983f). Papavizas and Lewis (Papavizas and Lewis, 1979; Lewis and Papavizas, 1980) developed another approach for integrated pest management. They prevented colonization of bean plant residues by *R. solani* by plowing to a depth of 20–25 cm, which reduced root rot of beans and fruit rot of cucumber. Better control was achieved by combining *Trichoderma* with plowing than by each method alone.

Data from most experiments (Grinstein et al., 1979) indicate that integration of low doses of fungicides with *Trichoderma* improves control.

Application of the biocontrol agent, following soil fumigation or solarization, significantly reduced soil reinfestation. These facts may lead to an effective control with low doses of the fungicide or fumigant. Since *Trichoderma*, when applied with a foodbase or as a seed coating can survive for long periods of time and even propagate in soil (Harman et al., 1980), its combination with chemical, cultural, or physical methods (Katan et al., 1976; Chet et al., 1982) can achieve a long-term controlling effect on soilborne plant pathogenic fungi.

5. ECOLOGY OF *TRICHODERMA* AND ITS SURVIVAL IN SOIL

According to soil samples taken from agricultural regions, the natural population of *Trichoderma* is rather low, usually not exceeding 10^2 CFU/g soil.

Henis et al. (1978) succeeded in generating soil suppressiveness to *R. solani* by a successive monoculture of radishes. Liu and Baker (1980) reported that soil suppressiveness was accompanied by a significant increase in *T. harzianum* propagule density. Inoculum density of *R. solani* was inversely proportional to the density of *Trichoderma* propagules following radish monoculture.

Naturally suppressive soil, from the vicinity of Bogota, Colombia, was analyzed and compared with conducive soil from Fort Collins, Colorado (Chet and Baker, 1981). The most pronounced difference was in the number of fungi. Whereas the Fort Collins soil contained a total of 10^3 *Trichoderma* propagules per gram, propagule density in the Colombian soil was extremely high (10^8 propagules/g). A significant difference in the density of the *Trichoderma* population was observed. Only 10^2 propagules/g were found in the Fort Collins clay loam, whereas an unusually high number of *Trichoderma* propagules (8×10^5/g) was detected in the Colombian organic soil. The pH of the Colombian soil was 5.1; that of Fort Collins soil—8.1. The low pH, apparently enhances the propagation of fungi in general and *Trichoderma* in particular. As a consequence, it was favorable for the development of suppressiveness. These results confirm a former study indicating that acidification of soil could induce suppressiveness by *Trichoderma* (Chet and Baker, 1981).

Later, Marshall (1982) found a significant reduction in damping-off of snap bean when seeds were inoculated with *T. harzianum* in a soil with a pH of 3.5. Moreover, acidic pH levels enhanced *in vitro* growth of *T. harzianum* and stimulated its conidiophore formation and conidial germination (Chet and Baker, 1980). Liu and Baker (1980) reported that in moist soil *Trichoderma* survived longer than in dry soil. However, certain isolates can grow even at −80 bars metric potential (Cook and Baker, 1983). At low temperatures the growth rate of *T. harzianum* in culture was greater than that of *S. rolfsii* (Elad et al., 1980a). Results show that *T. hamatum* was more efficient than *T. harzianum* (45.4 and 25.0% *R. solani* disease reduction in cotton, respectively) at low temperatures; at higher temperatures, *T. harzianum* was superior (78% disease reduction compared with 66% with *T. hamatum*) (Elad et al., 1982c). This corroborates the data obtained by Harman et al. (1980) showing that *T. hamatum* was active between 17 and 34°C. These are *Trichoderma* spp. isolated from moderate or warm regions, however, Tronsmo and Dennis (1977) found *Trichoderma* isolates active at temperatures as low as 2°C. These results suggest the possibility of applying the right isolate or a mixture of isolates (Ricard, in *Trichoderma* Workshop, Beltsville, Maryland, 1984) according to the growth conditions. In many of our experiments we found that the minimal effective amount of *Trichoderma* is around 1×10^6 CFU/g soil (Chet and Baker, 1980). When soil samples were assayed 1 yr after carnation planting, the density of *Trichoderma* population was 1.1×10^4 and 4.5×10^2 CFU/g soil in *Trichoderma*-treated and control plots, respectively (Elad et al., 1981c). Papavizas (1981) reported that *T. harzianum* did not survive well in the rhizosphere of bean and pea seedlings when seeds were coated with conidia of the fungus. It also did not increase in the rhizosphere of pea seedlings, from conidia applied directly to soil 1 day before planting. Infusion of pea seed with the fungicide metalaxyl before coating it with conidia of *T. harzianum*, improved the survival of conidia in the rhizosphere. As already described, *Trichoderma* can tolerate many fungicides, such as methyl bromide, PCNB, captan and maneb (Ruppel et al., 1983). This means that the biocontrol agent must have a relative advantage to survive in agricultural fields, where residues of various pesticides are rather common.

Trichoderma spp. can also produce chlamydospores on natural substrates, such as oat kernels placed in sterile and natural soils. These structures may play an important role in the survival of this genus in soil (Henis and Papavizas, 1983).

6. INCREASED GROWTH RESPONSE OF PLANTS

During recent years, as a result of intensive work carried out with *Trichoderma*, several independent research groups have noticed that the addition of *Trichoderma* conidia induced the growth of various seedlings. Baker and his co-workers in Fort Collins, Colorado (Baker et al., 1984 and unpublished data) found that when *T. harzianum* was added to propagative medium or to greenhouse soil mix, at a rate higher than 10^5 CFU/g soil, significantly increased growth responses were observed, as measured by increases in weight, height and branch and flower production (compared with nontreated controls) for chrysanthemums and petunias.

Similarly, Kleifeld and Chet (unpublished data) found an earlier germination and increase in plant length, leaf area and dry weight upon adding *T. harzianum* conidia to melon, tomato, cucumber, pepper, radish and bean seedlings. Both groups noticed earlier flowering in some plants, a valuable parameter both in ornamentals and vegetables. Using a selective medium (Elad and Chet, 1983), Chang et al. (1985) succeeded in showing that live *Trichoderma* propagules are present in the roots of responding plants. Recently, an increased growth response was also confirmed by Harman at Geneva, New York, as a result of seed coating by *Trichoderma* (personal communication).

The mechanism of this phenomenon has not been thoroughly studied. *Trichoderma* may affect minor pathogens in the soil but it may, also, directly affect the plant by excreting a regulating hormone which may, in turn, increase the growth rate or the efficiency of nutrient uptake.

This positive effect of the antagonist on plant growth and flower production can serve as a valuable factor in using *Trichoderma* as a biocontrol agent.

7. POSSIBLE MECHANISMS INVOLVED IN *TRICHODERMA* ANTAGONISM

Trichoderma spp. are common fungi in almost every soil and are antagonistic to other fungi. Many isolates produce volatile and nonvolatile antibiotics (Dennis and Webster, 1971a,b). The ability to produce antifungal substances varies with the isolate, even within the same species-aggregates (Dennis and Webster, 1971a). From the literature, it appears that the most effective antagonists belong to the species *T. harzianum*. When their activity was studied, no typical antibiotics could be detected (Hadar et al., 1979a; Chet and Baker, 1980, Harman et al., 1980). Recently, Sivan et al. (1984) reported growing *Trichoderma*, which is antagonistic to *P. aphanidermatum*, on a cellophane membrane placed on agar. By removing the membrane and inoculating the agar with *Pythium* they showed that the growth of the pathogen was partially inhibited by substances produced by the antagonist. The same *Trichoderma* was stimulated by substances released by *Pythium*.

In other cases, especially when no effect of *Trichoderma* can be directly observed or detected but its controlling activity in natural soil is significant, the possibility of competition (Alexander, 1982) between the biocontrol agent and the pathogen should be considered (Cook and Baker, 1983).

In many experiments, showing successful biological control, the antagonistic *Trichoderma* was found to be a mycoparasite (Boosalis, 1964; Hadar et al.,

1979; Harman et al., 1980; Chet and Baker, 1981; Chet and Elad, 1982; Elad et al., 1983b).

From recent work, it appears that mycoparasitism is a complex process including several successive steps. The initial detectable interaction shows that the hypha of the mycoparasite grows directly toward its host (Chet et al., 1981). This phenomenon appears as a chemotropic growth of the *Trichoderma* to some stimuli in the host's hyphae or toward a gradient of chemicals excreted by the host (Chet and Elad, 1983) (Fig. 1). Chemotactic responses in a host–parasite relationship were previously found in lytic bacteria (Chet et al., 1971) or nematode-trapping fungi (Jansson and Nordbring-Hertz, 1979).

When the mycoparasite reaches the host, its hyphae often coil around it (Fig. 2) or are attached to it by forming hook-like structures. *Trichoderma hamatum* produced appressoria at the tips of short branches. Following these interactions, the mycoparasite sometimes penetrates the host mycelium, apparently, by partially degrading its cell wall (Elad et al., 1983b,c).

Dennis and Webster (1971c) studied the possible interaction of *Trichoderma* spp. with nylon threads with a diameter similar to that of *P. ultimum* hyphae. *Trichoderma* never coiled around the threads, suggesting that the coiling is not just a randomly occurring contact. Moreover, the antagonism of *Trichoderma* was found to be rather specific (Elad et al., 1980a; Chet and Baker, 1981; Sivan et al., 1984). The problem of specificity in fungal–fungal interaction and the possible role of agglutinins in mycoparasitism has only recently been studied.

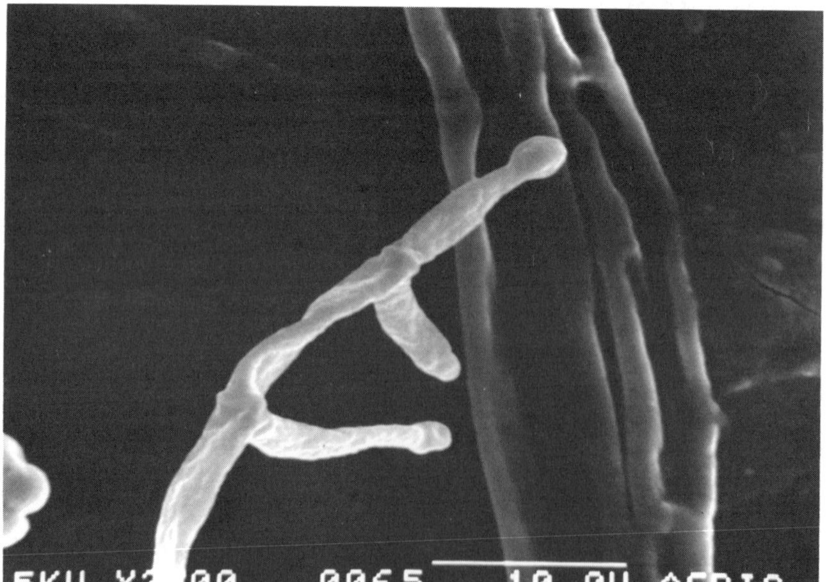

Fig. 1. Chemotropic growth of *Trichoderma harzianum* toward *Rhizoctonia solani* (×3000). (Chet, Sadovsky and Elad, unpublished data).

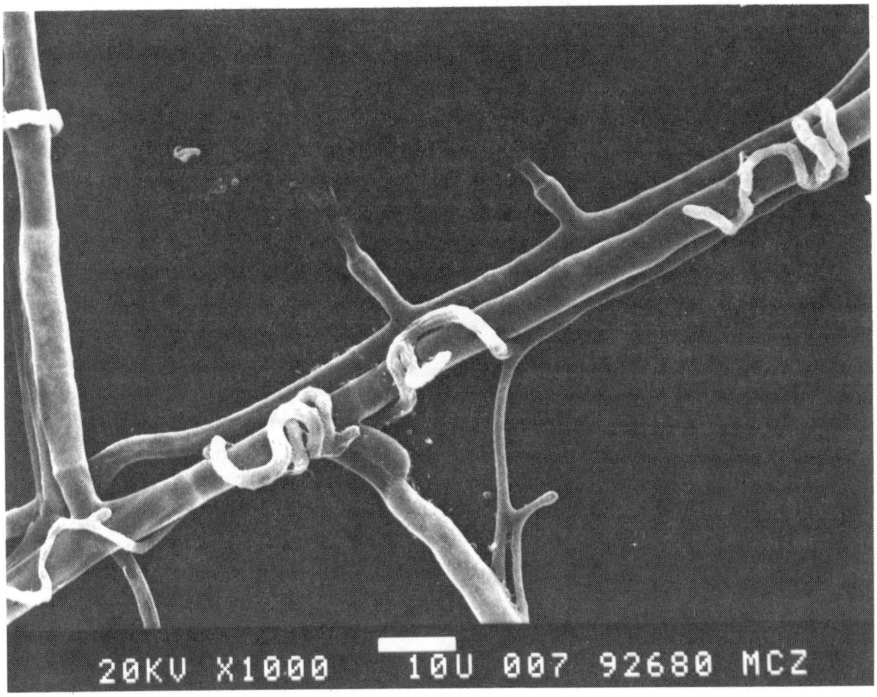

Fig. 2. Scanning electron micrograph of *Trichoderma* coils around *R. solani* hypha (×1800). (Chet, unpublished data).

Lectin activity in a host–mycoparasite relationship was demonstrated with *R. solani* and *T. harzianum*. Attachment of type O, but not type A and B, erythrocytes to hyphae occurred on *R. solani*, but not on its mycoparasite. A lectin, present in *R. solani* hyphae, binds to galactose residues on *Trichoderma* cell walls as well as to type O erythrocytes. This agglutinin may play a role in prey recognition by the predator. Moreover, since it does not distinguish among biological variants of the pathogen, it enables the *Trichoderma* species to attack various different *R. solani* isolates (Elad et al., 1983a). A recent study by Barak et al. (1985) indicates that a specific recognition between *Trichoderma* spp. and *S. rolfsii* may be due to the presence of lectins. The activity of this agglutinin, which was inhibited by D-glucose and D-mannose, depends on the presence of the cations Mn^{2+} and Ca^{2+}. It was purified on a Sepharose column and its separation on polyacrylamide gels showed two bands indicating two subunits. The agglutinin is apparently a glycoprotein with lectin characteristics (Mirelman et al., 1975).

Three *Trichoderma* isolates were tested for their ability to attack the mycelium of *S. rolfsii* cultures grown on agar plates. *Trichoderma hamatum* grew over the *S. rolfsii* colony and degraded its mycelium, whereas *T. harzianum* (No. 110) reached the *S. rolfsii* colony but did not invade it. Another isolate of *T.*

harzianum (No. 233) stopped growing before reaching the *S. rolfsii* colony, forming a clear zone. Washed conidia of *T. hamatum* were agglutinated by *S. rolfsii* lectin, whereas very slight and no agglutination was observed with *T. harzianum* No. 110 and No. 233, respectively. Moreover, suspended conidia of *T. hamatum* poured onto colonies of the pathogen were immediately agglutinated to the hyphae (Barak et al., 1985).

It appears that recognition may be an essential step in the specificity of mycoparasitism.

After recognition and attachment, *Trichoderma* grew along or coiled around the hyphae of the phytopathogenic fungus (Elad et al., 1983c; Harman et al., 1980; Henis and Chet, 1975; Wells et al., 1972). Sites of interaction were photomicrographed with infrared film. Bright regions indicating infrared irradiation were observed, apparently due to the high parasitic activity occurring there (Elad et al., 1983e). The microscopic observations (Weindling, 1932; Dennis and Webster, 1971c; Hadar et al., 1979a,b; Liu and Baker, 1980) lead to the possibility that mycolytic enzymes are excreted by *Trichoderma* spp. The successful biocontrol agent against *S. rolfsii* and *R. solani* did, indeed, produce extracellular β-1,3-glucanase and chitinase when grown on the cell wall of *R. solani* (Chet et al., 1979; Hadar et al., 1979a,b,; Chet and Baker, 1981). Later, Elad et al. (1982a) studied the enzymatic basis of this interaction in culture and soil.

Trichoderma harzianum excreted β-1,3-glucanase and chitinase into the medium when grown on laminarin and chitin, respectively. *Trichoderma harzianum* also showed high activity of both enzymes when grown on homogenized *S. rolfsii* sclerotia. A high lytic activity was detected in a wheat-bran culture of the fungus and in soil inoculated with this culture. In the presence of cycloheximide, antagonism was prevented and enzymatic activity was inhibited (Elad et al., 1983c). Isolates of *T. harzianum* differ in the levels of hydrolytic enzymes produced when mycelia of *S. rolfsii, R. solani* and *P. aphanidermatum* are attacked in soil. This phenomenon correlates with the ability of each of the *Trichoderma* isolates to control the respective soilborne pathogens. However, the specificity of *Trichoderma* could not be explained by the different activity of these enzymes, since nonantagonistic *Trichoderma* isolates also produce lytic enzymes, although fewer (Elad et al., 1982a). This may, therefore, support the idea that recognition is an important factor in the specificity of *Trichoderma*.

The activity of the enzymes was tested not only in liquid mixed cultures, but their specific effect was also observed using different microscopic techniques. Interaction sites were stained by fluorescein isothiocyanate-conjugated lectins or Calcofluor White M2R New. The appearance of fluorescence indicated the presence of localized cell wall lysis at points of interaction between the antagonist and its host (Elad et al., 1983c).

Electron micrographs show that during the interaction of *Trichoderma* spp. with either *S. rolfsii* or *R. solani* hyphae of the parasite contacted their host and then enzymatically digested their cell walls (Figs. 3 and 4). Extracellular fibrillar material was deposited between the interacting cells. Parasite organelles (e.g.,

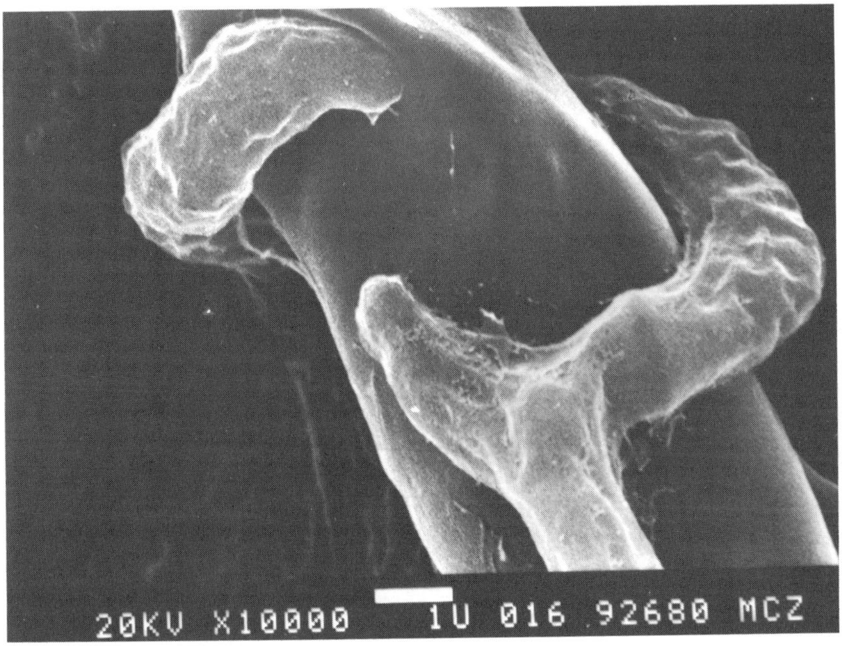

Fig. 3. Hypha of *T. harzianum* coiling around and penetrating one of *R. solani* (×8300). (Elad et al. (1983). *Phytopathology*, **73**:85–88).

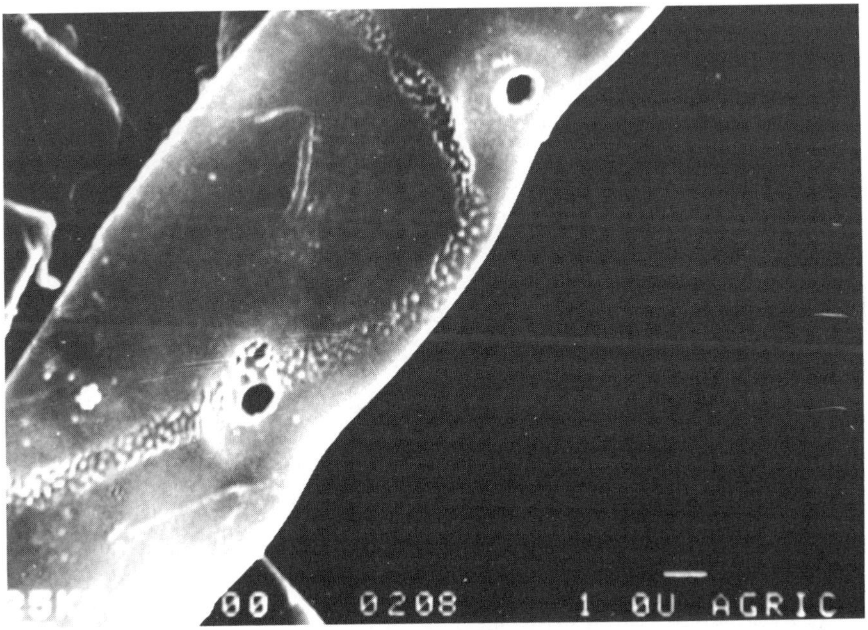

Fig. 4. Hypha from *S. rolfsii* from which a coiling hypha of *T. harzianum* was removed, showing digested zone with penetration sites caused by the antagonist (×5500). (Elad et al. (1983). *Phytopathology*, **73**:85–88).

mitochondria, vesicles and dark osmiophilic inclusions) accumulated in the parasitizing cells. In response to the invasion, the host produced a sheath matrix which encapsulated the penetrating hyphae and the host cells became empty of cytoplasm (Elad et al., 1983b).

It can be assumed that *T. harzianum* attacks the pathogenic mycelium first by dissolving its cell wall in certain locations, followed, sometimes, by hyphal penetration (Chet et al., 1981); it then uses other extracellular enzymes, for example, lipase and protease (Elad et al., 1982a). The susceptible host hyphae, when attacked, showed rapid vacuolation, collapse and disintegration (Chet et al., 1981). Similarly, *T. harzianum* isolates were also able to attack *S. rolfsii* hyphae and sclerotia (Henis et al., 1983; Artigues et al., 1984). Electron microscopy shows that the mycoparasite degraded sclerotial cell walls and that the attacked cells lost their cytoplasmic content. It is assumed that *T. harzianum* utilizes sclerotial cell content, thus enabling it to sporulate on sclerotial surfaces and inside the digested regions (Elad et al., 1984; Chet and Henis, 1985).

Findings, reported within the last few years, indicate that mycoparasitism is one of the main mechanisms involved in the antagonism of *Trichoderma* as a biocontrol agent. The process apparently includes (1) chemotropic growth of *Trichoderma*, (2) recognition of the host by the mycoparasite, (3) excretion of extracellular enzymes and (4) lysis of the host.

Some isolates of *Trichoderma* may indeed excrete inhibitory substances, but evidence of their importance in biological control is, as yet, insufficient. Competition may play a role in this microbial interaction; it may be an independent phenomenon; or it may be connected with antibiosis or parasitism. However, further studies are still needed to elaborate and clarify this question.

8. CONCLUSIONS AND EXPECTATIONS FOR THE FUTURE

The antagonistic ability of *Trichoderma* was discovered more than 50 years ago (Weindling, 1932). The potential of the fungus to serve as a biocontrol agent was already suggested at that early stage. However, only during the last few years has a worldwide effort been carried out to develop the fungus as a commercial preparation.

According to scientific reports, *Trichoderma* spp. can effectively control the following soilborne fungi: *S. rolfsii, R. solani, Pythium* spp., *Fusarium* spp., *Aspergillus niger* (Elad et al., 1983d; Chet and Henis, 1985), *Sclerotium cepivorum* (De Oliveira et al., 1984) and others. Moreover, *Trichoderma* has been successfully sprayed against *Botrytis* sp. in grapes (Dubos et al., 1978) and strawberries (Tronsmo and Dennis, 1977) and against *Septoria tritici* in wheat (Levi et al., unpublished data). Since its activity is rather specific, an alternate target for future research is the selection of isolates or mutants with a wide host range.

One of the main limitations of developing a biological preparation based on *Trichoderma* is the large amount needed for broadcast application. The economic analysis is much more promising today, owing to new application techniques that show significant reduction in disease incidence using much smaller amounts of the product. Broadcast application, can, for example, be replaced by furrow application with similar results.

Trichoderma was also successfully loaded into the root zone of seedlings in the nursery before transplanting them to infested fields. Growing the seedlings in trays (often called "speedlings") enables the grower to transfer them with their surrounding peat mixed with *Trichoderma*, thus forming a "protective layer." Moreover, conidial coating of seeds and bulbs is a very efficient technique, especially against pre- and post-emergence damping-off diseases, using very small amounts of the biocontrol agent. The minimal effective amount of *Trichoderma* is about 10^6 CFU/g soil. This level can be kept in the soil for several months, depending on the foodbase applied with the agent, type of soil, pH, temperature and moisture conditions. Under all experimental conditions the number of *Trichoderma* propagules dropped within a year by 1–2 orders of magnitude. It does not seem likely, therefore, that an irreversible imbalance in soil microflora may occur as a result of the use of *Trichoderma* for biological control.

It is quite obvious today that no biocontrol agent and no fungicide can totally kill the pathogen under practical conditions. But, in many cases, *Trichoderma* could successfully replace the common fungicides. Moreover, while many chemicals are degraded after a short time, *Trichoderma* survives and even propagates in the soil.

When considering integrated pest management, *Trichoderma* can be a very useful and efficient component. Today, large amounts of various fungicides are applied to the soil, however, concern is mounting because of their toxicity and environmental impact. Integration of resistant isolates of *Trichoderma* with low doses of pesticide can lead to a synergistic effect resulting from suppression of competitive soil microflora. *Trichoderma* can be applied together with or following various soil treatments for long-term control of soilborne plant pathogens.

Controlled environmental conditions, such as plastic tunnels, greenhouses and nurseries, where temperatures and moisture levels can be predicted and the right isolate chosen, will undoubtedly improve the chances for successful biological control. Nevertheless, many field experiments during the last few years have shown that significantly good control can also be obtained in commercial field plots.

Recent findings that *Trichoderma* directly affects plants and can live in their roots, open a new field for study. *Trichoderma* enhances plant germination, growth and flower production and may serve as a growth promotor. Moreover, its effect on minor pathogens in the soil is yet to be elaborated. There is no doubt in my mind, that, in spite of all these achievements, more effort will be put into selections, mutations and genetic manipulations.

Genetic engineering provides a way to isolate the gene coding for the lytic functions. Cloning of these genes into bacteria or yeasts will enable the manipulation of gene expression in the new host. The isolated genes may then be reintroduced into the lytic antagonist, to produce a superactive biocontrol agent. This future work depends on thorough physiological-ecological research, which will reveal the essential mechanisms enabling a specific isolate of *Trichoderma* to be an efficient and successful biocontrol agent.

The ability to use the potential of *Trichoderma* to control soilborne plant pathogens in different crops will only be successful by carrying out interdisciplinary research and development. This study should combine both basic and applied work related to the entire system—from the laboratory through the scaling-up and formulation processes, to application in the greenhouse and finally the field.

ACKNOWLEDGMENT

I am indebted to all my students and colleagues, who were always enthusiastic about biological control and without whose help most of this work could not have been accomplished.

REFERENCES

Alexander, M. (1982). *Introduction to Soil Microbiology.* Wiley, New York, pp. 467.

Artigues, M., P. Davet, and C. Roure (1984). Comparison des aptitudes parasitaires de clones de *Trichoderma* vis-a-vis de quelques chapignons a sclerotes. *Soil Biol. Biochem.*, **16:**413-417.

Backman, P. A. and R. Rodriguez-Kabana (1975). A system for the growth and delivery of biological control agents to the soil. *Phytopathology*, **65:**819-821.

Baker, K. F. and J. R. Cook (1982). *Biological Control of Plant Pathogens.* American Phytopathologica Society, 433 pp.

Baker, R., Y. Elad, and I. Chet (1984). The controlled experiment in the scientific method with special emphasis on biological control. *Phytopathology*, **74:**1019-1021.

Barak, R., Y. Elad, D. Mirelman, and I. Chet (1985). Lectins: a possible basis for specific recognition in *Trichoderma-Sclerotium rolfsii* interaction. *Phytopathology*, **75:**458-462.

Boosalis, M. G. (1964). Hyperparasitism. *Ann. Rev. Phytopathol.*, **2:**363-376.

Chang, Y., Y. A. Chang, R. Baker, O. Kleifeld, and I. Chet 1985. Increased growth of plants induced by the biological control and *Trichoderma harzianum. Plant Dis.* (in press).

Chet, I. and R. Baker (1980). Induction of suppressiveness to *Rhizoctonia solani* in soil. Phytopathol, **70:**994—998.

Chet, I. and R. Baker (1981). Isolation and biocontrol potential of *Trichoderma hamatum* from soil naturally suppressive of *Rhizoctonia solani. Phytopathology*, **71:**286-290.

Chet, I. and Y. Elad (1982). Prevention of plant infection by biological means. In *La Selection des Plantes, Bordeaux (France). Colloq. l'INRA*, **11:**195-204.

Chet, I. and Y. Elad (1983). Mechanism of mycoparasitism. *Colloq. l'INRA*, **18:**35-40.

Chet, I. and Y. Henis (1985). *Trichoderma* as a biocontrol agent against soilborne root pathogens. Proceedings of the 4th International Congress of Plant Pathology, Melbourne, Australia.

Chet, I., S. Fogel, and R. Mitchell (1971). Chemical detection of microbial prey by predator. *J. Bacteriol.*, **106:**863-867.

Chet, I., Y. Hadar, Y. Elad, J. Katan, and Y. Henis (1979). Biological control of soil-borne plant pathogens by *Trichoderma harzianum.* In *Soil-Borne Plant Pathogens* (B. Schippers and W. Gams, eds.). Academic, London, pp. 585-592.

Chet, I., G. E. Harman, and R. Baker (1981). *Trichoderma hamatum*: Its hyphal interactions with *Rhizoctonia solani* and *Pythium* spp. *Microb. Ecol.*, **7**:29–38.

Chet, I., Y. Elad, A. Kalfon, Y. Hadar, and J. Katan (1982). Integrated control of soil-borne and bulb-borne pathogens in iris. *Phytoparasitica*, **10**:229–231.

Cook, R. S. and K. F. Baker (1983). *The Nature and Practice of Biological Control of Plant Pathogens.* The American Phytopathological Society, St. Paul, MN, 539 pp.

Curl, E. A., E. A. Wiggind and S. C. Anders (1977). Interactions of *Rhizoctonia solani* and *Trichoderma* spp. with PCNB and herbicides affecting cotton seedling disease. Proc. American Phytopath Soc. *3*, St. Paul Minn., No. *75*.

Davet, P. (1979). Technique pour l'analyse des populations de *Trichoderma* et de *Gliocladium virens* dans le sol. *Ann. Phytopathol.*, **11**:529–533.

Davet, P., M. Artiguez, and C. Martin (1981). Production condition non aseptiques d'inoculum de *Trichoderma harzianum* Rifai des essais de latte biologique. *Agronomie*, **1**:933–936.

Dennis, C. and J. Webster (1971a). Antagonistic properties of species-groups of *Trichoderma*. I. Production of nonvolatile antibiotics. *Trans. Br. Mycol. Soc.*, **57**:25–39.

Dennis, C. and J. Webster (1971b). Antagonistic properties of species-groups of *Trichoderma*. II. Production of volatile antibiotics. *Trans. Br. Mycol. Soc.*, **57**:41–48.

Dennis, C. and J. Webster (1971c). Antagonistic properties of species-groups of *Trichoderma*. III. Hyphal interaction. *Trans. Br. Mycol. Soc.*, **57**:363–369.

De Oliveira, V. L., M. M. Bellei, and A. C. Borges (1984). Control of white rot of garlic by antagonistic fungi under controlled environmental conditions. *Can. J. Microbiol.*, **30**:884–889.

Dubos, B., J. J. Guillaumin, and M. Schubert (1978). Action du *Trichoderma viride* Pers., apporte avec divers substrats organiques, sur l'initiation et la croissance des rhizomorphes d'*Armillariella mellea* (Vahl.) Karst. dans deux types de sols. *Ann. Phytopathol.*, **10**:187–196.

Dunn, R. T., S. A. Lewis, and G. C. Papavizas (1983). Production and formulation of two biological control agents from liquid fermentation. *Phytopathology*, **73**:165 (Abstract).

Elad, Y. and I. Chet (1983). Improved selective medium for isolation of *Trichoderma* or *Fusarium* spp. *Phytoparasitica*, **11**:55–58.

Elad, Y., I. Chet, and J. Katan (1980a). *Trichoderma harzianum*: a biocontrol effective against *Sclerotium rolfsii* and *Rhizoctonia solani. Phytopathology*, **70**:119–121.

Elad, Y., J. Katan, and I. Chet (1980b). Physical biological and chemical control integrated for soil-borne diseases in potatoes. *Phytopathology*, **70**:418–422.

Elad, Y., I. Chet, and Y. Henis (1981a). Biological control of *Rhizoctonia solani* in strawberry fields by *Trichoderma harzianum. Plant Soil*, **60**:245–254.

Elad, Y., I. Chet, and Y. Henis (1981b). A selective medium for improving quantitative isolation of *Trichoderma* spp. from soil. *Phytoparasitica*, **9**:59–67.

Elad, Y., Y. Hadar, E. Hadar, I. Chet, and Y. Henis (1981c). Biological control of *Rhizoctonia solani* by *Trichoderma harzianum* in carnation. *Plant Dis.*, **65**:675–677.

Elad, Y., I. Chet, and Y. Henis (1982a). Degradation of plant pathogenic fungi by *Trichoderma harzianum. Can. J. Microbiol.*, **28**:719–725.

Elad, Y., Y. Hadar, I. Chet, and Y. Henis (1982b). Prevention with *Trichoderma harzianum* Rifai aggr. of reinfestation by *Sclerotium rolfsii* Sacc. and *Rhizoctonia solani* Kuhn of soil fumigated with methyl bromide and improvement of disease control in tomatoes and peanuts. *Crop Prot.*, **1**:199–211.

Elad, Y., A. Kalfon, and I. Chet (1982c). Control of *Rhizoctonia solani* in cotton by seed-coating with *Trichoderma* spp. spores. *Plant Soil*, **66**:279–281.

Elad, Y., R. Barak, and I. Chet (1983a). Possible role of lectins in mycoparasitism. *J. Bacteriol.*, **154**:1431–1435.

Elad, Y., R. Barak, I. Chet, and Y. Henis (1983b). Ultrastructural studies of the interaction between *Trichoderma* spp. and plant pathogenic fungi. *Phytopath. Z.*, **107**:168–175.

Elad, Y., I. Chet, P. Boyle, and Y. Henis (1983c). Parasitism of *Trichoderma* spp. on *Rhizoctonia solani* and *Sclerotium rolfsii.* Scanning electron microscopy and fluorescence microscopy. *Phytopathology*, **73:**85–88.

Elad, Y., Y. Hadar, and I. Chet. (1983d). The potential of *Trichoderma harzianum* as a biocontrol agent under field conditions. *Colloq. l'INRA*, **18:**305–310.

Elad, Y., Z. Sadovski, and I. Chet (1983e). Detection of mycoparasitism by infrared photomicrography. *Microb. Ecol.*, **9:**185–187.

Elad, Y., A. Sivan, O. Zeidan, O. Kleifeld, and I. Chet (1983f). Biological and integrated control of soil pathogens as a means for reduction of fungicides. In *Developments in Ecology and Environmental Quality* (Shuval, H. ed.). Balaban International Science Service Israel, pp. 587–594.

Elad, Y., R. Barak, and I. Chet (1984). Parasitism of *Sclerotium rolfsii* sclerotia by *Trichoderma harzianum. Soil Biol. Biochem.* **16:**381–386.

Grinstein, A., Y. Elad, J. Katan, and I. Chet (1979). Control of *Sclerotium rolfsii* by means of a herbicide and *Trichoderma harzianum. Plant Dis. Rep.*, **63:**823–826.

Hadar, Y., I. Chet, and Y. Henis (1979a). Biological control of *Rhizoctonia solani* damping-off with wheat bran culture of *Trichoderma harzianum. Phytopathology*, **69:**64–68.

Hadar, E., Y. Elad, S. Ovadia, Y. Hadar, and I. Chet (1979b). Biological and chemical control of *Rhizoctonia solani* in carnation. *Phytoparasitica*, **7:**55.

Hadar, Y., G. E. Harman, and A. G. Taylor (1984). Evaluation of *Trichoderma koningii* and *T. harzianum* from New York soils for biological control of seed rot caused by *Pythium* spp. *Phytopathology*, **74:**106–110.

Harman, G. E. (1982). Techniques to increase the biocontrol exerted by *Trichoderma* spp. applied to seeds. *Phytoparasitica*, **10:**120 (Abstract).

Harman, G. E., I. Chet, and R. Baker (1980). *Trichoderma hamatum* effects on seed and seedling disease induced in radish and pea by *Pythium* spp. or *Rhizoctonia solani. Phytopathology*, **70:**1167–1172.

Henis, Y. and I. Chet (1975). Microbial control of plant pathogens. *Adv. Appl. Microbiol.*, **19:**85–111.

Henis, Y. and G. C. Papavizas (1983). Factors affecting germinability and susceptibility to attack of sclerotia of *Sclerotium rolfsii* by *Trichoderma harzianum* in field soil. *Phytopathology*, **73:**1469–1474.

Henis, Y., G. Ghaffar, and R. Baker (1978). Integrated control of *Rhizoctonia solani* damping-off of radish: Effect of successive plantings. PCNB and *Trichoderma harzianum* on pathogen disease. *Phytopathology*, **68:**900–907.

Henis, Y., P. B. Adams, J. A. Lewis, and G. C. Papavizas (1983). Penetration of sclerotia of *Sclerotium rolfsii* by *Trichoderma* spp. *Phytopathology*, **73:**1043–1046.

Henis, Y., S. A. Lewis, and G. C. Papavizas (1984). Interactions between *Sclerotium rolfsii* and *Trichoderma* spp. relationship between antagonism and disease control. *Soil Biol. Biochem.* **16:**391–395.

Hoffman, G. M. and H. P. Malkomes (1974). Bromide residues in vegetable crops after soil fumigation with methyl bromide. *Agric. Environ.*, **1:**321–328.

Huang, C. S., R. C. V. Tenente, F. C. C. Da Silva, and I. A. R. Lara (1981). Effect of *Crotalaria spectabilis* and two nematocides on numbers of *Meloidogyne inognita* and *Helicotylenchus dihystera. Nematologia*, **27:**1–5.

Jackson, C. R. and D. K. Bell (1969). Diseases of peanuts (groundnut) caused by fungi. University of Georgia College of Agriculture Experimental Research Bulletin, No. 56, 137 pp.

Jansson, H. B. and B. Norbring-Hertz (1979). Attraction of nematode to living mycelium of nematophagous fungi. *J. Gen. Microbiol.*, **112:**89–93.

Katan, J., A. Greenberger, H. Alon, and A. Grinstein (1976). Solar heating by polyethylene mulching for the control of diseases caused by soil-borne pathogens. *Phytopathology*, **66:**683–688.

Kelley, W. D. (1976). Evaluation of *Trichoderma harzianum*—impregnated clay granules as a biocontrol for *Phytophthora cinnamomi* causing damping-off of pine seedlings. *Phytopathology*, **66:**1023–1027.

Lewis, J. A. and G. C. Papavizas (1980). Integrated control of *Rhizoctonia* fruit rot of cucumber. *Phytopathology*, **70:**85–89.

Lewis, J. A. and G. C. Papavizas (1984). Chlamydospore formation by *Trichoderma* spp. in natural substrates. *Can. J. Microbiol.*, **30:**1–7.

Liu, S. and R. Baker (1980). Mechanism of biological control in soil suppressive to *Rhizoctonia solani*. *Phytopathology*, **70:**404–412.

Marois, J. J. and D. J. Mitchell (1981a). Effects of fungal communities on the pathogenic and saprophytic activities of *Fusarium oxysporum* f. sp. *radicis-lycopersici*. *Phytopathology*, **71:**1251–1256.

Marois, J. J. and D. J. Mitchell (1981b). Effects of fumigation and fungal antagonists on the relationships of inoculum density to infection incidence and disease severity in Fusarium crown rot of tomato. *Phytopathology*, **71:**167–170.

Marois, J. J., D. J. Mitchell, and R. M. Sonoda (1981). Biological control of *Fusarium* crown rot of tomato under field conditions. *Phytopathology*, **71:**1257–1260.

Marshall, D. S. (1982). Effect of *Trichoderma harzianum* seed treatment and *Rhizoctonia solani* inoculum concentration on damping-off of snap bean in acidic soils. *Plant Dis.*, **66:**788–789.

Mirelman, D., E. Galun, N. Sharon, and M. Lotan (1975). Inhibition of fungal growth by wheat germ agglutinin. *Nature*, **256:**414–416.

Munnecke, D. E., M. J. Kolbezen, and W. D. Wilbur (1973). Effect of methyl bromide or carbon disulfide on *Armillaria* and *Trichoderma* growing on agar medium and relations to survival of *Armillaria* in soil following fumigation. *Phytopathology*, **63:**1352–1357.

Munnecke, D. E., M. J. Kolbezen, and H. D. Ohr (1981). Interactions involved in controlling *Armillaria mellea*. *Plant Dis.*, **65:**384–389.

Okon, Y., I. Chet, and Y. Henis (1973). Effect of lactose, ethanol and cycloheximide on the translocation pattern of radioactive compounds and of sclerotium formation in *Sclerotium rolfsii*. *J. Gen. Microbiol.*, **74:**251–258.

Papavizas, G. D. (1973). Status of applied biological control of soil-borne plant pathogens. *Soil Biol. Biochem.*, **5:**709–720.

Papavizas, G. C. (1981). Survival of *Trichoderma harzianum* in soil and in pea and bean rhizospheres. *Phytopathology*, **71:**121–125.

Papavizas, G. C. and J. A. Lewis (1979). Integrated control of *Rhizoctonia solani*. In *Soil-Borne Plant Pathogens* (B. Schippers and W. Gams eds.). Academic, New York, pp. 415–425.

Papavizas, G. C. and R. D. Lumsden (1982). Improved medium for isolation of *Trichoderma* spp. from soil. *Plant Dis.*, **66:**1019–1020.

Papavizas, G. C., J. A. Lewis, and T. H. Abd-El Moity (1982). Evaluation of new biotypes of *Trichoderma harzianum* for tolerance to benomyl and enhanced biocontrol capabilities. *Phytopathology*, **72:**126–132.

Ricard, J. L. (1981). Commercialization of a *Trichoderma* based mycofungicide. Some problems and solutions. *Biocontrol News Inform.*, **2:**95–98.

Ricard, J. L. (1983). Field observation on the biocontrol of Dutch elm disease with *Trichoderma viride* pellets. *Eur. J. For. Pathol.*, **13:**60–62.

Rifai, M. (1969). A revision of the genus *Trichodera*. *Mycol. Pap.* **116.** Commonw. Mycol. Inst., Assoc. Appl. Biologists, Kew, Surrey, England.

Ruppel, E. G., R. Baker, G. E. Harman, J. P. Hubbard, R. J. Hecker, and I. Chet (1983). Field tests of *Trichoderma harzianum* Rifai aggr. as a biocontrol agent of seedling disease in several crops and *Rhizoctonia* root rot of sugar beet. *Crop. Prot.*, **2:**399–408.

Sivan, A. and I. Chet (1982). Biological control of *Pythium* by *Trichoderma. Phytoparasitica*, **10:**118 (Abstract).

Sivan, A. and I. Chet (1985). Biological control of *Fusarium* spp. in cotton, wheat and musk melon by *Trichoderma harzianum* Photophath 2 (in press).

Sivan, A., Y. Elad, and I. Chet (1984). Biological control effects of a new isolate of *Trichoderma harzianum* on *Pythium aphanidermatum. Phytopathology*, **74:**498–501.

Strashnov, Y., Y. Elad, A. Sivan, Y. Rudich, and I. Chet (1985a). Control of *Rhizoctonia solani* fruit rot of tomatoes by *Trichoderma harzianum. Crop Prot.* **4:** 359–364.

Strashnov, Y., Y. Elad, A. Sivan, and I. Chet (1985b). Integrated control of *Rhizoctonia solani* Kühn by methyl bromide and *Trichoderma harzianum* Rifai Aggr. *Plant Pathol.* **34:**146–151.

Tronsmo, A. and C. Dennis (1977). The use of *Trichoderma* species to control strawberry fruit rot. *Neth. J. Pl. Path.*, **83:**449–455.

Webster, J. and N. Lomas (1964). Does *Trichoderma viride* produce gliotoxin and viridin? *Trans. Br. Mycol. Soc.*, **47:**535–540.

Wells, H. D., D. K. Bell, and C. A. Jawarski (1972). Efficacy of *Trichoderma harzianum* as a biological control for *Sclerotium rolfsii. Phytopathology*, **62:**442–447.

Weindling, R. (1932). *Trichoderma lignorum* as a parasitic of other soil fungi. *Phytopathology*, **22:**837–845.

Weindling, R. (1934). Studies on a lethal principle effective in the parasitic action of *Trichoderma lignorum* on *Rhizoctonia solani* and other soil fungi. *Phytopathology*, **24:**1153–1179.

Weindling, R. and O. H. Emerson (1936). The isolation of a toxic substance from the culture of a *Trichoderma. Phytopathology*, **26:**1068–1070.

Weindling, R. and H. S. Fawcett (1936). Experiment in the control of *Rhizoctonia* damping-off of citrus seedlings. *Hilgardia*, **10:**1–16.

Wilhelm, S. and A. O. Poulus (1980). How soil fumigation benefits the California strawberry industry. *Plant Dis.*, **64:**264–270.

Zamir, D. and I. Chet (1985). Application of enzyme electrophorsis for identification of isolates in *Trichoderma harzianum* Ann. of Microbiol. **31:**578–580.

7

ALTERNATIVE METHODS OF CONTROLLING WHITE ROT DISEASE OF *ALLIUM*

JOHN R. COLEY-SMITH
Department of Plant Biology and Genetics, University of Hull, Hull, United Kingdom

Although soilborne sclerotium-forming fungi cause some of the world's most important and destructive diseases of plants, there are no effective chemical control methods for a number of them. Diseases caused by *Phymatotrichum, Sclerotinia, Sclerotium* and, until recently, *Rhizoctonia* remain uncontrolled either because there are no effective fungicides available or for some other reason. One of the main causes for the lack of effective control measures is that soil-acting fungicides represent an underexploited area and, for economic reasons, it is highly likely that in the future they will receive only limited attention (Brooks, 1984). Undoubtedly many modern fungicides used in the control of foliar diseases are intrinsically highly toxic to soilborne sclerotium-forming fungi, but they are inadequate for use in soil for a number of reasons. One of the most important is failure to reach the target organism either because of poor placement of the fungicide in soil or the lack of phloem mobility. In addition, although acquired fungicide resistance has not yet proved to be particularly important with soilborne fungi, the potential for its development undoubtedly exists and the increasing use of highly active site-specific toxophores suggests that it may become much more important in the future (Dekker, 1982). An example of what could happen was recently demonstrated with the fungicide tolclofos-methyl, a site-specific type (Ohtsuki and Fujinami, 1982), which, at present, gives good control of diseases caused by *Rhizoctonia* in a number of crops. Although no instances of loss of control in the field with this fungicide have yet occurred, the development of stable acquired resistance of *Rhizoctonia solani* in the laboratory was reported (Van Bruggen and Arneson, 1984). Admittedly most of the resistant isolates possessed accompanying features, such as reduction in growth rates and certain morphological changes, which suggested that they might have reduced ecological fitness, but because resistance was induced with such ease suggests that the emergence of fungicide resistant and ecologically fit types can by no means be ruled out.

1. FUNGICIDES AND WHITE ROT DISEASE OF *ALLIUM*

A wide variety of fungicides has been used, with varying degrees of success, to control white rot disease of onions and other species of *Allium* (*Sclerotium cepivorum*). The first successful material was mercurous chloride (calomel), which was applied to the drills as a dust (Booer, 1945, 1946), to transplants as a slurry, or to seeds with different types of sticker (Croxall et al., 1953; Harrison, 1956; Fletcher, 1964). In the United Kingdom calomel seed treatment usually gives some control of white rot in salad onions during the summer, but is less effective later in the season (Entwistle and Munasinghe, 1973). The material has now fallen out of favor because of both environmental implications and cost (Fletcher et al., 1971). Quintozene and dicloran have subsequently been used against white rot (Lafon and Bugaret, 1966), but neither compound has lived up to its early promise (Locke, 1969). More recently, the use of benomyl for root-dusting treatments showed that this chemical is to some extent phytotoxic (Ryan and Kavanagh, 1976a,b). Nevertheless benomyl, thiabendazole and thiophanate-methyl have all been used to some extent for white rot control (Entwistle and Munasinghe, 1973; Maloy and Machtmes, 1974).

The 3,5-dichlorophenyl-*N*-cyclic imide fungicides, usually called dicarboximides, are, at present, the most effective against white rot in the United Kingdom and the most commonly used (Entwistle, 1983). Successful control of white rot in salad onions has been achieved with seed and stem base treatments with iprodione (Entwistle and Munasinghe, 1980a). These treatments delayed the appearance and reduced the incidence of stem base symptoms and root infection with *S. cepivorum* in autumn- and spring-sown salad onions (Entwistle and Munasinghe, 1980b). In dry-bulb onions and direct drilled leeks the dicarboximides have generally been less successful than in salad onions, probably because less fungicide is introduced into soil at lower seeding rates. Even in salad onions there is little cause for complacency about the effectiveness of the dicarboximides. Reliance on materials with a related chemical basis, particularly ones with a site-specific mode of action, suggests that dicarboximide-resistant strains might eventually become a problem. At present there is no evidence that resistant strains have developed in the field, but laboratory induction of iprodione resistance in *S. cepivorum* occurs readily (Entwistle, 1983).

During 1981 and 1982 iprodione failed to control white rot on a number of farms on which it had previously proved successful and in 1983 loss of control was also reported with vinclozolin (Entwistle and Hawling, 1984). In 1984 almost complete loss of control of white rot in garlic was experienced in France with the previously effective iprodione (Lafon, personal communication).

The loss of control with iprodione has now been attributed to the accelerated degradation of the fungicide in previously treated soils, probably as a result of changes in the microbial populations therein (Walker et al., 1984). In view of the enormous range of microorganisms in soils, the development of such microbial populations is hardly surprising. The failure of dicloran to control white rot in fields previously treated with this chemical in the United States was shown to have resulted from enhanced degradation (Groves and Chough, 1970).

It is quite clear that with *S. cepivorum*, as with many other soilborne fungi, alternatives to control with fungicides need to be sought and it is my purpose in this chapter to suggest some of these alternatives. Any such alternative methods will certainly depend upon a thorough understanding of white rot disease and of the biology of the organism that causes it.

2. THE IMPORTANCE OF WHITE ROT DISEASE

With the development of a satisfactory seed treatment (Maude and Presley, 1977a,b) for controlling the disease neck rot (*Botrytis allii*), white rot now poses the most serious threat to onion growing on a world scale. In the United Kingdom the pattern of events that often occurs (Evendon, personal communication) with continuous onion growing in Lincolnshire, the main area cultivated with dry bulb onions, appears to be:

Year 1. No white rot or very low levels.
Year 2. 5% infection.
Year 3. 30% infection.
Year 4. Uneconomic levels of infection.

In this area some growers have consequently stopped growing onions because of white rot.

Similar serious levels of infection have also been recorded in a number of countries including Australia (Somerville and Merriman, 1983), Canada (Oudemans and Edgington, 1983), Egypt (Georgy, 1983), Finland (Osara, 1983), Italy (Garibaldi, 1983), The Netherlands (Koert, 1979) and the United States (Adams, 1983).

The potentially serious nature of white rot disease is perhaps best demonstrated by the present state of the dry bulb onion crop in Egypt. Before 1965 onions were one of the major export crops of this country, with an area of 20,000 ha providing 160,000 tons of onions for export. By 1978 the area of land under onion cultivation had fallen to 8,000 ha and only 30,000 tons were available for export (Abd-El-Moity, 1979). This decline reflects a serious increase in the contamination of soils suitable for onion growing in the Nile valley. White rot was first recorded in Egypt in 1929 (Nattrass, 1931) and subsequently spread slowly, apparently in the dung of grazing sheep and goats (Mikhail et al., 1974). During this time most of the soil used for onion production in Upper Egypt was flooded by the river Nile for a period of 3 months in every year. This annual inundation deposited a thick layer of uncontaminated soil on top of that used in the previous year. In addition the prolonged flooding may have reduced the numbers of viable sclerotia in the soil. Following the construction of the high dam at Aswan in the 1960s the waters of the Nile were regulated and new soil was no longer deposited. This has resulted in a disastrous increase in the spread of white rot and at the present time over 70% of the land suitable for growing onions in Upper Egypt is infested with *S. cepivorum.*

Although attention is often focused on white rot in onions it should be remembered that *S. cepivorum* also causes serious disease in garlic (*Allium sativum*) and in *Allium fistulosum.* Leeks (*Allium porrum*) may also be infected but appear to be less susceptible than the other major edible species of the genus (Coley-Smith and Esler, 1983a).

3. ETIOLOGY OF WHITE ROT DISEASE

Sclerotium cepivorum is confined as a pathogen to members of the genus *Allium* (Esler and Coley-Smith, 1984). It is a typical root-infecting fungus and is a poor competitor in soil in which it survives as sclerotia in the absence of host plants. The sclerotia are normally maintained in an inactive state by the mycostatic influence of soil, but will germinate in the presence of *Allium* species. This is a specific triggering effect (Coley-Smith, 1960; Coley-Smith and Holt, 1966); little or no germination takes place in the absence of host plants. The effect of *Allium* plants results from the exudation by their roots of nonvolatile alkyl- and alkenyl-cysteine sulphoxides (Coley-Smith and Cooke, 1971), which are metabolized by the soil microflora to yield a range of volatile thiols and sulphides which activate the dormant sclerotia (Coley-Smith and King, 1969; King and Coley-Smith, 1969). The relationships between sclerotia, host plants, and the soil microflora are shown in Fig. 1.

The sclerotia of *S. cepivorum* germinate eruptively (Coley-Smith, 1960; New et al., 1984) and emerging hyphae infect the roots of host plants. Hyphae grow up through the roots and may eventually reach the developing bulb on which there is often a profuse growth of white mycelium. Later this mycelium dies down to form a mat in which masses of small black, nearly spherical sclerotia are formed (Mordue, 1976; Georgy and Coley-Smith, 1982). Infected bulbs develop a characteristic soft rot and the tissues eventually decay, releasing the sclerotia and replenishing the population of these in the soil. An important feature of germination in *S. cepivorum* is that the sclerotia only germinate once and usually expend their entire contents during this process. Although they have been observed to form secondary sclerotia either within the parent sclerotium prior to

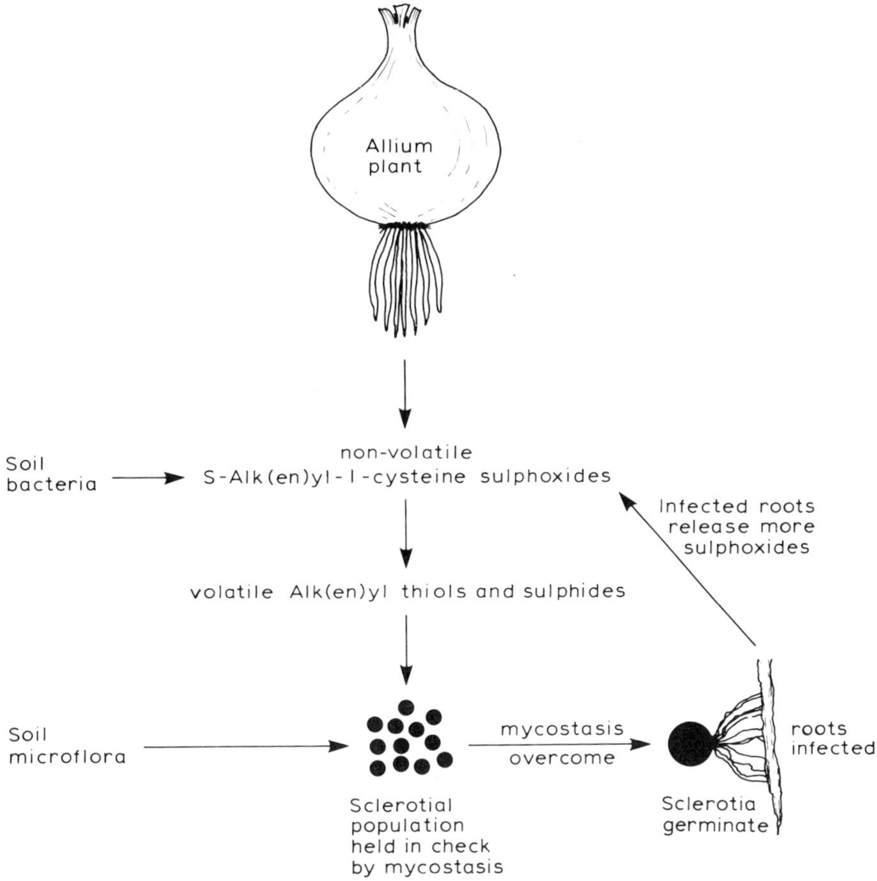

Fig. 1. The effect of *Allium* plants and soil microorganisms on sclerotia of *Sclerotium cepivorum*.

germination or outside it from germinating hyphae (Entwistle and Munasinghe, 1981), they do not appear to do so frequently. After germination they are usually left as broken fragments of rind which are incapable of further activity.

Mycelium from germinating sclerotia usually persists for about 10–14 days, depending on temperature, and then lyses (Coley-Smith, 1960). Growth of mycelium in the soil is negligible unless the fungus is sustained by a foodbase such as a sclerotium or an infected plant (Scott, 1956a; Sadasivan and Subramanian, 1960). Close proximity of roots and stem bases appears to be necessary for infection of adjacent plants and secondary spread by this means can be considerable among closely spaced plants, such as salad onions (Entwistle and Munasinghe, 1977). In plants that are spaced more widely, such as dry bulb onions, secondary spread is much less important (Scott, 1956b; Ryan and Kavanagh, 1976c; Coley-Smith and Esler, 1983a), and in this situation most infections arise directly from germinating sclerotia.

A large proportion of sclerotia produced in artificial culture survived for 15 years under field conditions in the United Kingdom (Coley-Smith, 1983) in the absence of host plants. That sclerotia possess the capacity for long-term survival is also supported by outbreaks of white rot disease in soils that remained free from *Allium* crops for at least 20 years (Croxall et al., 1953). In spite of this evidence it has been claimed that many sclerotia decay soon after release into soil from infected bulbs (Entwistle and Munasinghe, 1978). In the Fraser valley of British Columbia (Leggett and Rahe, 1985), sclerotia decay fairly rapidly and this appears to result from the near continuous saturation of the soils of this area during the winter months. Hence, although sclerotia are very resistant structures, they may not always survive for long periods under all conditions. Survival in natural populations of sclerotia has not been studied in most of the countries in which white rot is a problem and this aspect of the biology of *S. cepivorum* is certainly in need of further attention.

Since the sclerotia are the only perennating structures possessed by *S. cepivorum*, they are clearly primary targets for alternative control strategies. Unfortunately there is still insufficient knowledge about the size of sclerotial populations in infested soils in most countries; nor is there much information about the relationship between the density of sclerotia in the soil and the incidence of disease. Densities of 1000–1500 sclerotia/kg of soil are apparently common in Egypt (Georgy and Coley-Smith, 1982), but in most countries densities are probably considerably lower. Available data (Crowe et al., 1980; Hall and Somerville, 1983) suggest that populations of sclerotia would have to be reduced to densities of less than 10/kg of soil to reduce the incidence of white rot significantly. This means of course that with very high levels of infestation 99% of the sclerotia might have to be eliminated, at least in the upper layers of the soil.

4. POSSIBLE ALTERNATIVE METHODS OF CONTROL

4.1. Disease Avoidance

Although white rot disease has been reported in many countries, there are some in which it has never become a serious problem. In certain cases this appears to result from climatic conditions that are unsuitable for *S. cepivorum*. White rot is a cool weather disease (Walker, 1926; Adams and Papavizas, 1971) and is undoubtedly restricted by soil temperatures outside the range 10–20°C. This feature can sometimes be put to good use, as demonstrated in certain areas in the United States where spring-planted crops could be safely grown in fields where white rot caused complete loss in fall-planted crops (Adams and Springer, 1977). In Egypt, soil temperatures are too high for active growth of *S. cepivorum* for most of the year. Onions are usually grown from transplants as a winter crop and are relatively slow to mature. Promising levels of control have recently been obtained (El Gammal and Ibrahim, 1983) by growing early maturing varieties from sets. These are harvested sufficiently early to avoid much of the period during which *S. cepivorum* is active. Unfortunately it would not be possible to use such methods in many areas where white rot is a problem.

4.2. Soil Partial-Sterilization

Several soil treatments give some degree of control of white rot disease. One of the most promising is the use of soil partial-sterilants, such as metham sodium. This material gave a good degree of control in the United States when applied through irrigation lines at a rate of 234 l a.i./ha in 2.5 cm of water 4–5 weeks before planting leeks and bunching onions (Adams and Johnston, 1983). This system is now being tested in a number of countries, and in, small-scale trials in the United Kingdom (Coley-Smith, 1985), it was found that, with double the concentration of metham given above, a large proportion of sclerotia were killed to a depth of 10 cm and there was considerable death even at 20 cm. Unfortunately the cost of these treatments is high and in most countries it is doubtful whether they could be considered for use in the dry bulb onion crop. However, on farms with a few isolated outbreaks of white rot such treatment might be viable because the risk of reintroduction would be low (Entwistle and Munasinghe, 1978). On farms where outbreaks are more numerous the risk of reintroduction of the disease by machinery contaminated with infested soil or by wind movement of sclerotia would limit the value of such treatments.

A second promising soil partial-sterilization treatment is solarization (Katan, 1981), which involves mulching the soil with transparent polyethylene sheets. This has been tested for its effectiveness in controlling white rot in Egypt and in Australia. In Egypt (El-Yamani et al., 1983) a 50–67% reduction in disease, but without any corresponding increase in yield, was obtained, whereas in Australia (Porter, 1983) the reduction in plant deaths resulted in a significant increase in yield. Thus this method has considerable potential for use in certain countries but is of little value in countries such as those of northern Europe, where there is usually insufficient sunshine during the summer to sustain high soil temperatures for long periods. For a country such as Egypt the cost of treating large areas with the sheeting would probably be prohibitive and, where such a large proportion of the land is infested, there would be a serious risk of reintroduction of the pathogen either from neighboring fields or from viable sclerotia brought up by ploughing from lower layers in the soil.

4.3. Plant Resistance

Within the genus *Allium* there is little sign of resistance to *S. cepivorum*. Adams and Papavizas (1971) tested 16 species and only found evidence of a significant level of resistance in *A. caeruleum*. More recently, a larger number of species was tested for the ability to trigger germination of sclerotia and for resistance (Esler and Coley-Smith, 1983, 1984). The majority were highly stimulatory, but six species, commonly grown as ornamentals, either failed to trigger germination of sclerotia altogether or only had a weak ability to do so. These species contained *S*-methyl-L-cysteine sulfoxide as their principle flavor and odor precursor and they had low overall flavor and odor levels. In the presence of sclerotia that were stimulated to germinate by the addition of garlic extract, the roots of nonstimulatory species were infected as readily as those of stimulatory types. Lack of ability to trigger germination of sclerotia operating through differences in flavor and odor type would clearly be unacceptable, even if it were possible to transfer such characteristics to edible species. In the work described above it was also found that certain wild species, that are thought to be the wild progenitors of the common cultivated species of *Allium*, were both highly stimulatory and very susceptible to infection.

It is commonly held that among edible species of *Allium* garlic is particularly susceptible to white rot and leek cultivars are more resistant than those of onion. Recent experimental work from the United Kingdom supports these views (Coley-Smith and Esler, 1983a). Leeks have now been shown to trigger germination of sclerotia less than cultivars of onion (Coley-Smith and Esler, 1983b) and this appears to be the major reason for the observed differences in the incidence of disease. Levels of stimulatory activity appear to be very similar among different cultivars of a particular species.

The search for resistant cultivars of onion has been pursued for a long time with some workers finding little evidence of any differences (Semb et al., 1978; Rondomanski and Doruchowski, 1978; Coley-Smith and Esler, 1983a) and others claiming substantial differences (Utkhede and Rahe, 1978; Rahe and Utkhede, 1983; Van Der Meer et al., 1983). Differences in the degree of resistance to white rot have even been claimed to occur among different seed lots of the same cultivar (Utkhede et al., 1982). In an international trial undertaken to examine the stability of resistance to white rot (Utkhede et al., 1982), it was found that although significant differences did occur among the cultivars tested, the ranking of these in terms of resistance varied according to location. This emphasizes one of the major difficulties facing workers in their search for resistance among a wide range of onion types. There is usually very little genetic diversity among cultivars grown in a particular region and apparent differences in resistance, discovered in trials involving cultivars from different regions, may result from the poor growth brought about by environmental influences. Differences in growth may also explain why resistance can apparently be detected in large-scale field experiments, but it is usually difficult or impossible to demonstrate in laboratory tests. Rahe (1981) considers that resistance among onion cultivars observed in field trials may result from differential stimulation of sclerotium germination. Although there is as yet no direct evidence to support this idea, there is certainly none either for the possibility that differences in tissue resistance may be involved.

4.4. The Use of Antagonistic Microorganisms

There have been a number of attempts to control white rot disease biologically by the use of antagonistic microorganisms. Among those which are claimed to have some effect are *Penicillium nigricans* (Ghaffar, 1969), *Bacillus subtilis* (Utkhede and Rahe, 1980), and *Trichoderma harzianum* (Abd-El-Moity and Shatla, 1978).

The work on *P. nigricans* has not been followed up in the field and the potential of this organism as a disease control agent for *S. cepivorum* is impossible to evaluate at present. In the case of *B. subtilis*, four isolates were shown in Canadian trials to provide season-long protection, when applied to seeds (bacterization), with the partially resistant onion cultivar Festival, and significant protection even with the susceptible cultivar Autumn Spice. The levels of protection with some of the isolates were as good as those given by chemical treatments. Unfortunately, although positive results were obtained with seed bacterization in two successive years, the method failed to give adequate protection in the following two years (Rahe, 1983).

A similar pattern of unreliability has now emerged with *T. harzianum*. The initial results in Egypt with this fungus were extremely promising. In 1980–1981 the incidence of disease in treated plots was 10–15%, compared with 85% in untreated plots. In 1981–1982 more than 100 ha were treated and the overall level of control was 82%, which was similar to that given by iprodione. In 1982–1983 about 100 ha were again treated, with an overall level of control of about 87% (Abd-El-Moity, 1983). Despite these apparent successes, high levels of infection were observed in some treated fields and these failures have been attributed to a number of factors, including soil temperature and type, irrigation patterns, and dense weed populations. Attempts to use *Trichoderma* in the control of white rot in the United Kingdom have resulted in failure. Recently the effect of methyl bromide soil treatment as an aid to the establishment of *Trichoderma* has been studied in the United Kingdom in overwintered salad onions (Entwistle et al., 1984). The treatments were applied during the autumn of 1982. In May 1983 white rot incidence in soil sterilized with methyl bromide was 6% in *Trichoderma*-treated plots and 29% in untreated plots; in nonsterilized soil, infection in the *Trichoderma*-treated and untreated plots was 53% and 49%, respectively. By midsummer the effect of methyl bromide on *Trichoderma* establishment had disappeared and there was little difference between the treatments.

It is apparent from these results that antagonistic microorganisms can reduce white rot disease under certain circumstances. The potential for disease control has been demonstrated, but the erratic nature of the results obtained suggests that much more research is required. The mechanisms by which these antagonists control *S. cepivorum* are not known and this makes the proper manipulation of such organisms extremely difficult.

4.5. Mycoparasitism

It has been pointed out (Papavizas and Lumsden, 1980) that mycoparasitism has not been very well explored and is poorly understood in relation to the survival of fungal propagules in soil. This is certainly true in the case of *S. cepivorum*. Sclerotia of this fungus are obviously very resistant to microbial attack; nevertheless, there are a number of organisms capable of attacking them. The most important are probably *Coniothyrium minitans* and the more recently discovered *Sporidesmium sclerotivorum* (Uecker et al., 1978) and *Teratosperma oligocladum* (Uecker et al., 1980). *Coniothyrium minitans* is a well-known mycoparasite of a number of sclerotium-forming fungi (Turner and Tribe, 1976). Seed dressing or soil dusting with a pycnidial preparation of this fungus protected onion plants from white rot in glasshouse tests (Ahmed and Tribe, 1977), giving 57–61% disease control, which was similar to that achieved with calomel. In spite of this apparent success no large-scale field trials appear to have been undertaken with this fungus. All the evidence available at present suggests that *C. minitans* is a more efficient mycoparasite of *Sclerotinia* species than of *S. cepivorum*, but it would certainly be worth testing on a field scale against the latter.

Sporidesmium sclerotivorum and *T. oligocladum* possess a number of features that make them particularly attractive candidates for biological control of some sclerotium-forming fungi, including *S. cepivorum* (Ayers and Adams, 1981; Papavizas, 1984). These dematiaceous mycoparasites were thought to occur only in the United States but have now been reported from soils in a number of countries (Parfitt et al., 1983; Adams and Ayers, 1985). The macroconidia of these two fungi possess the ability to germinate rapidly in soil containing sclerotia of *Sclerotinia* species and at distances up to almost a centimeter from a sclerotium. They also spread through soil from one sclerotium to another, producing new macroconidia along a network of branching hyphae. Adams and Ayers (1982) showed that application of *S. sclerotivorum* to field plots infested with *Sclerotinia minor* in May 1978 reduced the population of sclerotia by 94% in 6 months and gave 63–83% control of lettuce drop in two consecutive crops in 1979 and 65–82% control in 1980. *Sporidesmium sclerotivorum* is also associated with the natural decline of populations of sclerotia both of *Sclerotinia* species and of *S. cepivorum* in the United States (Adams and Ayers, 1981). There is some evidence however that both *S. sclerotivorum* and *T. oligocladum* are more efficient mycoparasites of species of *Sclerotinia* than of *S. cepivorum*. In the United Kingdom over 150 soils from *Sclerotinia* and *S. cepivorum* sites have now been tested for the presence of the two mycoparasites. *Sporidesmium sclerotivorum* has so far only been found in a single sample (Sansford, personal communication) and, although *T. oligocladum* is more common, it still appears to be absent from most soils. If this pattern is true for other countries, there does seem to be ample opportunity for introducing these organisms into soils that do not already contain them. Initially, this might result in substantial oscillations in the levels of *S. cepivorum* and its mycoparasite (Deacon, 1983); in the long term a steady-state condition might be achieved with plant pathogen and mycoparasite in a rough state of balance, but it is hoped that the plant pathogen would be at a lower level than without its mycoparasite.

4.6. Sclerotium Germination Stimulants

The use of sclerotium germination stimulants is based upon the specific response of sclerotia of *S. cepivorum* to species of *Allium* and to their flavor and odor compounds (Coley-Smith and King, 1969; King and Coley-Smith, 1969). Trials undertaken in Australia (Merriman et al., 1981) with artificial onion oil showed that in the Werribee area of southern Victoria this material reduced the population of sclerotia from 44 to 13/kg soil and reduced white rot in dry bulb onions from 57 to 13%. In the Colac area of western Australia similar treatments reduced the sclerotial population from 187 to 87/kg soil, but had no effect upon the incidence of disease. In recent trials in the Colac area some measure of disease control was achieved using diallyl disulfide (Merriman, 1983), but it was not as good as that obtained with certain treatments using iprodione.

Trials involving onion oil have also been undertaken in Canada (Utkhede and Rahe, 1982). In one trial the oil was incorporated into soil 1 month prior to seeding. This gave a 50% reduction in sclerotial numbers and reduced white rot by 70%. In the following year (Rahe, 1983) the oil was applied as a furrow drench at seeding, but no significant effect on the incidence of white rot was observed.

It is clear from these results that neither artificial onion oil nor diallyl disulfide have given reliable control of white rot under field conditions. In view of this a study has been started in this department to investigate the potential of germination stimulants for controlling white rot disease. The main emphasis, so far, has been the evaluation, under ideal conditions, of the maximum possible effects such materials might have. Should these maximum effects prove to be inadequate, there would be little reason to undertake further large-scale field trials.

All the major constituents of artificial onion oil have now been tested for their effects on germination of sclerotia and, of these, diallyl disulfide has proved to be the most active (Coley-Smith et al., 1981). Diallyl disulfide has subsequently been shown to be as effective as the complete artificial onion oil (Coley-Smith and Parfitt, 1983). In laboratory experiments (Entwistle et al., 1982) a clear concentration effect with diallyl disulfide was demonstrated and there was evidence that at high concentration (20%) the material was inhibitory.

During the last 3 years small-scale field trials have been undertaken here in which the effects of diallyl disulfide, injected directly onto sclerotia buried in soil in known numbers in polyester fabric bags, were carefully monitored (Coley-Smith, 1985). Field samples of *S. cepivorum* sclerotia taken from different parts of the country showed little variation in response, but a marked seasonal effect in the reaction of sclerotia was discovered. Treatments made during June, July and August were usually much less effective than treatments made at other times of the year. During the winter months the response of sclerotia was very slow and March, April and October appear to be the most favorable months for such applications (Fig. 2). Gas chromatographic analyses of treated soils showed that the poor response of sclerotia was associated with the rapid disappearance of diallyl disulfide during the summer months. Although many of the experiments were done with a relatively high concentration of diallyl disulfide (5%), it has subsequently been shown that concentrations down to 0.02% are also very effective. In most of the experiments germination of sclerotia varied between 95 and 100%. This work has revealed that application of artificial germination stimu-

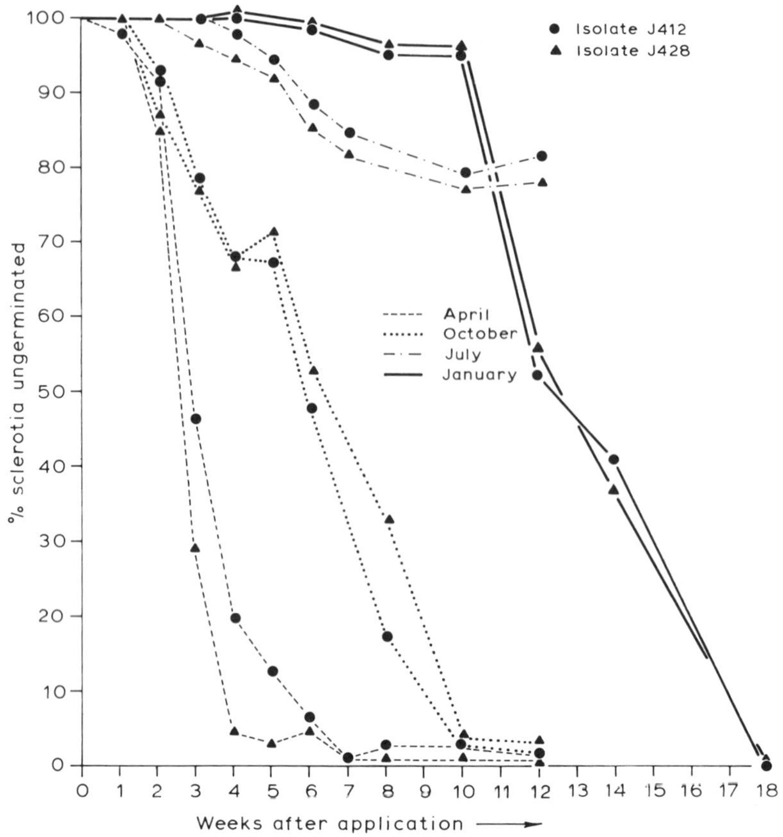

Fig. 2. The effect of applications of 5% diallyl disulfide applied at different times of the year on sclerotia of two isolates of *Sclerotium cepivorum*.

lants can have substantial effects on populations of sclerotia under ideal conditions. Efforts are now being directed toward using the materials in ways that would be possible on a large scale on farms.

5. CONCLUSIONS

Among the possible alternatives to fungicides for the control of white rot disease, soil partial-sterilants, mycoparasites and sclerotium germination, stimulants appear to offer the best chances of success, at least in the foreseeable future. Soil partial-sterilants are reasonably effective, but their use will probably be restricted by their cost in many countries, at least in the dry bulb onion crop.

Mycoparasites such as *S. sclerotivorum* and *T. oligocladum* also show some promise, but they have not received sufficient attention, so far, to allow judgments about whether they will be sufficiently effective against *S. cepivorum*.

With the notable exception of research being done in the United States, work on these mycoparasites is in its infancy.

The outlook for the use of sclerotium germination stimulants seems fairly promising after a rather erratic beginning. There is mounting evidence that considerable reductions in sclerotial populations may be achieved by the use of these materials, but in this case not only cost but also problems of pollution and nuisance might be involved. Present indications are that the cost might be considerably lower than that of soil partial-sterilants and the nuisance factor might be alleviated to some extent with diallyl disulfide, if it could be encapsulated.

The problems involved in attempting to control *S. cepivorum*, by methods other than the use of fungicides, are those with soilborne fungi in microcosm. The evaluation of alternative strategies is, inevitably, a slow and laborious activity and it may take many years to explore the various possibilities completely.

REFERENCES

Abd-El-Moity, T. H. (1979). The situation of white rot disease in Egypt. *Proceedings of the First International Workshop on* Allium *White rot.* Wageningen, The Netherlands, pp. 14-15.

Abd-El-Moity, T. H. (1983). Biological control of white rot of onion in the field. *Proceedings of the Second International Workshop on* Allium *White Rot.* Beltsville, MD, pp. 81-82.

Abd-El-Moity, T. H. and M. N. Shatla (1978). Biological control of white rot disease of onion (*Sclerotium cepivorum* Berk.) by *Trichoderma harzianum* Rifai. 3rd International Congress on Plant Pathology, Munchen, p. 197 (Abstract).

Adams, P. B. (1983). *Allium* white rot in the United States. *Proceedings of the Second International Workshop on* Allium *White Rot.* Beltsville, MD, pp. 1-4.

Adams, P. B. and W. A. Ayers (1981). *Sporidesmium sclerotivorum.* Distribution and function in natural biological control of sclerotial fungi. *Phytopathology*, **71**:90-93.

Adams, P. B. and W. A. Ayers (1982). Biological control of *Sclerotinia* lettuce drop in the field by *Sporidesmium sclerotivorum. Phytopathology*, **72**:485-488.

Adams, P. B. and W. A. Ayers (1985). The world distribution of the mycoparasites *Sporidesmium sclerotivorum, Teratosperma oligocladum* and *Laterispora brevirama. Soil Biol. Biochem.*, **17**:583-584.

Adams, P. B. and S. A. Johnston (1983). Factors affecting efficacy of metham applied through sprinkler irrigation for control of *Allium* white rot. *Plant Dis.*, **67**:978-980.

Adams, P. B. and G. C. Papavizas (1971). Effect of inoculum density of *Sclerotium cepivorum* and some soil environmental factors on disease severity. *Phytopathology*, **61**:1253-1256.

Adams, P. B. and J. K. Springer (1977). Time of infection of fall-planted onions by *Sclerotium cepivorum. Plant Dis. Rep.*, **61**:722-724.

Ahmed, A. H. M. and H. T. Tribe (1977). Biological control of white rot of onion (*Sclerotium cepivorum*) by *Coniothyrium minitans. Plant Pathol.*, **26**:75-78.

Ayers, W. A. and P. B. Adams (1981). Mycoparasitism and its application to biological control of plant diseases. In *Biological Control In Crop Production* (G. C. Papavizas, ed.). Allanheld, Osmun, Totowa, NJ, pp. 91-103.

Booer, J. R. (1945). Experiments on the control of white rot (*Sclerotium cepivorum* Berk.) in onions. *Ann. Appl. Biol.*, **32**:210-213.

Booer, J. R. (1946). Further experiments on the control of white rot (*Sclerotium cepivorum* Berk.) in onions, shallots and leeks. *Ann. Appl. Biol.*, **33**:413-419.

Brooks, D. H. (1984). Is there a future for soil-acting pesticides? In *Soils and Crop Protection Chemicals* (R. J. Hance, ed.). Monograph No. 27, British Crop Protection Council, pp. 3-13.

Coley-Smith, J. R. (1960). Studies of the biology of *Sclerotium cepivorum* Berk., IV. Germination of sclerotia. *Ann. Appl. Biol.*, **48:**8-18.

Coley-Smith, J. R. (1983). Integrated control of *Sclerotium cepivorum. Proceedings of the Second International Workshop on* Allium *White Rot.* Beltsville, MD, pp. 87-98.

Coley-Smith, J. R. (1985). Methods for the production and use of sclerotia of *Sclerotium cepivorum* in field germination studies. *Plant Pathol.* **34:**380-384.

Coley-Smith, J. R. and R. C. Cooke (1971). Survival and germination of fungal sclerotia. *Annu. Rev. Phytopathol.*, **9:**65-92.

Coley-Smith, J. R. and G. Esler (1983a). Infection of cultivars of onion, leek, garlic and *Allium fistulosum* by *Sclerotium cepivorum. Plant Pathol.*, **32:**373-376.

Coley-Smith, J. R. and G. Esler (1983b). The nature of resistance in *Allium* species and cultivars. *Proceedings of the Second International Workshop on* Allium *White Rot.* Beltsville, MD, pp. 53-68.

Coley-Smith, J. R. and R. W. Holt (1966). The effect of species of *Allium* on germination in soil of sclerotia of *Sclerotium cepivorum* Berk. *Ann. Appl. Biol.*, **58:**273-278.

Coley-Smith, J. R. and J. E. King (1969). The production by species of *Allium* of alkyl sulphides and their effect on germination of sclerotia of *Sclerotium cepivorum* Berk. *Ann. Appl. Biol.*, **64:**289-301.

Coley-Smith, J. R. and D. Parfitt (1983). Response of sclerotia to artificial stimulants at different times of the year. *Proceedings of the Second International Workshop on* Allium *White Rot.* Beltsville, MD, pp. 99-106.

Coley-Smith, J. R., G. A. Esler, and C. M. New (1981). Possibilities for biological and integrated control of white rot disease of *Allium. Proceedings 1981 British Crop Protection Conference—Pests and Diseases*, pp. 459-465.

Crowe, F. J., D. H. Hall, A. S. Greathead, and K. G. Baghott (1980). Inoculum density of *Sclerotium cepivorum* and the incidence of white rot of onion and garlic. *Phytopathology*, **70:**64-69.

Croxall, H. E., R. W. Sidwell, and J. E. E. Jenkins (1953). White rot (*Sclerotium cepivorum*) of onions in Worcestershire with special reference to control by seed treatment with calomel. *Ann. Appl. Biol.*, **40:**166-175.

Deacon, J. W. (1983). *Microbial Control of Plant Pests and Diseases.* Van Nostrand Reinhold, United Kingdom, 88 pp.

Dekker, J. (1982). Can we estimate the fungicide resistance hazard in the field from laboratory and greenhouse tests. In *Fungicide Resistance in Crop Protection* (J. Dekker and S. E. Georgopoulos, eds.). Centre for Agricultural Publications and Documentation, Wageningen, pp. 128-138.

El Gammal, M. M. and M. Y. Ibrahim (1983). Avoiding white rot disease through growing onion from sets. *Proceedings of the Second International Workshop on* Allium *White Rot.* Beltsville, MD, pp. 115-117.

El-Yamani, T., M. F. Abdel-Rahim, K. Y. Michael, A. Grinstein, and J. Katan (1983). Soil solarization—a potential method for controlling white rot and other soilborne diseases in onions. *Proceedings of the Second International Workshop on* Allium *White Rot.* Beltsville, MD, p. 126.

Entwistle, A. R. (1983). Needs and outlook for chemical control. *Proceedings of the Second International Workshop on* Allium *White Rot.* Beltsville, MD, pp. 20-22.

Entwistle, A. R. and S. L. Hawling (1984). Changes in response of white rot to vinclozolin and iprodione. *Report of the National Vegetable Research Station, Wellesbourne for 1983*, p. 67.

Entwistle, A. R. and H. L. Munasinghe (1973). White rot disease of salad onions. *Report of the National Vegetable Research Station, Wellesbourne for 1972*, pp. 94-95.

Entwistle, A. R. and H. L. Munasinghe (1977). White rot disease of onions. *Report of the National Vegetable Research Station, Wellesbourne for 1976*, pp. 96–98.

Entwistle, A. R. and H. L. Munasinghe (1978). Epidemiology and control of white rot disease of onions. In *Plant Disease Epidemiology* (P. R. Scott and A. Bainbridge, eds.). Blackwell Scientific Publications, Oxford, pp. 187–191.

Entwistle, A. R. and H. L. Munasinghe (1980a). The effect of seed, furrow and stem base spray treatment with iprodione on white rot disease (*Sclerotium cepivorum*) in spring sown salad onions. *Ann. Appl. Biol.*, **94:**215–224.

Entwistle, A. R. and H. L. Munasinghe (1980b). The effect of iprodione on sclerotium germination, root infection and mycelial spread of *Sclerotium cepivorum* in salad onions. *Ann. Appl. Biol.*, **95:**329–339.

Entwistle, A. R. and H. L. Munasinghe (1981). Formation of secondary sclerotia by *Sclerotium cepivorum. Trans. Br. Mycol. Soc.*, **77:**432–434.

Entwistle, A. R., P. R. Merriman, H. L. Munasinghe, and P. Mitchell (1982). Diallyl-disulphide to reduce the numbers of sclerotia of *Sclerotium cepivorum* in soil. *Soil Biol. Biochem.*, **14:**229–232.

Entwistle, A. R., S. L. Hawling, and J. G. White (1984). Evaluation of *Trichoderma* spp. to control white rot. *Report of the National Vegetable Research Station, Wellesbourne for 1983*, p. 66.

Esler, G. and J. R. Coley-Smith (1983). Flavour and odour characteristics of species of *Allium* in relation to their capacity to stimulate germination of sclerotia of *Sclerotium cepivorum. Plant Pathol.*, **32:**13–22.

Esler, G. and J. R. Coley-Smith (1984). Resistance to *Sclerotium cepivorum* in *Allium* and other genera. *Plant Pathol.*, **33:**199–204.

Fletcher, J. T. (1964). The control of onion white rot using a pure calomel seed dressing and a new sticker. *Plant Pathol.*, **13:**182–184.

Fletcher, J. T., B. C. Knight, E. Bate, and I. A. Cragg (1971). The control of white rot (*Sclerotium cepivorum* Berk.) in salad onions with dicloran. *Plant Pathol.*, **20:**88–92.

Garibaldi, A. (1983). The economic importance of *Allium* white rot in Italy. *Proceedings of the Second International Workshop on* Allium *White Rot.* Beltsville, MD, p. 15.

Georgy, N. I. (1983). *Allium* white rot: Geographic distribution and economic importance (Egypt). *Proceedings of the Second International Workshop on* Allium *White Rot.* Beltsville, MD, pp. 16–17.

Georgy, N. I. and J. R. Coley-Smith (1982). Variation in morphology of *Sclerotium cepivorum* sclerotia. *Trans. Br. Mycol. Soc.*, **79:**534–536.

Ghaffar, A. (1969). Biological control of white rot of onion. II. Effectiveness of *Penicillium nigricans* (Bain) Thom. *Mycopathol. Mycol. Appl.*, **38:**113–127.

Groves, K. and K. S. Chough (1970). Fate of the fungicide 2,6-dichloro-4-nitroaniline (DCNA) in plants and soils. *J. Agric. Food Chem.*, **18:**1127–1128.

Hall, D. H. and P. A. Somerville (1983). Effect of inoculum potential on control of white rot by fungicides. *Proceedings of the Second International Workshop on* Allium *White Rot.* Beltsville, MD, pp. 113–114.

Harrison, D. E. (1956). Onion white rot: Further control experiments. *J. Agric. Victoria*, **54:**17–20.

Katan, J. (1981). Solar heating (solarization) of soil for control of soilborne pests. *Annu. Rev. Phytopathol.*, **19:**211–236.

King, J. E. and J. R. Coley-Smith (1969). Production of volatile alkyl sulphides by microbial degradation of alliin and alliin-like compounds in relation to germination of sclerotia of *Sclerotium cepivorum* Berk. *Ann. Appl. Biol.*, **63:**303–314.

Koert, J. L. (1979). The white rot situation in the Netherlands. *Proceedings of the First International Workshop on* Allium *White Rot.* Wageningen, The Netherlands, pp. 18–19.

Lafon, R. and Y. Bugaret (1966). Étude sur l'efficacité du dicloran contre la pourriture blanche de l'ail (*Sclerotium cepivorum* Berk.). *Phyt. Phytopharm.*, **3:**165–170.

Leggett, M. E. and J. E. Rahe (1985). Factors affecting the survival of sclerotia of *Sclerotium cepivorum* in the Fraser Valley of British Columbia. *Ann. Appl. Biol.* **106:**255–263.

Locke, S. B. (1969). Botran tolerance of *Sclerotium cepivorum* isolates from fields with different botran treatment histories. *Phytopathology*, **59:**13 (Abstract).

Maloy, O. C. and R. Machtmes (1974). Control of onion white rot by furrow and root-dip application of fungicides. *Plant Dis. Rep.*, **58:**6–9.

Maude, R. B. and A. H. Presley (1977a). Neck rot (*Botrytis allii*) of bulb onions. I. Seed-borne infection and its relationship to the disease in the onion crop. *Ann. Appl. Biol.*, **86:**163–180.

Maude, R. B. and A. H. Presley (1977b). Neck rot (*Botrytis allii*) of bulb onions. II. Seed-borne infection in relationship to the disease in store and the effect of seed treatment. *Ann. Appl. Biol.*, **86:**181–188.

Merriman, P. R. (1983). Effect of diallyl disulphide and iprodione on sclerotia of *Sclerotium cepivorum* and incidence of white rot in dry bulb onions. *Proceedings of the Second International Workshop on* Allium *White Rot.* Beltsville, MD, pp. 107–112.

Merriman, P. R., S. Isaacs, R. R. MacGregor, and G. B. Towers. (1981). Control of white rot in dry bulb onions with artificial onion oil. *Ann. Appl. Biol.*, **96:**163–168.

Mikhail, S., D. M. Stewart, M. K. H. El-Haggagy, and R. E. Wilkinson (1974). The role of grazing animals in the dissemination of the onion white rot pathogen in Egypt. *Food Agric. Org. Plant Prot. Bull.*, **22:**37–41.

Mordue, J. E. M. (1976). *Sclerotium cepivorum.* Commonwealth Mycological Institute. Description of Pathogenic Fungi and Bacteria, No. 512.

Nattrass, R. M. (1931). The occurrence of the white rot of the onion (*Sclerotium cepivorum* Berk.) in Egypt. *Bulletin, Ministry of Agriculture, Egypt; Technical and Scientific Service (Plant Protection Section)* No. 107, 9 pp.

New, C. M., J. R. Coley-Smith, and N. I. Georgy (1984). Scanning electron microscopy of sclerotial germination in *Sclerotium cepivorum. Trans. Br. Mycol. Soc.*, **83:**690–693.

Osara, K. (1983). The distribution of *Allium* white rot in Finland. *Proceedings of the Second International Workshop on* Allium *White Rot.* Beltsville, MD, pp. 13–14.

Ohtsuki, S. and A. Fujinami (1982). *Rizolex (tolclofos-methyl). Jpn. Pest. Inf.*, **41:**21–25.

Oudemans, P. and L. V. Edgington (1983). Distribution of white rot of onions in Ontario. *Proceedings of the Second International Workshop on* Allium *White Rot.* Beltsville, MD, p. 10.

Parfitt, D., J. R. Coley-Smith, and T. M. Jeves (1983). *Teratosperma oligocladum*, a mycoparasite of fungal sclerotia. *Plant Pathol.*, **32:**459–460.

Papavizas, G. C. (1984). Soilborne plant pathogens: new opportunities for biological control. *Proceedings 1984 British Crop Protection Conference—Pests and Diseases*, pp. 371–378.

Papavizas, G. C. and R. D. Lumsden (1980). Biological control of soilborne fungal propagules. *Annu. Rev. Phytopathol.*, **18:**389–413.

Porter, I. J. (1983). Soil solarization in Victoria for control of white rot in onions caused by *Sclerotium cepivorum. Proceedings of the Second International Workshop on* Allium *White Rot.* Beltsville, MD, pp. 118–125.

Rahe, J. E. (1981). Lack of correlation between field and laboratory tests for resistance with special reference to white rot of onions. In *Plant Disease Control—Resistance and Susceptibility* (R. C. Staples and G. H. Toenniessen, eds.). Wiley, New York, pp. 193–200.

Rahe, J. E. (1983). Biological control. *Proceedings of the Second International Workshop on* Allium *White Rot.* Beltsville, MD, pp. 71–73.

Rahe, J. E. and R. S. Utkhede (1983). Sources and nature of resistance to *Allium* white rot in *Allium cepa. Proceedings of the Second International Workshop on* Allium *White Rot.* Beltsville, MD, pp. 69–70.

Rondomanski, W. and R. W. Doruchowski (1978). Introductory searches on onion resistance to white rot. *Biul. Warzywniczy*, **22:**99–103.

Ryan, E. W. and T. Kavanagh (1976a). White rot of onion (*Sclerotium cepivorum*). I. Control by fungicidal pelleting of onion seed. *Ir. J. Agric. Res.*, **15:**317–323.

Ryan, E. W. and T. Kavanagh (1976b). White rot of onion (*Sclerotium cepivorum*). 2. Control by fungicidal dusting of onion sets. *Ir. J. Agric. Res.*, **15:**325–329.

Ryan, E. W. and T. Kavanagh (1976c). White rot of onion (*Sclerotium cepivorum*). 3. Epidemiology. *Ir. J. Agric. Res.*, **16:**57–63.

Sadasivan, T. S. and C. V. Subramanian (1960). Interaction of pathogen, soil, other microorganisms in the soil and host. In *Plant Disease: An Advanced Treatise*, Vol. 2 (J. G. Horsfall and E. B. Cowling, eds.). Academic, New York, pp. 273–313.

Scott, M. R. (1956a). Studies of the biology of *Sclerotium cepivorum* Berk. 1. Growth of the mycelium in soil. *Ann. Appl. Biol.*, **44:**576–583.

Scott, M. R. (1956b). Studies of the biology of *Sclerotium cepivorum* Berk. 2. The spread of white rot from plant to plant. *Ann. Appl. Biol.*, **44:**584–589.

Semb, L., J. Vik, and A. Hirvonen-Semb (1978). Resistance of onion (*Allium cepa*) and other *Allium* spp. to the white rot fungus (*Sclerotium cepivorum* Berk.). *Biul. Warzywniczy*, **22:**105–113.

Somerville, P. and P. Merriman (1983). Distribution and economic importance of white rot of onions in Australia. *Proceedings of the Second International Workshop on* Allium *White Rot.* Beltsville, MD, pp. 11–12.

Turner, H. J. and H. T. Tribe (1976). On *Coniothyrium minitans* and its parasitism of *Sclerotinia* species. *Trans. Br. Mycol. Soc.*, **66:**97–105.

Uecker, F. A., W. A. Ayers, and P. B. Adams (1978). A new hyphomycete on sclerotia of *Sclerotinia sclerotiorum. Mycotaxon*, **7:**275–282.

Uecker, F. A., W. A. Ayers, and P. B. Adams (1980). *Teratosperma oligocladum*, a new hyphomycete on sclerotia of *Sclerotinia sclerotiorum, S. trifoliorum* and *S. minor. Mycotaxon*, **10:**421–427.

Utkhede, R. S. and J. E. Rahe (1978). Screening commercial onion cultivars for resistance to white rot. *Phytopathology*, **68:**1080–1083.

Utkhede, R. S. and J. E. Rahe (1980). Biological control of onion white rot. *Soil Biol. Biochem.*, **12:**101–104.

Utkhede, R. S. and J. E. Rahe (1982). Treatment of muck soil with onion oil to control onion white rot. *Can. J. Plant Pathol.*, **4:**79–80.

Utkhede, R. S., J. E. Rahe, J. R. Coley-Smith, Q. P. Van Der Meer, J. G. Brewer, and V. Criscola (1982). Genotype-environment interactions for resistance to onion white rot. *Can. J. Plant Pathol.*, **4:**269–271.

Van Bruggen, A. H. C. and P. A. Arneson (1984). Resistance in *Rhizoctonia solani* to tolclofos-methyl. *Neth. J. Plant Pathol.*, **90:**95–106.

Van Der Meer, Q. P. (1983). Review of literature with respect to screening of *Allium* for resistance to white rot. *Proceedings of the Second International Workshop on* Allium *White Rot.* Beltsville, MD, pp. 45–49.

Van Der Meer, Q. P., J. L. Van Bennekom, and A. C. Van Der Giessen (1983). Screening for resistance to white rot caused by *Sclerotium cepivorum* Berk. in onions (*Allium cepa* L.) and leeks (*Allium porrum* L.). *Euphytica*, **32:**697–701.

Walker, A., A. R. Entwistle, and N. J. Dearnaley (1984). Evidence for enhanced degradation of iprodione in soils treated previously with this fungicide. In *Soils and Crop Protection Chemicals* (R. J. Hance, ed.). Monograph No. 27. British Crop Protection Council, pp. 117–123.

Walker, J. C. (1926). The influence of soil temperature and soil moisture upon white rot of *Allium. Phytopathology*, **16:**697–710.

8

VESICULAR-ARBUSCULAR MYCORRHIZAL FUNGI AND THE CONTROL OF FUNGAL ROOT DISEASES

N. C. SCHENCK
Plant Pathology Department, University of Florida, Gainesville, Florida

There are several recent reviews on the nature of vesicular-arbuscular (VA) mycorrhizal fungi (Hayman, 1983; Harley and Smith, 1983; Gianinazzi-Pearson and Gianinazzi 1981; Gianinazzi-Pearson and Diem, 1982; Rhodes, 1981; Tinker and Gildon, 1982). Therefore, no details are summarized in this chapter in regard to their form or function unless they pertain directly to some aspect of a host-mycorrhiza-pathogen interaction that is being discussed. There are also several recent reviews summarizing the effect of VA mycorrhizal fungi on root diseases or their causal organisms (Dehne, 1982; Hussey and Roncadori, 1982; Schenck, 1981). Considerable additional research on mycorrhizae-disease interactions has been reported since these reviews. Although the intent of this chapter is not to review these recent papers, several of the more pertinent ones are discussed.

Emphasis is placed on the relationships between VA mycorrhizal fungi and fungal plant pathogens causing root diseases, since root disease fungi are the most common pathogens evaluated in interaction studies. However, other microorganisms affecting plant root health are also mentioned. The objectives of this chapter are (1) to examine in some detail the procedures used in recent studies of mycorrhizae-disease interactions, (2) to make recommendations of procedures for future studies, (3) to discuss some of the suggested means by which VA mycorrhizal fungi affect root diseases and (4) to indicate how VA mycorrhizal fungi could be used in systems of nonconventional root-disease control.

1. EARLY STUDIES

The study of the effects of VA mycorrhizal fungi on soilborne diseases is relatively new, with the first report in 1968 (Safir, 1968). However, there are over 75 papers on this subject at present. Early research involving mycorrhizal fungi and root pathogens was summarized by Schenck and Kellam (1978). In many of these early studies, detailed information on the inoculation procedures with the pathogen was lacking or the pathogen was added to the host in amounts or forms that normally did not occur in natural soils. Thus it was difficult or impossible to evaluate the significance of the conclusions reached or to determine if the observed effects were operative in a natural soil. With few exceptions, these studies were conducted in fumigated, steamed or irradiated soils only. This procedure was necessary to remove the naturally occurring mycorrhizal and pathogenic fungi so that one host, one mycorrhizal fungus and one pathogen could be evaluated.

Most interaction studies utilized plants that were precolonized or exposed to the mycorrhizal fungus before being challenged by a soilborne pathogen. The rationale behind this procedure, in many cases, was to assure the presence of the VA mycorrhizal fungus in the root at the time of inoculation with the pathogen. For most species of mycorrhizal fungi, several days are required for spore germination, hyphal growth through soil and host root penetration (van Nuffelen and Schenck, 1984). Additional time is required for establishment of the mycorrhizal fungus in the host (Abbott and Robson, 1981). Although prior colonization of the host with VA mycorrhizal fungi may seem like biasing the outcome in favor of the symbiont, it has some merit. This procedure may be justified in preliminary evaluations of the effectivity of a mycorrhizal fungus in a disease interaction. In addition, many crops are grown in a nursery before being used in the field and these crops could be precolonized easily with mycorrhizal fungi before exposure to a pathogen in the field. In this instance, mycorrhizal precolonization would be a normal procedure before challenge with a pathogen. However, with most annual, direct-seeded crops, this would not be an acceptable procedure for estimating what might occur in the field.

Many of these early studies were executed in soils that had low levels of available phosphorus. Most research on VA mycorrhizal plants have demonstrated that they are capable of absorbing greater amounts of phosphorus than nonmycorrhizal plants. This is the result of hyphal extension by the VA mycorrhizal fungi beyond the phosphorus depletion zone, which occurs near the root as a result of phosphorus absorption by the plant. The hyphae of VA mycorrhizal fungi, which extend beyond this zone, absorb phosphorus and translocate it to the plant (Harley and Smith, 1983). Thus plants with mycorrhizae grow better than plants without VA mycorrhizal fungi when available phosphorus is limiting. This increased vigor and growth could result in plants with larger root systems, which are better able to withstand the invasion and subsequent damage from pathogens. Unfortunately, many host–mycorrhiza–pathogen systems are not evaluated with adequate phosphorus controls to determine if this increased tolerance to the presence of a pathogen results merely from phosphorus-induced growth increases or from some other factor(s).

Until 1978, 11 different host–soilborne plant pathogen combinations were affected by the presence of VA mycorrhizal fungi (Schenck and Kellam, 1978). Despite the shortcomings mentioned above, these data indicated that mycorrhizal fungi, for whatever reason, could possibly result in beneficial effects to the host, at least with certain soilborne diseases.

2. SPECIFICITY

Vesicular–arbuscular mycorrhizal fungi have wide host ranges, establishing symbioses with numerous plants. A mycorrhizal fungus obtained from a perennial plant such as sweetgum (*Liquidambar styraciflua* L.) can also readily establish on an annual plant such as soybean [*Glycine max* (L.) Merr.], bahiagrass (*Paspalum notatum* Flugge) or sorghum (*Sorghum vulgare* Pers.). Because of this capability, isolates from several different hosts can be maintained in pot culture on a single host species. However, the effects of a VA mycorrhizal fungus on various hosts can differ. Mosse (1972) noted that different species and isolates of mycorrhizal fungi can have different effects on plant growth response. McGraw and Schenck (1980) noted several different effects of five species of VA mycorrhizal fungi on seven plant species including citrus, vegetable and ornamental crops. In some instances host biomass of the various plants was affected, in others flowering, fruit set and yields. Others found differences in response to species of mycorrhizal fungi among plant cultivars in soybean (Carling and Brown, 1980; Schenck et al., 1975) and citrus (Nemec, 1978). Edriss et al. (1984) found that five citrus rootstocks varied in growth response to four isolates of mycorrhizal fungi, but differences among these fungi were consistent across rootstocks.

In a similar manner, effects of VA mycorrhizal fungi on interactions with root-infecting plant pathogens can vary with host and species of mycorrhizal fungi. Schenck et al. (1975) observed differences between soybean cultivars and the effect of three species of VA mycorrhizal fungi on root-knot nematode [*Meloidogyne incognita* (Kofoid and White) Chitwood]. With the root-knot susceptible cultivar Pickett, plants colonized by *Glomus macrocarpum* Tul. and Tul. had the highest shoot biomass and yield, yet the nematode numbers were appreciably less than around those plants associated with *Gigaspora margarita* Becker and Hall and *G. heterogama* (Nicol. and Gerd.) Gerd. and Trappe. With the resistant soybean cultivar Forrest, *G. heterogama* was the most effective species in stimulating growth response and reducing nematode numbers. Davis and Menge (1981) also observed differences between VA mycorrhizal fungi and their effects on two citrus rootstock cultivars (sweet orange and Troyer citrange) infected with *Phytophthora parasitica* Dast. Of the five species of VA mycorrhizal fungi evaluated in their tests, plants inoculated with *Glomus mosseae* (Nicol. and Gerd.) Gerd. and Trappe and *Gigaspora margarita* were not significantly reduced in total plant or root dry weights when concomitantly inoculated with *P. parasitica. Glomus fasciculatum* (*G. deserticola* Trappe, Bloss and Menge) produced the greatest growth response on these rootstock cultivars, but the percentage reduction in total plant dry weight from *P. parasitica* was the same or more than the control plants without mycorrhizae.

Thus the presence of mycorrhizal fungi on the roots of a plant does not insure a positive effect on disease and their effects are very specific on the host and the disease. Therefore, mycorrhizal fungi in a control system may need to be very specifically tailored for each host–pathogen combination.

3. METHODOLOGY

As already indicated, little effort has been made to use natural systems in evaluating VA mycorrhizal fungi and their effects on soilborne diseases. For this reason more than any other, results from the studies reported to date are confusing, difficult to interpret and impossible to relate to field situations. If we look at a series of studies for one host–pathogen combination in detail, perhaps the importance of the procedures as they affect results can be better appreciated. We will examine in some detail a series of papers by Davis and others on *P. parasitica* on citrus, since much research on mycorrhizal interactions with pathogens in the genus *Phytophthora* have been reported.

Davis et al. (1978) evaluated the effect of the mycorrhizal fungus *Glomus fasciculatum* (*Glomus deserticola*) on *P. parasitica* on citrus [sweet orange = *Citrus sinensis* (L.) Osbeck, cultivar Pineapple]. The method of sweet orange inoculation with mycorrhizae was to precolonize the roots with *G. deserticola* 8 wk prior to inoculating them with *P. parasitica*. Inoculation with *P. parasitica* consisted of submerging the colonized sweet orange roots in 600 ml of a spore suspension of zoospores (1000/ml) for 1 hr. After 11 wk growth in autoclaved sand, the plants were examined. The growth stimulation obtained by *G. deserticola* (968% increase in dry weight) was negated when mycorrhizal plants were inoculated with *P. parasitica*. However, the visual examination of roots for discoloration and decay (evidence for effects of the pathogen) indicated that mycorrhizal plants inoculated with *P. parasitica* had significantly more "healthy" roots than the nonmycorrhizal roots inoculated with *P. parasitica*, despite the rather "harsh" inoculation method with the pathogen.

In another study (Davis and Menge, 1980) with citrus-*G. deserticola*-*P. parasitica*, several variations from the procedures described above were employed. Split-root procedures were used in which sweet orange was inoculated with both the mycorrhizal fungus and the pathogen on the same or different sides of the plant root system. After 11 wk, plants precolonized with *G. deserti-*

cola on one or both sides of the root were inoculated with *P. parasitica* by submersion into a 600-ml suspension of zoospores (500/ml) for 1 hr. This was half the concentration of zoospores used in the previous study. The presence of *P. parasitica*, again, significantly reduced plant-growth stimulation resulting from *G. deserticola* inoculation; however, a local effect of the pathogen on *G. deserticola* was noted. The greatest reduction in shoot dry weight resulted when *P. parasitica* and *G. deserticola* occurred on the same side of the split-root system. When they were on opposite sides of the split root, no significant reduction in root weight was seen. The same was true for the percentage of healthy roots (this time determined using tetrazolium stain). Thus additional information was gained using the split-root method to evaluate the interaction, but the overall benefit from mycorrhizae was again reduced by the pathogen. In this same publication, additional tests were reported in which the effect of phosphorus on the severity of *P. parasitica* was noted. The system used was similar to the previous experiment, except that plant roots were not split and phosphorus treatments (6, 56 and 600 μg/g soil) were included. The added phosphorus (at 56 μg/g soil) significantly reduced the disease produced by the pathogen. This reduction in disease was equal to that of the mycorrhizal effect obtained at 6 μg/g of phosphorus, indicating that the probable benefit of VA mycorrhizal fungi in this system, was the result of increased uptake of phosphorus by the mycorrhizal plant. In both the split root and phosphorus experiments, the 500 zoospore/ml inoculum significantly reduced the mycorrhizal effect on plant growth but did not completely negate it, as occurred with the higher inoculum level (1000 zoospore/ml level) used in a previously published study.

In another attempt to evaluate the host–mycorrhiza–pathogen system for slightly different aspects, Davis and Menge (1981) used chlamydospores of *P. parasitica* as inoculum (0, 20 and 50 chlamydospores/g soil), and in another experiment (using 500 zoospores/ml inoculum as above) evaluated several species and isolates of mycorrhizal fungi for their effect on *P. parasitica*. Using the soil–chlamydospore method of *P. parasitica* inoculation, this pathogen significantly reduced the total and root dry weight of mycorrhizal and nonmycorrhizal sweet orange. However, the mycorrhizal plant had 2–2.5 times as much total dry weight as the pathogen-inoculated nonmycorrhizal plant and was superior always to the nonmycorrhizal, noninoculated plant in dry root weight. In the evaluation of different species and isolates of VA mycorrhizal fungi, *G. deserticola*, the fungus producing the greatest growth response, was not the best in reducing the effects of the pathogen on citrus. *Glomus mosseae* and *Gigaspora margarita* had the least reduction in root and shoot weight by *P. parasitica*. One wonders what the results of this series of three papers might have been if *G. mosseae* or *Gi. margarita* had been utilized instead of *G. deserticola*. As indicated previously, mycorrhizal fungi are specific in their effects on plants and the selection of a species or isolate may influence the results obtained in mycorrhiza–disease interaction studies.

An evaluation of the composite effects reported in the three papers is of interest. In the first and second papers (Davis et al., 1978; Davis and Menge, 1980), the mycorrhizal dependency values were over 1000%, whereas in the experiments presented in the third paper (Davis and Menge, 1981) these values were 250–320%, one-fourth to one-third as high. The reason for these wide differences in mycorrhizal dependency is not apparent from reading the papers, since the host, pathogen, mycorrhizal fungus, soil types, phosphorus levels and other conditions were apparently similar. In spite of these differences in dependency, the overall effects on disease by the presence of the mycorrhizal fungus were similar. In all tests mentioned above (Davis et al., 1978; Davis and Menge, 1980; and three tests in Davis and Menge, 1981, i.e., at 20 and 50 chlamydospores/g soil and with the dip inoculum at 500 zoospores/ml) there were more healthy roots in plants inoculated with mycorrhizae and the pathogen than those inoculated with pathogen alone. In four out of five of these tests, there were significantly more healthy roots in the mycorrhiza-inoculated plants. Although there was, consistently, a greater reduction in total dry weight by the pathogen with mycorrhizal rather than nonmycorrhizal plants, total plant weight, after pathogen inoculation of mycorrhizal plants was always two to four times greater than that of nonmycorrhizal plants. These differences among dry plant weights were statistically significant in four of the five tests. Only with the dip inoculation, using 1000 zoospores of *P. parasitica* per milliliter, were these differences insignificant. Thus, in these evaluations, the methods of inoculation did not drastically change the results, except for the highest inoculum level used.

4. ADDITIONAL STUDIES

Few papers have studied the effects of mycorrhizal fungi in natural soils or involving several pathogens on one host. Zambolim and Schenck (1983) noted that *G. mosseae* reduced the effects of three pathogens [*Macrophomina phaseoli* (Tassi) Goid., *Rhizoctonia solani* Kuhn, and *Fusarium solani* (Mart.) App. and Wr. emend. Snyd. and Hans.] individually on soybean in autoclaved soil. In this study the pathogens were added to soil in amounts and forms that were frequently encountered in field soils in Florida. The propagules were distributed throughout the soil by using a mixer to provide a distribution of the pathogen similar to that in a natural soil. The mycorrhizal fungus was added as chlamydospores only in a single layer 5 cm below the soil surface (approximately 500 chlamydospores/2 kg soil) at the same time as exposure to the pathogen. In each instance the mycorrhizal plant inoculated with the pathogen was significantly superior to the nonmycorrhizal plant in growth at 45 days. The mycorrhizal plant was able to compensate for each pathogen's reductions in plant growth. The same effect occurred in natural field soil and resulted in yield increases that exceeded the yields of control plants (no additional mycorrhizal fungi or pathogens added). This natural soil had a relatively high level of available phosphorus (92 ppm, double acid extraction) and soybean plants grown in it had good mycorrhizal colonization (49–65%) from *G. etunicatum* and *Gi. margarita*. This is the first good evidence that the mycorrhizal effect on disease can occur in natural soils and against several pathogens on the same host.

McGraw (1983) was the first to evaluate a mycorrhizal-disease interaction introducing the mycorrhizal fungus into autoclaved soil in form (chlamydospores) and numbers similar to those found in natural soils. Previous studies of mycorrhizae and Fusarium wilt on tomato had used the conidial suspension dip method at transplanting as the method to inoculate tomato with *Fusarium oxysporum* Schlect f. sp. *lycopersici* (Sacc.) Snyd. and Hans. (Dehne 1977; Dehne and Schönbeck, 1975). In their study the tomato host had been inoculated with the mycorrhizal fungus *G. mosseae* several weeks prior to inoculation with the pathogen. With this system *G. mosseae* had a marked effect on wilt disease by reducing wilt symptoms, vascular invasion and sporulation of the pathogen. In McGraw's study, the pathogen and mycorrhizal fungus were added simultaneously, in a propagule form (chlamydospores), and at a rate comparable to that found in natural soils. Propagules of both organisms were dispersed in autoclaved soil to approach conditions in nature. Utilizing these methods, McGraw noted that early wilt symptoms occurred on mycorrhizal plants 8–10 days before symptoms on nonmycorrhizal plants. However, after 2 months, disease severity was significantly less on mycorrhizal plants than on nonmycorrhizal plants. Although disease symptoms were less on mycorrhizal plants inoculated with *G. etunicatum* than those inoculated with *G. mosseae*, both mycorrhizal fungi had less disease than nonmycorrhizal plants. In this instance, improving the methods so that they resembled a more natural system did not change or greatly modify the final outcome of results from that using less "natural" systems.

Although most mycorrhizae-disease studies are conducted without adequate phosphorus controls, several papers have reported the effects of various levels of phosphorus on mycorrhizae-disease interactions (Davis et al., 1978; Davis and Menge, 1980; Graham and Menge, 1982). The effects were found to vary with the pathogen-host-mycorrhizal fungus included in each study. The addition of the mycorrhizal fungus *G. deserticola* (*G. fasciculatum*) increased vascular browning and the incidence of the pathogen *Verticillium dahliae* Kelb on cotton. Although the presence of *V. dahliae* did not reduce the incidence of *G. deserticola* in the roots, application of phosphorus (300 μg/g soil) to the potting soil reduced mycorrhizal colonization by 45.3%. Since the addition of phosphorus increased the incidence of *V. dahliae* in the host, it was assumed that the increased disease from the presence of *G. deserticola* was the result of increased phosphorus levels in the host.

It was previously mentioned that the effect of the mycorrhizal fungus *G. deserticola* on Phytophthora root rot of citrus was the result of increased phosphorus provided to the host by the mycorrhiza (Davis and Menge, 1980). In a study with *G. deserticola* and the pathogen *Gaeumannomyces graminis* (Sacc.) von Arx and Oliver var. *tritici* Walker on wheat (*Triticum aestivum* L.), Graham and Menge (1982) noted that phosphorus reduced the quantity of lesioned roots. This reduction was equivalent to the reduction in disease obtained by the mycorrhizal fungus. In this study as in those mentioned above (Davis et al., 1980; Davis and Menge, 1981), inoculum levels of the pathogen influenced the results. Whereas *G. deserticola* reduced the incidence of disease at the low level of *G. graminis* inoculum (0.1% pathogen-infected wheat kernels in the potting mix, w/w), this depression in disease incidence was not apparent at the higher inoculum level (0.5% pathogen-infected wheat kernels in the potting mix).

Thus the level of phosphorus in the studies described above affected the interaction differently. With *V. dahliae*, increased host phosphorus levels worsened disease, hence, mycorrhizal hosts had greater disease than nonmycorrhizal hosts. However, with *G. graminis* on wheat, disease was lessened by increased host phosphorus levels and the mycorrhizal host had less disease, at least at the lower level of pathogen inoculum.

5. RECOMMENDATIONS FOR FUTURE INTERACTION STUDIES

Based on the results mentioned, some recommendations can be made to those considering evaluations of mycorrhizae-disease interactions. In regard to the pathogen, perhaps the most attention should be given to the level of inoculum used so that the effect (if any) of the mycorrhizal fungus is not "overwhelmed." Various inoculum levels of the pathogen should be evaluated with or without the mycorrhizal fungus present to determine an amount of the pathogen that does not rapidly kill the host but provides a moderate amount of disease. It would appear preferable to use pathogen inoculum in the form of propagules that are normal in soil (resting or overwintering propagules). However, Davis and Menge (1981) found that a root-dip procedure using 500 zoospores/ml of the pathogen had results similar to 50 chlamydospores/g soil dispersed throughout the potting medium. In addition, Dehne and Schönbeck (1975) noted a mycorrhizal effect with the Fusarium wilt pathogen using precolonized mycorrhizal plants and a conidial suspension root-dip inoculation procedure for the pathogen. McGraw (1983) observed a mycorrhizal effect using a more natural inoculum propagule for both the pathogen and mycorrhizal fungus. Although the end results were similar in these studies, despite the methods used, the use of a natural propagule (chlamydospores) at levels approximating those in the field would relate more directly to conditions in natural soil and be more easily transferred to a field situation.

In regard to the mycorrhizal fungus, it would be wise to screen several species or isolates to select one that has an effect on the disease to be considered. Also, it might be desirable to use several mycorrhizal fungi together, since it has been shown, in at least one incidence, that a mixed mycorrhizal population can have better results than a single species (Bartschi et al., 1981). Selection of indigenous mycorrhizal fungi may have an advantage of adaptation to the soils and climate in which the pathogen and host occur as opposed to nonindigenous mycorrhizal fungi. Natural population levels of the mycorrhizal fungus should be used, approximating those found in natural soils, and the mycorrhizal inoculum should be dispersed throughout the soil as occurs in nature.

In regard to the host, various cultivars should be evaluated for reaction to the pathogen and mycorrhizal fungus to determine if any differences in reaction among cultivars occur. Several host and disease parameters should be measured during the progress of the experiment to detect varying differences in response to the mycorrhiza–pathogen interaction. A series of phosphate applications in the test soil should be evaluated to determine if the mycorrhizal effect is simply a response to additional amounts of phosphorus supplied to the plant by the mycorrhizal fungus. The phosphorus level desired is one that allows adequate growth of the nonmycorrhizal host but also enables the mycorrhiza, on the mycorrhiza inoculated host plant, to colonize the root adequately (G. Smith, personal communication). In a recent study by Caron et al. (1985) the substrate used affected the interaction results. A calcined montmorillonite clay (Turface) was found to be best for detection of a mycorrhizal effect on *Fusarium oxysporum* Schlect f. sp. *radicis-lycopersici* Jarvis and Shoemaker on tomato. However, despite the effects of the medium on the interaction results, it probably would be desirable to evaluate the interaction in field soil to determine the likelihood of its occurring in nature. Conforming to these suggestions will not assure success in determining a mycorrhizal effect on a specific disease, but it could markedly reduce some of the variability and increase the reliability of results.

6. NATURE OF THE MYCORRHIZAL EFFECT ON DISEASE

The specific effect of the mycorrhizal fungus on disease is not well known. Most studies suggesting a means of action of the mycorrhizal fungus on disease have relied on observed differences between the mycorrhizal and nonmycorrhizal plant for chemical or morphological features. The rationale in these studies would implicate features present in the mycorrhizal plant as the possible cause, if they were absent or occurred at a lower degree in the nonmycorrhizal plant.

The suggested possible mycorrhizal effects on disease can be categorized as either chemical or morphological. In the chemical category, the indirect effect on disease, resulting from increased phosphorus uptake by mycorrhizal plants, has already been mentioned. The increase of amino acids (Baltruschat and Schönbeck, 1975), sugars (Safir, 1968), phenolic compounds (Krishna and Bagyaraj, 1983; Dehne and Schönbeck, 1979) and enzyme activity (Dehne et al., 1978) in the roots of mycorrhizal plants has been observed and suggested as the possible reason for the reduction of disease on mycorrhizal plants. Morphological phenomena noted in mycorrhizal plants that have been suggested as factors influencing disease include (1) a greater morphogenic response in cell walls to the pathogen's presence on mycorrhizal roots (Becker and Gerdemann, 1977), (2) increase in wound barrier formation (Wick and Moore, 1984), and (3) increase in lignification of the cell wall (Dehne and Schönbeck, 1979; Becker, 1976).

7. POTENTIAL USES IN AGRICULTURE

In several instances, VA mycorrhizal fungi can be detrimental to plants. Jones and Hendrix (1985) noted a suppression of tobacco in the field due to the presence of *Glomus macrocarpum*. Kiernan et al. (1983) reported several VA mycorrhizal fungi that inhibited sweetgum growth under conditions of high fertility. Mycorrhizal fungi increased the severity of several diseases by foliar pathogens (Schönbeck and Dehne, 1979). Schenck and Kellam (1978) discuss other instances reported in the literature of reduced plant growth from mycorrhizal fungi. In addition, mycorrhizal fungi cannot be readily cultured at present to produce inoculum in large quantities for extensive field inoculation.

With these attributes, why is there any interest in the use of mycorrhizal fungi as biocontrol agents for soilborne diseases? Root diseases are some of the most difficult to manage or control. Chemicals used to control these diseases are expensive to purchase and apply. Their use often contributes to soil pollutants that possibly can lead to ground-water contamination. There is considerable interest in using alternative techniques, especially nonpolluting biological methods to control root diseases in plants. Mycorrhizal fungi can not only reduce root diseases they can also confer many other attributes to plants (Rhodes, 1981), especially the ability to absorb relatively immobile elements (phosphorus, copper and zinc) from soil. As a result, they aid the growth of plants in soil in which these (and possibly other) elements are not readily available. In addition, lower amounts of fertilizer are needed to produce adequate growth and yields. Hence, fertilizer amounts in the soil can be reduced thus minimizing potential soil fertility pollutants, where this may be a problem. In developing countries where soils with low phosphorus availability may be one of the major constraints to increased agricultural production, the use of mycorrhizal fungi to alleviate this problem is a possibility. In these same countries, the use of pesticides to control root disease is not a viable option because of economics. Again, mycorrhizal fungi could possibly meet this need. The same biological agent has the potential to alleviate several problems.

Although at present the availability of large amounts of mycorrhizal inoculum for use in commercial agriculture is a constraint to their widespread application, there is still the possibility that these fungi can be cultured. Other obligate fungi have been grown in culture. Also, new techniques have been developed recently for their culture on the roots of plants. Mosse and Thompson (1984) and Elmes and Mosse (1984) recently reported on a method (nutrient flow procedure) for growing VA mycorrhizal fungi in relatively large quantities and the inoculum produced in this manner can be used effectively in the field (Elmes et al., 1983). Menge et al. (1977) developed a means to increase these fungi in modified pot culture and this system has been utilized commercially (Menge, 1983). Sufficient inoculum of mycorrhizal fungi could be produced by either of these methods for use with crops grown in nurseries and containers.

Although additional research is needed to demonstrate the use and effectiveness of mycorrhizal fungi, it is not necessary to understand all aspects of their life cycle or how they function before they can be utilized in agriculture. Some suggested areas for further study can be listed. Nursery or container-grown plants with root diseases that limit production or markedly reduce yields would be candidates for evaluation with VA mycorrhizal fungi. The recommendation for future studies mentioned in the previous section of this chapter should be considered, especially screening for the most effective isolates and evaluating them eventually in natural systems. Where mycorrhizal fungi are being evaluated for field inoculation, and there are several areas in the world where this is being done, consideration should be given to isolates that not only increase nutrient uptake but might also be effective in reducing the effects of predominant soilborne pathogens.

Using VA mycorrhizal fungi, we have the potential to increase plant growth and vigor under a number of stress conditions, including the presence of some root pathogens. At a time when alternative methods of disease control are needed, particularly for those organisms affecting the below-ground plant tissues, we should not overlook any possible means, and that includes mycorrhizal fungi. I am confident they can be useful in agriculture and in some manner be effective as biocontrols for root diseases. However, further study and research are required to achieve this goal.

REFERENCES

Abbott, L. K. and A. D. Robson (1981). Infectivity and effectiveness of vesicular arbuscular mycorrhizal fungi: Effect of inoculum type. *Austral. J. Agric. Res.*, **32:**631–639.

Bartschi, H., V. Gianinazzi-Pearson, and I. Vegh (1981). Vesicular-arbuscular mycorrhiza formation and root rot disease (*Phytophthora cinnamomi*) development in *Chamaecyparis lawsoniana. Phytopath. Z.*, **102:**212–218.

Baltruschat, H. and F. Schönbeck (1975). The influence of endotrophic mycorrhiza on the infestation of tobacco by *Thielaviopsis basicola. Phytopath. Z.*, **84:**172–188.

Becker, W. N. (1976). Quantification of onion vesicular-arbuscular mycorrhizae and their resistance to *Pyrenochaeta terrestris.* Ph.D. Dissertation, University of Illinois, Urbana, 72 pp.

Becker, W. N. and J. W. Gerdemann (1977). The resistance of vesicular-arbuscular onion mycorrhizae to *Pyrenochaeta terrestris.* Abstract of the 3rd North American Conference Mycorrhizae. Athens, Georgia.

Carling, D. E. and M. F. Brown (1980). Relative effect of vesicular-arbuscular mycorrhizal fungi on the growth and yield of soybean. *Soil Sci. Soc. Am. J.*, **44:**528–531.

Caron, M., J. A. Fortin, and C. Richard (1985). Influence of substrate on the interaction of *Glomus intraradices* and *Fusarium oxysporum* f. sp. *radicis-lycopersici* on tomatoes. *Plant and Soil*, **87:**233–239.

Davis, R. M. and J. A. Menge (1980). Influence of *Glomus fasciculatus* and soil phosphorus on Phytophthora root rot of citrus. *Phytopathology*, **70:**447–452.

Davis, R. M. and J. A. Menge (1981). *Phytophthora parasitica* inoculation and intensity of vesicular-arbuscular mycorrhizae on citrus. *New Phytol.*, **87:**705–715.

Davis, R. M., J. A. Menge, and G. A. Zentmyer (1978). Influence of vesicular-arbuscular mycorrhizae on Phytophthora root rot of three tree crop plants. *Phytopathology*, **68:**1614–1617.

Dehne, H.-W. (1977). Untersuchungen uber den Einfluss der endotrophen Mykorrhiza auf die Fusarium-Welke, Tomate and Gurge. Ph.D. Dissertation, Universitat Bonn, 150 pp.

Dehne, H.-W. (1982). Interaction between vesicular-arbuscular mycorrhizal fungi and plant pathogens. *Phytopathology*, **72:**1115–1119.

Dehne, H.-W. and F. Schönbeck (1975). Investigation on the influence of endotrophic mycorrhiza on Fusarium wilt of tomato. *Zeit. Pflanzenkr. Pflanzensch.*, **82:**630–632.

Dehne, H.-W. and F. Schönbeck (1979). The influence of endotrophic mycorrhiza on plant diseases. II. Phenol metabolism and lignification. *Phytopath. Z.*, **95:**210–216.

Dehne, H.-W., F. Schönbeck, and H. Baltruschat (1978). Investigations on the influence of endotrophic mycorrhiza on plant diseases. III. Chitinase activity and ornithine-cycle. *Phytopath. Z.*, **95:**666–678.

Edriss, M. H., R. M. Davis, and D. W. Burger (1984). Increased growth response of citrus by several species of mycorrhizal fungi. *Hort. Sci.*, **68:**537–539.

Elmes, R. P., C. M. Hepper, D. S. Hayman, and J. O'Shea (1984). The use of vesicular-arbuscular mycorrhizal roots grown by the nutrient film technique as inoculum for field sites. *Ann. Appl. Biol.*, **104:**437–441.

Elmes, R. P. and B. Mosse (1984). Vesicular-arbuscular endomycorrhizal inoculum production. II. Experiments with maize (*Zea mays*) and other hosts in nutrient flow culture. *Can. J. Bot.*, **62:**1531–1536.

Gianinazzi-Pearson, V. and H. G. Diem (1982). Endomycorrhizae in the tropics. In *Microbiology of Tropical Soils and Plant Productivity* (Y. R. Dommergues and H. G. Diem, eds.). M. Nijhoff/ W. Junk Publishers, The Hague, pp. 209–261.

Gianinazzi-Pearson, V. and S. Gianinazzi (1981). Role of endomycorrhizal fungi in phosphorus cycling in the ecosystem. In *The Fungal Community, Its Organization and Role in the Ecosystem* (D. T. Wicklow and G. C. Carrol, eds.). Marcel Dekker, Basel, pp. 637–652.

Graham, J. H. and J. A. Menge (1982). Influence of vesicular-arbuscular mycorrhizae and soil phosphorus on take-all disease of wheat. *Phytopathology*, **72:**95–98.

Harley, J. L. and S. E. Smith (1983). *Mycorrhizal Symbiosis*. Academic, London, 483 pp.

Hayman, D. S. (1983). The physiology of vesicular-arbuscular endomycorrhizal symbiosis. *Can. J. Bot.*, **61:**944–963.

Hussey, R. S. and R. W. Roncadori (1982). Vesicular-arbuscular mycorrhizae may limit nematode activity and improve plant growth. *Plant Dis.*, **66:**9–14.

Jones, K. and J. W. Hendrix (1985). Inhibition of root production as the mechanism of pathogenicity of *Glomus macrocarpum* to tobacco. Proceedings of the 6th North American Conference on Mycorrhizae. Bend, Oregon, p. 332.

Kiernan, J. M., J. W. Hendrix, and D. M. Maronek (1983). Fertilizer-induced pathogenicity of mycorrhizal fungi to sweetgum seedlings. *Soil Biol. Biochem.*, **15:**257–262.

Krishna, K. R. and D. J. Bagyaraj (1983). Interaction between *Glomus fasciculatum* and *Sclerotium rolfsii* in peanut. *Can. J. Bot.*, **61:**2349–2351.

McGraw, A.-C. (1983). The influence of inoculum density of vesicular-arbuscular mycorrhizal fungi on their development and on Fusarium wilt of tomato. Ph.D. Dissertation University of Florida, Gainesville, 131 pp.

McGraw, A.-C. and N. C. Schenck (1980). Growth stimulation of citrus, ornamental, and vegetable crops by select mycorrhizal fungi. *Proc. Florida State Hort. Soc.*, **93:**201–205.

Menge, J. A. (1983). Utilization of vesicular-arbuscular mycorrhizal fungi in agriculture. *Can. J. Bot.*, **93:**1015–1024.

Menge, J. A., H. Lembright, and E. L. V. Johnson (1977). Utilization of mycorrhizal fungi in citrus nurseries. *Proc. Int. Soc. Citricult.*, **1:**129–132.

Mosse, B. (1972). The influence of soil type and *Endogone* strain on the growth of mycorrhizal plants in phosphorus deficient soils. *Rev. Ecol. Biol. Soc.*, **9**:529–537.

Mosse, B. and J. P. Thompson (1984). Vesicular-arbuscular endomycorrhizal inoculum production. I. Exploratory experiments with beans (*Phaseolus vulgaris*) in nutrient flow culture. *Can. J. Bot.*, **62**:1523–1530.

Nemec, S. (1978). Response of six citrus rootstocks to three species of *Glomus*, a mycorrhizal fungus. *Proc. Florida State Hort. Soc.*, **91**:10–14.

Rhodes, L. H. (1981). The use of mycorrhizae in crop production systems. *Outlook Agric.*, **10**:275–281.

Safir, G. (1968). The influence of vesicular-arbuscular mycorrhiza on the resistance of onion to *Pyrenochaeta terrestris*. M. S. Thesis. University of Illinois, Urbana, 36 pp.

Schenck, N. C. (1981). Can mycorrhizae control root disease? *Plant Dis.*, **65**:230–234.

Schenck, N. C. and M. K. Kellam (1978). The influence of vesicular arbuscular mycorrhizae on disease development. *Florida Agric. Exp. Sta. Bul.*, **798,** 16 pp.

Schenck, N. C., R. A. Kinloch, and D. W. Dickson (1975). Interaction of endomycorrhizal fungi and root-knot nematode on soybean. In *Endomycorrhizas* (F. E. Sanders, B. Mosse, and P. B. Tinker, eds.). Academic, London, pp. 605–617.

Schönbeck, F. and H.-W. Dehne (1979). The influence of endotrophic mycorrhiza on plant diseases. 4. Fungal parasites on aerial plant parts, *Olpidium brassicae*, TMV. *Z. Pflanzenkr. Pflanzensch.*, **86**:103–112.

Tinker, P. B. and A. Gildon (1982). Mycorrhizal fungi and ion uptake. In *Metals and Micronutrients, Uptake and Utilization by Plants* (D. A. Robb, and W. S. Pierpoint, eds.). Academic, London, pp. 21–32.

van Nuffelen, M. and N. C. Schenck (1984). Spore germination, penetration, and root colonization of six species of vesicular-arbuscular mycorrhizal fungi on soybean. *Can. J. Bot.*, **62**:624–628.

Wick, R. L. and L. D. Moore (1984). Histology of mycorrhizal and nonmycorrhizal *Ilex crenata* "Helleri" challenged by *Thielaviopsis basicola. Can. J. Plant Pathol.*, **6**:146–150.

Zambolim, L. and N. C. Schenck (1983). Reduction of the effects of pathogenic root-infecting fungi on soybean by the mycorrhizal fungus, *Glomus mosseae. Phytopathology*, **73**:1402–1405.

9

GENETIC MANIPULATION TO IMPROVE THE EFFECTIVENESS OF BIOCONTROL FUNGI FOR PLANT DISEASE CONTROL

G. C. PAPAVIZAS

Soilborne Diseases Laboratory, Plant Protection Institute, USDA–ARS Beltsville, Maryland

1. INTRODUCTION

A recent report to the U. S. Department of State, compiled by a panel of scientists, stated that genetic engineering is "one of four major scientific revolutions of this century, on a par with unlocking the atom, escaping the earth's gravity, and the computer revolution" (U. S. Department of State, 1981). Since the elucidation of the molecular structure of DNA (Watson and Crick, 1953), the science of molecular genetics has rapidly evolved through molecular biology, recombinant DNA technology, and the function and expression of plant and microbial genomes into genetic engineering. Gene cloning, gene transfer, and other new techniques (e.g., recombinant protoplast techniques) have been used successfully by several branches of plant science and microbiology (especially those dealing with industrial and pharmaceutical microbiology) to engineer new microorganisms or improve existing qualities in old ones (Ball, 1984a,b). For instance, cells of the yeast *Saccharomyces cerevisiae* have been genetically altered to produce the antiviral protein interferon (Hitzeman et al., 1983).

Although wild-type strains of various fungal antagonists (*Aspergillus, Chaetomium, Dendrostilbella, Gliocladium, Penicillium, Pythium, Sporidesmium, Talaromyces, Teratosperma,* and *Trichoderma*) have shown considerable promise for biocontrol of plant diseases, especially soilborne diseases (Papavizas, 1985; Papavizas and Lewis, 1981), plant pathologists and mycologists dealing with disease suppression with biocontrol fungi have not delved into the new area of genetic engineering with the same enthusiasm and determination expressed by industrial microbiologists. For instance, the realization of techniques for protoplast fusion and transformation has resulted in the further development of new technologies for gene manipulation of industrial fungi (Ball, 1984a; Peberdy, 1983). None of these approaches has been used yet with biocontrol fungi to improve their effectiveness or their other desirable characteristics (tolerance to pesticides, survival ability in the environment, ability to colonize the rhizosphere, tolerance to adverse environmental conditions such as extremes of moisture and temperature, vigorous growth and sporulation in deep-tank fermentation, and long shelf life). With the exception of mutagenesis and conventional genetics, nothing was found in the literature indicating that any efforts have been made to improve the effectiveness of biocontrol fungi by gene-splicing, cloning, and transformations.

2. INDUCTION OF NEW BIOTYPES OF BIOCONTROL FUNGI BY EXPOSURE TO FUNGICIDES AND MUTAGENS

2.1. Exposure to Fungicides

Research performed during the past 10–15 years on resistance developed by fungi as a result of exposure to modern fungicides, confined almost exclusively to plant-pathogenic fungi, has been summarized adequately (Dekker, 1976; Delp, 1980; Georgopoulos, 1977). Despite the abundance of work performed on this subject, no research was done until 1980 on the tolerance of biocontrol fungi to fungicides or on purposeful induction of tolerance in biocontrol agents for use in integrated control of plant pathogens.

Abd-El Moity et al. (1982) presented evidence that it was feasible to develop strains of *Trichoderma harzianum* tolerant to fungicides by prolonged and repeated exposure to fungicides. Exposure of four wild-type strains of *T. harzianum* to the fungicides chlorothalonil, iprodione, procymidone, and vinclozolin resulted in the selection of several isolates tolerant to these fungicides (Abd-El Moity et al., 1982; Papavizas, 1980). None of the isolates developed tolerance to benomyl by repeated culture to media containing the fungicide. Several tolerant isolates differed from their respective wild-type strains, not only in their tolerance to the fungicides, but also in morphological characteristics, growth habit, and sporulation. A few tolerant isolates showed enhanced ability to suppress selected soilborne plant pathogens. For instance, the fungicide-tolerant isolate Th-1(procy-2M), derived from the wild-type strain Th-1, reduced white rot of onion caused by *Sclerotium cepivorum* more effectively than did Th-1 or other fungicide-tolerant isolates. Also, a combination of iprodione and an iprodione-tolerant isolate gave significantly higher control of white rot in the field than did iprodione alone (Abd-El Moity et al., 1982).

Tolerance to fungicides in fungi may be due to nongenetic adaptation (training) (Delp, 1980) or to permanent changes by fungicides. Hastie (1979) suggested that the "genetic effects" of some modern fungicides may include genetic mutation, chromosome breakage, and mitotic recombinations. In the experiments by Abd-El Moity et al. (1982) and Papavizas (1980) both nongenetic and genetic changes occurred. Several fungicide-tolerant isolates lost their ability to tolerate the fungicides after repeated transfers to fungicide-free culture media. A few isolates remained stable. In assessing the potential for using fungicide-tolerant isolates of biocontrol fungi, whose tolerance was the result of "training" to the fungicides, it is at once fairly obvious that the prospects for blending such isolates in integrated pest management systems may not be very promising for agricultural application. A practical approach to remedy the situation would involve "preserving" tolerance by maintaining cultures on media containing the appropriate fungicide (Abd-El Moity et al., 1982). However, such an empirical approach would raise several formidable problems to industrial developers, discussion of which is beyond the scope of this review.

2.2. Mutagenesis and Selection

Although there are very few published reports on the fundamental aspects of mutagenesis performed with industrial fungi, mutants giving rise to increased production of various metabolites or catabolic enzymes have been produced (Ball, 1973, 1980, 1984a; Picataggio, 1983). Improved strains of *Trichoderma reesei* were isolated by mutation and selection, and these produced more cellulase than did wild strains (Mandels et al., 1971; Montenecourt and Eveleigh, 1977, 1979; Picataggio, 1983). Mandels et al. (1971) irradiated *T. viride* with high-energy electrons from a linear accelerator and produced a new strain that secreted twice as much cellulase as its wild parent. A further approach to gain enhanced cellulase yields was to select mutants that were resistant to polyene antibiotics (Schimenti et al., 1983). Since these antibiotics act through selective sterol binding by the plasma membrane, such polyene antibiotic-resistant mutants may include pleiotropic strains that synthesize modified membranes, resulting in enhanced protein secretions. Schimenti et al. (1983) obtained nystatin-resistant mutants of *T. reesei* that showed an increased production of cellulase.

Since the middle 1940s, microbiologists have also made great progress in developing biotypes of fungi by genome modification to increase antibiotic production (Alikhanian, 1962; Ball, 1980, 1984a; Burnett, 1975; Demain, 1973; Elander et al., 1976; Lowe and Elander, 1983; Sermonti, 1969). Mutations and selections, coupled with parasexual hybridization processes, have increased antibiotic production by fungi and streptomycetes nearly 100- to 500-fold (Alikhanian, 1962; Lowe and Elander, 1983). For instance, initial production of penicillin by *Penicillium* spp. at the time of discovery was 20 units/ml. Because of the development of new biotypes of *P. chrysogenum* by the use of mutagenic factors and selection, antibiotic production increased to 8000 units/ml (Alikhanian, 1962). However, genetic manipulation by mutagenesis to enhance the antagonistic activity of biocontrol fungi and to increase their tolerance to commonly used pesticides is mostly unexplored.

Since the development of biotypes of biocontrol fungi resistant to fungicides by nongenetic adaptation or "training" presents serious problems of instability (Abd-El Moity et al., 1982), and most of the major increases in metabolite production by *Trichoderma* spp. and *Penicillium* spp. have been achieved by selecting high metabolite-yielding mutants, development of new biotypes resistant to pesticides and with enhanced biocontrol capabilities by mutagenesis appears quite promising. The first such program was initiated at Beltsville in 1978 when it was decided to explore ultraviolet (UV) radiation as a simple mutagenetic method (Papavizas and Lewis, 1983; Papavizas et al., 1982). A major difficulty in using the induced mutagenesis-selection approach with *Trichoderma* spp. and *Gliocladium* spp. was the lack of visible or detectable markers associated with desired qualities in the two genera. Papavizas and Lewis (1983) and Papavizas et al. (1982) used the ability of *Trichoderma* spp. to tolerate fungicides of the MBC group [benomyl, thiabendazole (TBZ)] as a marker in a selection system following induced mutagenesis with UV light. Wild-type strains of the two biocontrol agents are very sensitive to these fungicides. Using this approach, Papavizas and Lewis (1983) and Papavizas et al. (1982) developed several genetic variants from wild strains of *T. viride* and *T. harzianum*, respectively. Several new biotypes tolerated high concentrations (up to 500 μg a.i./ml) of benomyl and other MBC fungicides, as indicated by growth and conidial germination in fungicide-amended culture media.

Ultraviolet radiation of several wild-type strains of *T. hamatum* and *Gliocladium virens* for up to 80 min yielded few survivors, but no mutants tolerant to MBC fungicides (Papavizas, unpublished). Recently, mutants of *T. hamatum* and *Talaromyces flavus* (teliomorph of *Penicillium vermiculatum*) were produced by exposing conidia and ascospores, respectively, to *N*-methyl-N^3-nitro-*N*-nitrosoguanidine (NTG) by the method of Hashiba and Yamada (1984). A few mutants could tolerate benomyl and TBZ up to 100 μg a.i./ml (Katan et al., 1984).

Several UV-induced biotypes, or biotypes induced by chemical mutagenesis, of *Trichoderma* spp. differed from their respective wild-type strains not only in tolerance to benomyl, but also in growth characteristics, sporulation, and survival in soil (Papavizas and Lewis, 1983; Papavizas et al., 1982). A few mutants of *T. harzianum* and *T. viride* that were tolerant to benomyl were also more effective than their respective wild-type strains in suppressing the competitive saprophytic activity of *Rhizoctonia solani* in soil, damping-off (*Pythium ultimum*) of peas, damping-off (*R. solani*) of cotton and radish (Table 1), white rot (*Sclerotium cepivorum*) of onion, and damping-off and blight of bean (*S. rolf-*

TABLE 1. Biological Control of Seedling Disease of Cotton and Damping-Off of Radish Caused by *Rhizoctonia solani*[a] with *Trichoderma harzianum* Strain WT-6 and with Selected UV-Induced Biotypes[b]

	Plant Stand at 3 Wk (%)[c]	
Biotype and Kind of Inoculum	*Cotton*	*Radish*
Conidia in water[d]		
None (infested control)	16 abc	31 bc
WT-6 (wild strain)	20 abc	26 ab
WT-6-1	24 abc	37 bcde
WT-6-17	50 de	52 efg
WT-6-24	53 de	55 fg
WT-6-6	56 de	50 efg
None (uninfested control)	67 e	61 g
Dry antagonist preparation[e]		
WT-6-1	11 a	14 a
None (infested control)	16 ab	41 bc
WT-6 (wild strain)	20 abc	33 bcd
WT-6-19	20 abc	49 defg
WT-6-17	40 cd	46 cdef
WT-6-24	48 de	53 efg
WT-6-5	50 de	52 efg
WT-6-6	54 de	37 bcde
None (uninfested control)	67 e	61 g

[a]Inoculum of *R. solani* R-35 was produced on chopped-potato-soil (CPS) medium; large granular propagules (>589μm) were separated and added to soil at 1% 1 wk before the antagonist, *T. harzianum*.

[b]Reprinted by permission from *Phytopathology*, **72**:126–132, 1982 (58).

[c]In each column, values followed by the same letter do not differ significantly ($P = 0.05$) according to Duncan's multiple range test.

[d]Conidia from V-8 juice agar plates added to soil to provide 4.4×10^5 conidia/gram of soil.

[e]Antagonists were grown on sand-cornmeal-bran-modified gliotoxin fermentation medium (SCB-GFM) for 3 wk at 25°C; preparations were dried, crumbled, and added to soil in amounts to provide 4.4×10^5 CFU/gram of soil.

sii). One UV-induced biotype of *T. viride* that possesses tolerance to benomyl effectively suppressed wilt of chrysanthemum caused by *Fusarium oxysporum* f. sp. *chrysanthemi* in greenhouse culture, when repeated use of benomyl is critical for the control of foliar pathogens (Locke et al., 1985). Biotype T-1-R9 (U. S. Patent No. 4489161) was able to recolonize the soil even in the presence of benomyl, prevent recolonization of soil by the pathogen, and bring about control equivalent to that produced by the current integrated control approach that includes the use of benomyl, lime, and nitrate nitrogen (Locke et al., 1985). Biotype T-1-R9 was also found recently to be effective against *R. solani* on potatoes (Beagle-Ristaino and Papavizas, 1985).

The type of mutants studied with industrial fungi, apart from those introduced empirically, were specific mutants of which the primary and secondary metabolism were affected (Elander et al., 1976). There are two examples in the sparse literature of biocontrol fungi in which production of specific metabolites, possibly involved in biocontrol mechanisms, was altered by mutagenesis. Howell and Stipanovic (1983) produced mutants of *Gliocladium virens* by UV radiation that differed from the wild-type strain in their ability to produce the diketopiperazine antibiotic gliovirin. Gliovirin is highly toxic to *Pythium ultimum,* the pathogen that caused damping-off of cotton. A mutant of *G. virens* deficient for gliovirin production did not protect cotton seedlings from damping-off in soil infested with the pathogen. In contrast, a mutant with enhanced gliovirin production provided good biocontrol, even though it did not grow as well as the wild-type strain. The second example comes from studies at Beltsville. Papavizas et al. (1982) found that several UV-induced mutants of *T. harzianum* produced two toxic metabolites in the gliotoxin fermentation medium of Brian and Hemming (1945), one was heat-labile and the other heat-stable. The wild-type strain did not produce the heat-labile metabolite. Four mutants did not produce any metabolites toxic to the assay organism (*Sclerotium cepivorum*) (Table 2). Gas chromatographic analysis showed that the heat-labile metabolite was identical to an authentic sample of gliotoxin (Abd-El Moity, unpublished). The biotype that produced the highest amount of both metabolites was also the most effective biocontrol agent against white rot of onion (*S. cepivorum*).

The feasibility of manipulating the genome of an antagonistic fungus to interact favorably in a biological–chemical control system is also exemplified by *Talaromyces flavus,* a homothallic ascomycete with promising biocontrol ability against *Verticillium dahliae* of eggplant (Marois et al., 1982) and potatoes (Fravel, unpublished). In many practical agricultural systems potatoes do not do well in the field unless the potato seedpieces are treated with fungicides (e.g., TBZ, captan, benomyl, maneb, and polyram) to prevent fungal diseases other than *V. dahliae.* If *T. flavus* is to be added to potato seedpieces to suppress Verticillium wilt, it must be compatible with these fungicides, especially TBZ. Thiabendazole is the most commonly used fungicide for seedpiece decay and most certified potato seed is treated with TBZ. Several mutants of *T. flavus* developed by NTG treatment tolerated high amounts (up to 100 μg a.i./ml) of TBZ and benomyl (Katan et al., 1984). Rhizosphere populations of *T. flavus* from the rhizosphere of potato plants that originated from seedpieces treated with TBZ and a wild-type strain of T. flavus were not different from those in the rhizosphere of plants from seedpieces that received ascospores alone (Fravel et al., 1985). Higher populations of T. *flavus* were recovered from the rhizosphere of plants from seedpieces treated with ascospores of mutants resistant to TBZ and the fungicide TBZ than from those treated with ascospores alone.

Considerable progress has been made in a very short period of time in developing new effective biotypes of biocontrol agents by adaptation and genome modification through mutagenesis. It is equally important that it is possible to induce new biotypes of biocontrol agents that differ from wild strains in many

TABLE 2. The effect of Fungitoxic Metabolite Produced in Culture by *Trichoderma harzianum* Strain WT-6 and by UV-Induced, Benomyl-Resistant Biotypes on the Radial Growth of *Sclerotium cepivorum*[a]

		Percentage of Radial Growth Inhibition[b,c] *with Indicated Sterilization*	
Stain or Biotype	*Group*[d]	*Filtration*[e]	*Heat*[f]
WT-6-10, -13, -18, -20	I	0 a	0 a
WT-6 (wild strain)	II	13 b	0 a
WT-6-9	II	25 b	0 a
WT-6-7, -12	II	41 bc	0 a
WT-6-5, -6, -8, -15, -16	II	50 c	0 a
WT-6-1, -4	II	63 c	0 a
WT-6-21	III	73 cd	38 bc
WT-6-17	III	75 cd	25 b
WT-6-19	III	75 cd	38 bc
WT-6-24	III	88 d	50 c
WT-6-11	III	100 e	75 d

[a]Reprinted by permission from Phytopathology, **72:**126–132, 1982 (58).

[b]Strain WT-6 and 19 biotypes were grown in the gliotoxin fermentation medium for 10 days in the dark; sterile culture filtrates were added to potato–dextrose agar and plates were inoculated with *S. cepivorum:* the percentage of radial growth inhibition was determined by measuring radial growth after 4 days of incubation.

[c]Values followed by the same letter do not differ significantly ($P = 0.05$) according to Duncan's multiple range test.

[d]Three arbitrary groups on the basis of fungitoxic metabolite production.

[e]Millipore filtration.

[f]Autoclaving at 121°C for 30 min.

respects, including the important parameter of compatibility with fungicides for integrated pest management programs. It may even be possible to develop multiple pesticide tolerances in strains of biocontrol agents that would be more desirable than single-fungicide tolerant strains. The potential importance of this research to the long-range improvement of disease control is direct and substantial. Development of effective, pesticide-resistant strains would also be an inducement to industry to assume responsibility for large-scale production and pilot testing. Tolerant strains may be used successfully in association with minimal amounts of the chemicals for disease control.

Even though much new information on mutagenesis of biocontrol fungi has accumulated in a short period of time, very few definite statements can be made of the intrinsic genetic stability of mutants and of production of blocked mutants affecting either primary metabolic pathways (auxotrophs) or mutants blocking secondary metabolic pathways. The aim in mutagenesis should be to produce not only new strains possessing a *de facto* enhanced biocontrol ability, but also auxotrophic mutants to be used in a genetic recombination program by conventional genetic approaches and by protoplast fusion and transformation or DNA cloning. Of especial interest would be the production of auxotrophic mutants characterized by an increased production of metabolites (antibiotics and enzymes) implicated in the mechanism of biocontrol. Such mutants can be used to produce specific recombinant clones following the introduction of specific gene encoding for auxotrophy and metabolite production into a recipient biocontrol fungus that possesses other desirable characteristics. Auxotrophic requirements would be excellent markers, especially if they are linked with the production of desirable metabolites, as was shown in experiments for increased production of the antibiotic cephalosporin C by *Cephalosporium acremonium* (Hamlyn and Ball, 1979).

3. SEXUAL AND PARASEXUAL HYBRIDIZATION

Intraspecific and interspecific hybrids have been produced by the sexual cycle and the parasexual cycle in breeding several industrial fungi, especially *Penicillium* and *Aspergillus* (Ball, 1984a,b), and in plant pathogens and mycotoxin producers, including such fungi as *Cephalosporium, Cochliobolus, Fusarium, Humicola, Neurospora, Sordaria,* and *Verticillium* (Hastie, 1981). The sexual cycle critically differs from the parasexual cycle in that it involves a regular alternation of karyogamy and meiosis. The products of meiosis are recovered as special types of spores. Although the frequency of genetic reassortment in the sexual cycle is higher than in the parasexual cycle, prime consideration in breeding industrial and pathogenic fungi was given to the parasexual cycle. However, the sexual cycle has been exploited in the hybridization of *Aspergillus nidulans* and *Neurospora crassa* as model systems (Ball, 1984b; Davis, 1966). Since many excellent review articles (Backus and Stauffer, 1955; Ball, 1973, 1980, 1984a,b; Davis, 1966; Hastie, 1981; Lloas, 1967; Pontecorvo, 1956; Roper, 1966; Sermonti, 1969) and several books (Ball, 1984b; Burnett, 1975; Smith et al., 1980) appeared that deal, either exclusively or in a large part, with the sexual and parasexual cycle of fungi, there is no need to elaborate further in this chapter.

Despite all the advances made on fungal hybridization during the last 50 years, genetic manipulation by exploiting the sexual cycle and parasexual cycle to enhance antagonistic activity of biocontrol fungi and increase their tolerance to commonly used pesticides is mostly unexplored. These approaches are not expected to produce useful results with such biocontrol fungi as *Gliocladium* and *Trichoderma* because the teliomorph (*Hypocrea*) is not well defined in these two genera (Rifai, 1969). In addition, under several conditions, these biocontrol agents may be incapable of anastomosis, a prerequisite to heterokaryon formation. Even when anastomosis leading to heterokaryon formation occurs, complementation may not take place (Bojňanská et al., 1980), or the heterokaryons may be unstable, reverting to the original nuclear condition (and parental phenotypes) during conidiation (Weinman-Greenshpan and Galun, 1969).

The first attempt to improve biocontrol ability and tolerance to fungicides by sexual and parasexual hybridization was done at Beltsville with *Talaromyces flavus* (Katan et al., 1984). The sexual cycle of *T. flavus* critically differs from the parasexual cycle in that nuclear fusion, transient diploid formation, and segregation take place in specialized structures (cleistothecia) that can be isolated readily. Katan et al. (1984) showed that it is feasible to induce morphological mutants and mutants resistant to benomyl and TBZ, to cross various strains, and to select new biotypes that possess tolerance to these fungicides. Crosses between mutants resistant to benomyl and sensitive wild types, as well as between different resistant isolates showed that the two levels of benomyl resistance encountered are conferred by different alleles at a single locus. More importantly, some of the double crosses and back crosses possessed superior ability to suppress damping-off (*R. solani*) of cotton compared to the wild types or to the mutants (Papavizas, unpublished). This study showed for the first time that genetic recombination in fungi obtained by using sexual and parasexual hybridization can be used successfully to improve biocontrol efficiency and other desirable characteristics of biocontrol fungi.

4. HYBRIDIZATION AND RECOMBINATION FOLLOWING PROTOPLAST FUSION

In a recent review Lowe and Elander (1983) stated that "Parasexual recombination has now been overshadowed by the development of protoplast fusion technology." Protoplast fusion is now regarded as an effective technique for studying the genetics and biochemistry of fungi and yeasts and for leading to new approaches in genetic manipulation (Ball, 1984b; Evans and Cocking, 1975; Groves and Oliver, 1984; Kevei and Peberdy, 1984; Peberdy, 1976, 1979, 1983; Picataggio, 1983; Toyama et al., 1984). With this technique, we can prepare cell-free extracts and organelles and generate novel intraspecific and interspecific hybrids in fungi, overcoming natural barriers to hybridization, and opening up new possibilities in the study of gene transfer in fungi (Ferenczy et al., 1975). The use of protoplasts permits metabolic processes to occur in which the tough cell wall normally impairs access of extraneous genetic material.

Fungal protoplasts or spheroplasts, osmotically sensitive spherical or subspherical bodies free of cell walls, can easily be prepared from spores or young hyphae by the enzymatic digestion of the cell walls in the presence of an osmotic stabilizer (Anné and Eyssen, 1978; Anné and Peberdy, 1975, 1976; Peberdy, 1983; Toyama et al., 1984). Hashiba and Yamada (1984) developed an effective method for obtaining uninucleate protoplasts from multinucleate filamentous cells of *Rhizoctonia solani.* Kawula and Ling (1984) produced protoplasts by exposing blastospores of *Beauvaria bassiana* to β-glucuronidase for 90 min and collecting the protoplasts by centrifugation. Toyama et al. (1984) also succeeded in preparing protoplasts of *Trichoderma reesei* from conidia by the use of cell-wall lytic enzymes of *T. viride.* Hyphal cells of *T. reesei* are multinucleate, but the conidia are uninucleate. Detailed reviews of typical digestive enzymes used in fungal protoplast formation, osmotic stabilizers, state of cells used for protoplast production, mycelial age, pretreatment of mycelium with various agents, secretions by protoplasts, metabolic properties, protoplast regeneration, and methods used to study regeneration frequencies can be found in the literature (Hamlyn et al., 1981; Kevei and Peberdy, 1984; Lowe and Elander, 1983; Moore and Peberdy, 1976; Peberdy, 1976, 1979, 1983).

Fusion of protoplasts can be fairly easily obtained in the presence of polyethylene glycol and Ca^{2+} or other stabilizers (Anné and Peberdy, 1975, 1976; Peberdy, 1983; Peberdy and Gibson, 1971). The frequency of protoplast fusion is measured by the complementation of nutritionally deficient mutants on minimal media (or media containing drugs or pesticides if different markers are used), and the nutritional complementation of auxotrophic mutants is accepted as a proof of fusion. Cell fusion may permit a limited degree of nuclear fusion and reassortment of chromosomes. Usually, there is a higher probability for reassortment with the protoplast fusion technique than mycelial anastomosis because the former permits parental nuclei to be brought into close proximity, especially with uninucleate protoplasts. In hyphal cells, juxtapositioning of nuclei is infrequent (Hamlyn and Ball, 1979). Success in hybrid formation has been most readily achieved when the parental strains belong in closely related fungal taxa, such as *Aspergillus nidulans* + *A. rugulosus* (Kevei and Peberdy, 1977, 1979), *Penicillium chrysogenum* + *P. cyaneofulvum* (Peberdy et al., 1977), or *P. cyaneofulvum* + *P. citrinum* (Anné and Eyssen, 1978). In all these crosses the hybrid had a phenotype totally different from the parental strains and exhibited heterozygosity for all the genetic markers used. In some instances, it was shown that the hybrid phenotype was a consequence of nuclear fusion (Kevei and Peberdy, 1979, 1984). Instability in the hybrids, either spontaneous or induced, resulted in the segregation of progeny with novel phenotypes, which in the *A. nidulans* + *A. rugulosus* hybrid was shown to be a consequence of random chromosome assortment (Kevei and Peberdy, 1979).

It appears from the previous discussion that protoplast fusion is an excellent alternative approach to the construction of fungal strains with novel gene combinations. Although segregants derived from hybrids as a result of protoplast fusion may represent a new mix of gene pools, and protoplast recombination can be used to blend several desirable properties together in a single biotype, this novel technology has not been used for genetic manipulation of biocontrol fungi. By employing the protoplast fusion technique, barriers preventing genetically unrelated fungal strains, even those at present assigned to different genera (e.g., *Gliocladium* and *Trichoderma*), from exchanging genetic components can be overcome. If mating specificity of potential biocontrol fungi could be completely overcome by partial or complete removal of cell walls, then protoplasts of any two unrelated strains, even those from two fungal genera, should be able to fuse and the fusion products should be stable in respect to parental properties and capable of proliferation. This would suggest again that the cell wall constituents may be the only important barrier preventing unrelated fungal taxa from combining their genomes. It would, therefore, be expected that owing to lack of specificity in the process of protoplast fusion (Peberdy, 1983), formation of fusion products between protoplasts of *Trichoderma* and *Gliocladium*, or *Penicillium vermiculatum* (*Talaromyces flavus*) and other *Penicillium* spp. will be possible. Ferenczy et al. (1975) described a simple technique that results in high-frequency interspecific protoplast fusion and postulated that among others "*Aspercillium*" and "*Penigillus*" strains, the results of *Aspergillus–Penicillium* protoplast fusion, would be especially interesting from both theoretical and practical points of view. In the same context "*Glioderma*" strains combining the enhanced biocontrol ability of *Gliocladium* and tolerance to benomyl from *T. viride* or *T. harzianum* mutants could theoretically be produced as a result of *Gliocladium* and *Trichoderma* protoplast fusion. However, attempts to develop intergeneric hybrids by this approach have been disappointing (Peberdy, 1983). For instance, protoplasts of *Saccharomyces lipolytica* and *Pichia guilliermondii* (Spata and Weber, 1980), and those of *S. cerevisiae* and *Schizosaccharomyces pombe* (Svoboda, 1980), fused successfully, but the hybrids were unstable.

Although no specific examples dealing with the use of the protoplast fusion technique to improve the performance of biocontrol fungi could be found in the literature, several published examples dealing with other classes of fungi can be used to illustrate the usefulness of the possible applicability of the technique to genetically manipulate biocontrol fungi. One result of hybridization by protoplast fusion, for instance, is the development of segregant classes in which metabolic pathways may interact to produce novel metabolites, or increased amounts of known metabolites. Hamlyn and Ball (1979) found that a strain of *Cephalosporium acremonium* that could synthesize the antibiotic cephalosporin C from inorganic sulfur yielded three times the amount of the antibiotic made by a distantly related strain. The former grew slowly without producing spores and the latter grew at about four times the rate of the other strain and sporulated abundantly. From crosses initiated by protoplast fusion made between these two strains, a prototroph was selected whose segregated progeny differed in growth rates and morphological characteristics. One segregant gave 40% more cephalosporin C than the higher titer prototrophic or genetically labeled parent and also grew faster and sporulated better than the prototroph. Pettey and Crawford (1984) recently showed that protoplast fusion can be used to construct acid-precipitable polymeric lignin (APPL)-overproducing strains of *Streptomyces*, which may be useful in the bioconversion of lignin to simpler chemicals. They used protoplast fusion as a technique to manipulate genetically two lignin-degrading strains, *S. viridosporus* and *S. setonii*. Four of 19 recombinants showed enhanced production of APPL. In another recent analysis, Kevei and Peberdy (1984), based on a pairwise analysis of genetic markers used in the generation of *A. nidulans* + *A. rugulosus* hybrids, showed the existence of homology between linkage groups of the two species. Also, in a study of the penicillin-producing capability of segregants from such hybrids, they showed that penicillin yield could be enhanced beyond that of the highest producing parent. Although these examples may appear to be somewhat remotely related to the genetic manipulation of biocontrol agents, the wider implications are that interspecific or intraspecific hybridizations by protoplast fusion might be an aid to metabolite yield improvement where one of the parents is a high-yielding strain. For instance, if one of four antibiotics is known to be produced by a given *T. flavus* strain sensitive to benomyl, and this particular antibiotic is implicated in the mechanism of action against *Verticillium dahliae* (Marois et al., 1982), protoplast fusions with another strain tolerant to benomyl could yield hybrids or segregants that would combine the advantageous properties together in a single strain.

The realization of techniques for protoplast isolation, regeneration, fusion, and transformation will undoubtedly be used as the basis for genetic manipulation of biocontrol fungi. The examples from other fungi justify considerable optimism for future successes. Johansen et al. (1984), however, expressed some doubts on the possibility of successfully using the protoplast fusion technique among anamorphic yeast strains in the genus *Kluyveromyces.* The fusion technique in this genus did not offer a reliable alternative to hybridization since the fusion products were unstable after a single passage through a complete medium. According to them, the cell walls may not be the only important barrier that prevents genomic recombination in unrelated yeast strains. Also, the genetic affinity may not always be improved by the removal of cell walls from cells of hybridizing and nonhybridizing strains.

5. PROTOPLAST TRANSFORMATIONS

Hinnen et al. (1978) defined transformation as "the process by which naked DNA is introduced into a cell, resulting in a heritable change." The process of transformation involves two basic steps, namely, partial or complete elimination of cell barriers (cell walls) to allow the uptake of DNA into the cell and the stable integration, maintenance, and self-replication of the DNA in the cell (Case, 1983). Isolated protoplasts or spheroplasts are excellent materials for the introduction of foreign or donor DNA. Uptake of DNA, TMV-RNA, proteins, DNA, polystyrene latex particles, bacterial plasmids, and DNA-protein complexes has been demonstrated (Hess, 1976).

Transformations can be done in at least three ways: (1) fusion of whole protoplasts from two different strains, species, or genera; (2) transfer of whole natural plasmids between microorganisms, for example, mitochondrial (mt) plasmids; and (3) transfer of DNA segments (autonomously replicating and integrated vectors). The first approach was discussed under protoplast fusion. The second technique (plasmid-mediated transformation) differs from the third (recombinant DNA technique) in that hybrids contain genetic materials only from the naturally interbreeding microbes from which the crossed strains are derived. The third method is based on the facility with which DNA segments can be inserted or attached to bacterial plasmids (vectors) for subsequent transfer to other cells (Case, 1983). In this approach, eukaryotic genes can first be cloned into bacteria (e.g, *Escherichia coli* and *Bacillus subtilis*). The rearranged sequences can then be taken out of the vectors to determine the impact of induced changes on the function and expression of the genes of the eukaryotic donor. The rearranged sequence can be introduced into the donor or into a new eukaryotic organism (Hinnen et al., 1978).

Our understanding of the structure, organization, and function of genes of *E. coli* and of some viruses has progressed far beyond our understanding of the genes in eukaryotes. The ability to maintain selected genes, regardless of sources in *E. coli,* could be used for transformation in fungi (Helling and Lomax, 1978; Hinnen et al., 1978). Procedures for cloning DNA segments in *E. coli* by insertion into a plasmid are now well established and have been used to isolate many prokaryotic and eukaryotic sequences. In the early 1970s, several attempts to transform fungi by cloning exogenous DNA failed to establish that the foreign DNA had become incorporated as an autonomously replicating genetic material (Hinnen et al., 1978). In the initial tests with yeast (*S. cerevisiae*) and *Neurospora crassa* the failure could have been due to the use of a non-self-replicating plasmid (Case, 1983) or because not one but many genes may govern cell characteristics. According to Hess (1976), "To introduce all of them [genes] one after the other seems to be impossible." The major breakthrough was accomplished by Hinnen et al. (1978), who succeeded in developing highly efficient cloning vectors by constructing plasmids containing three different types of autonomously replicating sequences from yeast. They showed that a stable leu-2^- yeast strain (double mutant strain) was transformed with a bacterial chimeric plasmid (Col El) containing yeast DNA (leu-2^+) as a source of highly enriched genes and as a molecular probe for the sequences introduced into recipient cells. The transformed strain was well characterized both for the inserted DNA location and for the presence of *E. coli* plasmid DNA sequences inserted into the receiver genome. Beggs (1978) also developed a highly efficient yeast transformation system with the use of yeast-bacterial hybrid plasmids.

Although cloning technology in other fungi is not yet as advanced as it is in the yeast *S. cerevisiae*, a few examples in the literature demonstrate that considerable progress had been made. An efficient transformation system was developed by Case et al. (1979), who used spheroplasts and hybrid plasmid DNA (pVK88). This plasmid is a recombinant *E. coli* plasmid carrying the *N. crassa* qa-2^+ gene which encodes for catabolic dehydrogenase. The qa-2^+ gene from *N. crassa* was ligated into a suitable plasmid vector (pBR322) to give the chimeric plasmid pVK88 (Case 1983; Case et al., 1979). The *N. crassa* qa-2^+ mutants were transformed with this plasmid, which was integrated into the fungal genome at several sites in a manner similar to that described by Hinnen et al. (1978) for yeast. Efficient transformation experiments were also performed recently with *Aspergillus nidulans* (Yelton et al., 1984). Yelton et al. (1984) isolated a trifunctional trpC gene from a wild-type strain of *A. nidulans* and constructed a chimeric plasmid, designated pHY201, by inserting trpC into the bacterial plasmid pBR329. Plasmid pHY201 replicated in *E. coli* and imparted resistance to the antibiotics ampicillin and chloramphenicol. The plasmid pHY201 was then used to transform a type trpC$^-$ strain of *A. nidulans* to trpC$^+$

at efficiencies similar to those observed for yeast (Hinnen et al., 1978) and *N. crassa* (Case et al., 1979). Plasmid pHY201 could be successfully recovered from *A. nidulans* transformants by restriction enzymes. Their research could be used as an excellent model to develop similar cloning approaches in biocontrol fungi.

So far, there have been no attempts to physically incorporate exogenous DNA into the genome of biocontrol fungi. The technique of DNA transformation, especially plasmid-mediated transformation, presents a great opportunity to clone selected desirable pieces of DNA molecules (genes) into biocontrol fungi. Before researchers delve in earnest into genetic cloning to improve effectiveness of biocontrol fungi by plasmid-mediated transformation that will involve DNA transfer into fungal protoplasts or spheroplasts, much related or peripheral work should be done. Good protoplast formation and generation methods should be developed with such fungi to match the progress that has already been made with yeasts. Transformation has been handicapped by poor quality protoplasts or spheroplasts (Murakami et al., 1983) or by the slow speed of their production and the expense required to obtain pure lytic enzymes (Peberdy, 1979). Many published reviews (Peberdy, 1976, 1979, 1983) indicate that the problems of methods should not be serious, especially with the fungal genera where most of the biocontrol fungi belong.

Attempts should also be made to find out whether circular or linear plasmids exist in biocontrol fungal strains and whether such extrachromosomal genetic elements are associated with mt DNA. Following the first discovery of plasmids in fungi (yeast) in 1967 (cited by Picataggio, 1983), and in *Podospora anserina* in 1978 (Esser et al., 1983; Stahl et al., 1982), circular or linear plasmids were also found in other fungi, such as wild strains of *N. crassa* (Collins et al., 1981). However, there is no evidence that a chimeric plasmid containing mt plasmid sequences can be used as a self-replicating vector in *Neurospora* (Case et al., 1979). Despite this, mt DNA, regardless of whether it originates from mt plasmids or from the mt "chromosome," can be used to construct autonomously replicating and integrated vectors in other fungi (Esser et al., 1983; Stahl et al., 1982). Mitochondrial plasmid-DNA isolated from *P. anserina* also has the potential to be used as a vector in the development of a transformation model in this fungus (Stahl et al., 1982). In fact, Stahl et al. (1982) showed that a bacterial-mt hybrid plasmid could replicate and express itself in *P. anserina*. Hybrid plasmids made up by the bacterial plasmid pBR322 and plasmid-like DNA (pl DNA) sequences from *P. anserina* replicated in *E. coli* and in the fungus. The fungus was able to express both the eukaryotic senescence and the prokaryotic β-lactamase. One might expect that the mt DNA replicon of other fungi could be used to develop fungal vectors. Strain improvement of biocontrol fungi can therefore be undertaken by using eukaryotic vectors consisting of their own mt DNA. In this way, the risk of instability of the transferred organism is expected to be minimal, because a compatibility is expected between the vector and host DNA. Unfortunately, nothing is known about plasmids in biocontrol fungi.

Genomic clones that encode for a variety of desirable characters are not known in biocontrol fungi. Because of the lack of information on the mechanism of action of most of these microorganisms, cloning and recombinant DNA research will be difficult to pursue. If we knew for instance, that *Trichoderma* spp. lyse the cell walls of pathogenic fungi with a specific enzyme [e.g., β-(1–3)-glucanase or chitinase], we could identify chromosomal or extrachromosomal genes that encode for the production of such metabolites. The genetic information could then be transferred to a prokaryotic vector to transform protoplasts of wild strains or mutants that possess other desirable characteristics (tolerance to pesticides, ability to colonize plant rhizosphere, good survival ability in the environment, good growth and sporulation, and long shelf life). Equally important with the identification of genes encoding for specific properties or products is the development of plasmid vector systems. This would be the most difficult research phase, especially for stable, self-replicating systems. The realization of techniques for efficient transformation with yeast, *N. crassa*, and *A. nidulans* will be useful in the further development of new technologies for general strain improvement as well as for novel strain development.

Specific guidelines for cloning and recombinant DNA research with biocontrol fungal genera are not completely lacking. There are a few recent specific examples of molecular cloning of major genes in such genera. Although such research was done with genes that encode for production of specific enzymes in industrial strains, such examples can be used as models for similar research with biocontrol strains. For instance, the chromosomal gene encoding for the major cellulase of *Trichoderma reesei*, exo-cellobiohydrolase I (CBHI), was successfully cloned and partially characterized (Shoemaker et al., 1983; Teeri et al., 1983). One of the clones contained a gene coding for CBHI of *T. reesei* (Teeri et al., 1983). The first prerequisite of this research was the successful identification of the gene coding for CBHI by hybrid selection of messenger RNA (Teeri et al., 1983) and confirmed by DNA sequence analysis (Shoemaker et al., 1983). Also, Picataggio et al. (1981, 1983) were able to clone the gene encoding for cellulase in *T. reesei* into *E. coli* HB101 via the plasmid pBR322. Picataggio (1983) also constructed gene banks of *T. reesei* DNA using the yeast cosmid pBTI-1. The gene banks were introduced into auxotrophic strains of *S. cerevisiae* and the transformants were screened for functional expression of the cloned *Trichoderma* DNA by complementation of auxotrophic mutants. The DNA hybridization and plasmid isolation of yeast transformants showed that the large recombinant plasmids in the receiver strains were subject to structural instability, a problem that can perhaps be solved in the future by additional research.

6. FUTURE RESEARCH PRIORITES

The brilliant genetic manipulation of bacteria and their viruses during the last three decades and the several advances on molecular cloning with yeast and some fungi, especially *N. crassa*, *A. nidulans*, *C. acremonium*, and *P. anserina*, are excellent models to be used for improving the effectiveness of biocontrol fungi. Although it is difficult to predict at this time the full extent to which genetic manipulation will be used with such microorganisms, it is anticipated that work with such systems will be pursued with increasing momentum, even though systems in eukaryotes are more complex than those in prokaryotes. It is expected that filamentous biocontrol fungi will be used more extensively than in the past for conventional genetic manipulation and as cloning hosts to develop efficient transformation systems, thus greately expanding the possibilities of manipulating the genome to enhance biocontrol effectiveness.

Before the time arrives when improved fungal strains become available commercially for biocontrol, major priorities will be included in the following research areas: (1) physical and chemical mutagenesis to increase the gene pool and develop badly needed, well-defined markers for transformation research; (2) expanded use of the sexual cycle and the parasexual cycle for conventional hybridization and recombination studies, especially with biocontrol fungi whose teliomorph can be produced readily in culture; (3) improvement of techniques for protoplast production, regeneration, and fusion; (4) unraveling the sequence of events that follow protoplast fusion, and control of such events to increase the frequency of fusion and widen the range of species, and even genera, that can be hybridized by this method; (5) unraveling the mechanisms of action of biocontrol fungi, especially if those are mediated by specific metabolites, and determining the genes that encode for such activities or metabolite production; (6) discovery of plasmids of biocontrol fungi and their possible use to construct eukaryotic vectors for transformation experiments; (7) adaptation to techniques for DNA transformation using prokaryotic vectors to modify the genotypes of biocontrol fungi; and (8) obtaining novel organisms via these approaches that will possess enhanced biocontrol ability combined with other desirable traits.

In anticipating an expansion of genetic manipulation research, the following difficulties should not be underestimated: (1) those that may be encountered even in studying a comparatively simple process such as DNA uptake, stability, and self-replication; (2) the fact that in eukaryotes, genes of related functions are seldom linked (Groves and Oliver, 1984), thus necessitating transferring all the genes that code for certain traits or metabolites; and (3) the inability of many genes to function when cloned in prokaryotic vectors (Hinnen et al., 1978).

REFERENCES

Abd-El Moity, T. H., G. C. Papavizas, and M. N. Shatla (1982). Induction of new isolates of *Trichoderma harzianum* tolerant to fungicides and their experimental use for control of white rot of onion. *Phytopathology,* **72**:396–400.

Alikhanian, S. I. (1962). Induced mutagenesis in the selection of microorganisms. *Adv. Appl. Microbiol.,* **4**:1–50.

Anné, J. and H. Eyssen (1978). Isolation of inter-species hybrids of *Penicillium citrinum* and *Penicillium cyaneo-fulvum* following protoplast fusion. *FEMS Microbiol. Lett.,* **4**:87–90.

Anné, J. and J. F. Peberdy (1975). Conditions for induced fusion of fungal protoplasts in polyethylene glycol solutions. *Arch. Microbiol.,* **105**:201–205.

Anné, J. and J. F. Peberdy (1976). Induced fusion of fungal protoplasts following treatment with polyethylene glycol. *J. Gen. Microbiol.*, **92:**413–417.

Backus, M. P. and J. F. Stauffer (1955). The production and selection of a family of strains in *Penicillium chrysogenum. Mycologia,* **47:**429–463.

Ball, C. (1973). The genetics of *Penicillium chrysogenum. Prog. Ind. Microbiol.*, **12:**47–72.

Ball, C. (1980). Genetic modification in filamentous fungi. In *Fungal Biotechnology* (J. E. Smith, D. R. Berry, and B. Kristiansen, eds.). Academic, London, pp. 43–54.

Ball, C. (1984a). *Genetics and Breeding of Industrial Microorganisms.* CRC, Boca Raton, FL, 203 pp.

Ball, C. (1984b). Filamentous fungi. In *Genetics and Breeding of Industrial Microorganisms* (C. Ball, ed.). CRC, Boca Raton, FL, pp. 159–188.

Beagle-Ristaino, J. and G. C. Papavizas (1985). Biological control of *Rhizoctonia* stem canker and black scurf of potato. *Phytopathology,* **75:**560–564.

Beggs, J. D. (1978). Transformation of yeast by a replicating hybrid plasmid. *Nature,* **275:**104–109.

Bojňanśka, M., M. Sipiczki, and L. Ferenczy (1980). Characterization of conidiation mutants in *Trichoderma viride* by hyphal anastomosis and protoplast fusion. *Acta Microbiol. Acad. Sci. Hung.*, **27:**305–307.

Brian, P. W. and H. G. Hemming (1945). Gliotoxin, a fungistatic metabolic product of *Trichoderma viride. Ann. Appl. Biol.*, **32:**214–220.

Burnett, J. H. (1975). *Mycogenetics.* Wiley, New York, 375 pp.

Case, M. E. (1983). Transformation in fungi. In *Genetic Engineering in Eukaryotes* (P. F. Lurquin and A. Kleinhofs, eds.). Plenum, New York, pp. 1–5.

Case, M. E., M. Schweizer, S. R. Kushner, and N. H. Giles (1979). Efficient transformation of *Neurospora crassa* by utilizing hybrid plasmid DNA. *Proc. Natl. Acad. Sci. USA,* **76:**5259–5263.

Collins, R. A., L. L. Stohl, M. D. Cole, and A. N. Lambowitz (1981). Characterization of a novel plasmid DNA found in mitochondria of *Neurospora crassa. Cell,* **24:**443–452.

Davis, R. H. (1966). Mechanisms of inheritance 2. Heterokaryosis. In *The Fungi, An Advanced Treatise,* Vol. II (G. C. Ainsworth and A. S. Sussman, eds.). Academic, New York, pp. 567–588.

Dekker, J. (1976). Acquired resistance to fungicides. *Annu. Rev. Phytopathol.*, **14:**405–428.

Delp, C. J. (1980). Coping with resistance to plant disease control agents. *Plant Dis.*, **64:**652–657.

Demain, A. L. (1973). Mutation and the production of secondary metabolites. *Adv. Appl. Microbiol.*, **16:**177–202.

Elander, R. P., C. J. Corum, H. De Valeria, and R. M. Wilgus (1976). Ultraviolet mutagenesis and cephalosporin synthesis in strains of *Cephalosporium acremonium.* In *Genetics of Industrial Microorganisms* (K. D. Macdonald, ed.). Academic, London, pp. 253–271.

Esser, K., U. Kuck, U. Stahl, and P. Tudzynski (1983). Clone vectors of mitochondrial origin for eukaryotes: A new concept in genetic engineering. *Cur. Genet.*, **7:**239–243.

Evans, P. K. and E. C. Cocking (1975). The techniques of plant cell culture and somatic cell hybridization. In *New Techniques in Biophysics and Cell Biology, Vol. 2* (R. H. Pain and B. J. Smith, eds.). Wiley, New York, pp. 127–158.

Ferenczy, L., F. Kevei, and M. Szegedi (1975). High-frequency fusion of fungal protoplasts. *Experientia,* **31:**1028–1029.

Fravel, D. R., J. J. Marois, M. T. Dunn, and G. C. Papavizas (1985). Compatibility of *Talaromyces flavus* with potato seedpiece fungicides. *Soil Biol. Biochem,* **17:**163–166.

Georgopoulos, S. G. (1977). Pathogens become resistant to chemicals. In *Plant Disease, An Advanced Treatise. Vol. I. How Disease is Managed* (J. G. Horsfall and E. B. Cowling, eds.). Academic, New York, pp. 327–345.

Groves, D. P. and S. G. Oliver (1984). Formation of intergeneric hybrids of yeast by protoplast fusion of *Yarrowia* and *Kluyveromyces* species. *Curr. Genet.*, **8:**49–55.

Hamlyn, P. F. and C. Ball (1979). Recombination studies with *Cephalosporium acremonium.* In *Genetics of Industrial Microorganisms* (O. K. Sebek and A. I. Laskin, eds.). *American Society of Microbiology,* Washington, DC, pp. 185–191.

Hamlyn, P. F., R. E. Bradshaw, F. M. Mellon, C. M. Santiago, J. M. Wilson, and J. F. Peberdy (1981). Efficient protoplast isolation from fungi using commercial enzymes. *Enzyme Microb. Technol.*, **3:**321–325.

Hashiba, T. and M. Yamada (1984). Intraspecific protoplast fusion between auxotrophic mutants of *Rhizoctonia solani. Phytopathology,* **74:**398–401.

Hastie, A. C. (1979). The genetic effects of fungicides. In *Proceedings of Symposia IX International Congress of Plant Protection, Vol. I* (T. Kommedahl, ed.). Washington, DC, August 5–12, 1979, pp. 238–241.

Hastie, A. C. (1981). The genetics of conidial fungi. In *Biology of Conidial Fungi, Vol. II* (G. T. Cole and B. Kendrick, eds.). Academic, New York, pp. 511–547.

Helling, R. B. and M. I. Lomax (1978). The molecular cloning of genes-general procedures. In *Genetic Engineering* (A. M. Chakrabarti, ed.). CRC, Boca Raton, FL, pp. 1–30.

Hess, D. (1976). Uptake and expression of foreign genetic material in plant protoplasts. In *Microbial and Plant Protoplasts* (J. F. Peberby, A. H. Rose, H. J. Rogers, and E. C. Cocking, eds.). Academic, London, pp. 125–144.

Hinnen, A., J. B. Hicks, and G. R. Fink (1978). Transformation of yeast. *Proc. Natl. Acad. Sci. USA,* **75:**1929–1933.

Hitzeman, R. A., D. W. Leung, L. J. Perry, W. J. Kohr, H. L. Levine, and D. V. Goeddel (1983). Secretion of human interferons by yeast. *Science,* **219:**620.

Howell, C. R. and R. D. Stipanovic (1983). Gliovirin, a new antibiotic from *Gliocladium virens,* and its role in the biological control of *Pythium ultimum. Can. J. Microbiol.*, **29:**321–324.

Johansen, E., L. Halland, and A. Opperman (1984). Protoplast fusion within the genus *Kluyveromyces* van der Walt emend. van der Walt. *Can. J. Microbiol.*, **30:**540–552.

Katan, T., M. T. Dunn, and G. C. Papavizas (1984). Genetics of fungicide resistance in *Talaromyces flavus. Can. J. Microbiol.*, **30:**1079–1087.

Kawula, T. H. and A. J. Ling (1984). Production of protoplasts from *Beauvaria bassiana* blastospores. *J. Invert. Path.*, **43:**282–284.

Kevei, F. and J. F. Peberdy (1977). Interspecific hybridization between *Aspergillus nidulans* and *Aspergillus rugulosus* by fusion of somatic protoplasts. *J. Gen. Microbiol.*, **102:**255–262.

Kevei, F. and J. F. Peberdy (1979). Induced segregation in interspecific hybrids of *Aspergillus nidulans* and *Aspergillus rugulosus* obtained by protoplast fusion. *Mol. Gen. Genet.*, **170:**213–218.

Kevei, F. and J. F. Peberdy (1984). Interspecies hybridization following protoplast fusion in *Aspergillus.* In *Fungal Protoplasts: Their Use in Physiology, Biochemistry and Genetics* (J. F. Peberdy, and L. Ferenczy, eds.). Marcel Dekker, New York.

Lloas, P. (1967). Genetic analysis by means of the parasexual cycle in *Aspergillus niger. Genet. Res., Camb.*, **10:**45–61.

Locke, J. C., J. J. Marois, and G. C. Papavizas (1985). Biological control of Fusarium wilt of greenhouse grown chrysanthemums. *Plant Dis.*, **69:**167–169.

Lowe, D. A. and R. P. Elander (1983). Contribution of mycology to the antibiotic industry. *Mycologia,* **75:**361–373.

Mandels, M., J. Weber, and R. Parizek (1971). Enhanced cellulase production by a mutant of *Trichoderma viride. Appl. Microbiol.*, **21:**152–154.

Marois, J. J., S. A. Johnston, M. T. Dunn, and G. C. Papavizas (1982). Biological control of Verticillium wilt of eggplant in the field. *Plant Dis.*, **66:**1166–1168.

Montenecourt, B. S. and D. E. Eveleigh (1977). Semiquantitative plate assay for determination of cellulase production by *Trichoderma viride*. *Appl. Microbiol.*, **33:**178–183.

Montenecourt, B. S. and D. E. Eveleigh (1979). Production and characterization of high yielding cellulase mutants of *Trichoderma reesei*. *Tech. Assoc. Pulp. Pap. Ind. Annu. Meet.*, 101–108.

Moore, P. M. and J. F. Peberdy (1976). Release and regeneration of protoplasts from the conidia of *Aspergillus flavus*. *Trans. Br. Mycol. Soc.*, **66:**421–425.

Murakami, T., C. Nojiri, H. Toyama, E. Hayashi, K. Katumata, H. Anzai, Y. Matsuhashi, Y. Yamada, and K. Nagaoka (1983). Cloning of antibiotic-resistance genes in *Streptomyces*. *J. Antibiot.*, **36:**1305–1311.

Papavizas, G. C. (1980). Induced tolerance of *Trichoderma harzianum* to fungicides. *Phytopathology*, **70:**691–692 (Abstract).

Papavizas, G. C. (1985). *Trichoderma* and *Gliocladium*: Biology, ecology, and the potential for biocontrol. *Annu. Rev. Phytopathol.*, **23:**23–54.

Papavizas, G. C. and J. A. Lewis (1981). Introduction and augmentation of microbial antagonists for the control of soilborne plant pathogens. In *Biological Control in Crop Production* (G. C. Papavizas, ed.). Allanheld, Osmun, Totowa, NJ, pp. 305–322.

Papavizas, G. C. and J. A. Lewis (1983). Physiological and biocontrol characteristics of stable mutants of *Trichoderma viride* resistant to MBC fungicides. *Phytopathology*, **73:**407–411.

Papavizas, G. C., J. A. Lewis, and T. H. Abd-El Moity (1982). Evaluation of new biotypes of *Trichoderma harzianum* for tolerance to benomyl and enhanced biocontrol capabilities. *Phytopathology*, **72:**126–132.

Peberdy, J. F. (1976). Isolation and properties of protoplasts from filamentous fungi. In *Microbial and Plant Protoplasts* (J. F. Peberdy, A. H. Rose, H. J. Rogers, and E. C. Cocking, eds.). Academic, London, pp. 39–50.

Peberdy, J. F. (1979). Fungal protoplasts: Isolation, reversion and fusion. *Annu. Rev. Microbiol.*, **33:**21–39.

Peberdy, J. F. (1983). Genetic recombination in fungi following protoplast fusion and transformation. In *Fungal Differentiation* (J. E. Smith, ed.). Marcel Dekker, New York, pp. 559–581.

Peberdy, J. F. and R. K. Gibson (1971). Regeneration of *Aspergillus nidulans* protoplasts. *J. Gen Microbiol.*, **69:**325–330.

Peberdy, J. F., H. Eyssen, and J. Anné (1977). Interspecific hybridization between *Penicillium chrysogenum* and *Penicillium cyaneo-fulvum* following protoplast fusion. *Mol. Gen. Genet.*, **157:**281–284.

Pettey, T. M. and D. L. Crawford (1984). Enchancement of lignin degradation in *Streptomyces* spp. by protoplast fusion. *Appl. Environ. Microbiol.*, **47:**439–440.

Picataggio, S. K. (1983). *Maintenance and Expression of Cloned Trichoderma reesei DNA in the Yeast Saccharomyces cerevisiae*. Ph.D. Thesis, Rutgers, The State University of New Jersey, 258 pp.

Picataggio, S. K., B. S. Montenecourt, and D. E. Eveleigh (1981). The cloning of *Trichoderma reesei* genomic DNA in *Escherichia coli* HB101. *Annu. Meet. Am. Soc. Microbiol.*, **81:**116 (Abstract).

Picataggio, S. K., J. Noti, D. Morris, A. Szalay, and D. E. Eveleigh (1983). Maintenance and expression of cloned *Trichoderma reesei* DNA in *Saccharomyces cerevisiae*. *Annu. Meet. Am. Soc. Microbiol.*, **83:**135 (Abstract).

Pontecorvo, G. (1956). The parasexual cycle in fungi. *Annu. Rev. Microbiol.*, **10:**393–400.

Rifai, M. A. (1969). A revision of the genus *Trichoderma*. *Commonw. Mycol. Inst. Mycol. Pap. 116*, 56 pp.

Roper, J. A. (1966). Mechanisms of inheritance 3. The parasexual cycle. In *The Fungi, An Advanced Treatise, Vol. II* (G. C. Ainsworth and A. S. Sussman, eds.). Academic, New York, pp. 589–617.

Schimenti, J., T. Garrett, B. S. Montenecourt, and D. E. Eveleigh (1983). Selection of hypercellulolytic mutants of *Trichoderma reesei* based on resistance to nystatin. *Mycologia,* **75:**876–880.

Sermonti, G. (1969). Mutation. In *Genetics of Antibiotic-Producing Microorganisms* (G. Sermonti, ed.). Wiley-Interscience, New York, pp. 55–109.

Shoemaker, S., V. Schweickart, M. Ladner, D. Gelfand, S. Kwok, K. Myambo, and M. Innis (1983). Molecular cloning of exo-cellobiohydrolase I derived from *Trichoderma reesei* strain L27. *Biotechnology,* **1:**691–696.

Smith, J. E., D. R. Berry, and B. Kristiansen, eds. (1980). *Fungal Biotechnology.* Academic, London, 308 pp.

Spata, L. and H. Weber (1980). A study on protoplast fusion and parasexual hybridization of alkane utilizing yeasts. In *Advances in Protoplast Research* (L. Ferenczy and G. L. Farkas, eds.). Akademiai Kaido, Budapest, pp. 131–138.

Stahl, U., P. Tudzynski, U. Kuck, and K. Esser (1982). Replication and expression of a bacterial-mitochondrial hybrid plasmid in the fungus *Podospora anserina. Proc. Natl. Acad. Sci. USA,* **79:**3641–3645.

Svoboda, A. (1980). Intergeneric fusion of yeast protoplasts: *Saccharomyces cerevisiae* and *Schizosaccharomyces pombe.* In *Advances in Protoplast Research* (L. Ferenczy and G. L. Farkas, eds.). Akademiai Kaido, Budapest, pp. 119–124.

Teeri, T., I. Salovuori, and J. Knowles (1983). The molecular cloning of the major cellulase gene from *Trichoderma reesei. Biotechnology,* **1:**696–699.

Toyama, H., K. Yamaguchi, A. Shinmyo, and H. Okada (1984). Protoplast fusion of *Trichoderma reesei,* using immature conidia. *Appl. Environ. Microbiol.,* **47:**363–368.

U. S. Department of State (1981). Report and Recommendations of the *Ad Hoc* Expert Panel on the Foreign Policy Implications of Advances in Genetic Engineering. U. S. Department of State, pp. 1–6.

Watson, J. D. and F. H. C. Crick (1953). A structure for deoxyribose nucleic acids. *Nature,* **171:**737.

Weinman-Greenshpan, D. and E. Galun (1969). Complementation in nonconidiating mutants of *Trichoderma. J. Bacteriol.,* **99:**802–806.

Yelton, M. M., J. E. Hamer, and W. E. Timberlake (1984). Transformation of *Aspergillus nidulans* by using a trpC plasmid. *Proc. Natl. Acad. Sci. USA,* **81:**1470–1474.

10

THE USE OF AGROCIN-PRODUCING BACTERIA IN THE BIOLOGICAL CONTROL OF CROWN GALL

JENNIFER A. THOMSON
CSIR Laboratory for Molecular and Cell Biology, University of the Witwatersrand, Johannesburg, South Africa

1. INTRODUCTION

It has been documented that many bacteria produce low molecular weight toxic compounds called bacteriocins, which are active against nonproducing strains of the same species, but the practical application of such antibiotics has been limited. The only example of the use of a bacteriocin-producing strain in the prevention of a plant disease is in the case of crown gall, whose causative agent is *Agrobacterium tumefaciens*. A bacteriocin produced by strains of *Agrobacterium* was first reported by Stonier (1960) who discovered a nonproteinaceous, dialyzable antibiotic he called Agrobacteriocin 1, which is active against a variety of *Agrobacterium* strains, both pathogens and nonpathogens.

This bacteriocin was not, however, suitable for biological control and it was not until 1972 that an effective controlling strain was isolated. New and Kerr (1972) reported that when 10-week old peach seedlings were individually potted in soil artificially inoculated with a 1:1 ratio of a pathogen and nonpathogen, both of biotype 2, complete inhibition of gall formation was achieved. They suggested that the nonpathogen, strain 84, could be used for biological control by dipping seeds or the roots of young plants in a suspension of this strain before planting in potentially infested soil. At that time, however, the mechanism of gall inhibition was obscure and a suggestion was made that strain 84 might carry a phage closely related to, and therefore interfering with, the induction of the temperate phage, which at that stage some workers considered to be the causative agent of crown gall (New and Kerr, 1972).

The mystery was cleared up in 1974 when Kerr and Htay reported that the mechanism of control was by the production by strain 84 of a bacteriocin and noted a correlation between resistance to bacteriocin 84 and loss of pathogenicity. This was confirmed by Roberts and Kerr (1974), who showed that conversion of a nonpathogen to a pathogen by means of conjugation *in planta* was accompanied by a change from bacteriocin resistance to sensitivity. Engler et al. (1975) soon showed that this correlation was not fortuitous and that sensitivity to agrocin 84, the term they coined for this bacteriocin, was encoded by the Ti plasmid recently shown to be responsible for the pathogenicity of an *A. tumefaciens* strain, C58. They did this by curing the pathogen of its Ti plasmid by growth at 37°C and correlating resistance to agrocin with the inability to induce tumors.

It was not long before strain 84 became widely used in countries where crown gall is a problem. *Agrobacterium tumefaciens* has been described as one of the top three agricultural pathogens in California (Schroth et al., 1971) and it is a major peach-tree pathogen elsewhere in America (Alconero, 1980). This was confirmed by Kennedy and Alcorn (1980) in an analysis of United States crop losses due to prokaryotic phytopathogens during the period 1975–1977. Elsewhere crown gall is of major concern to grape growers in Europe (Lehoczyky, 1978) and South Africa; to rose, fruit, and grape growers in Greece (Panagopoulos et al., 1983); to fruit, rose, and chrysanthemum growers in Italy (Bazzi, 1983); and to peach, almond, and plum growers in Australia (Kerr, 1969). It has been estimated that if the losses from crown gall could be prevented savings of up to 138 million dollars could be achieved (El-Fiki and Giles, 1981). Efficient biological control of crown gall by strain 84 has been reported from a number of countries (Moore and Warren, 1979). In Italy, control was reported to be 80–100% successful on stone fruit seedlings and rose cuttings, but not on chrysanthemums, probably because of their infection with biotypes 1 and 3 pathogens (Bazzi, 1983). In Greece there was nearly complete prevention of crown gall on almond and peach seedlings planted in naturally infested soils (Panagopoulos et al., 1983).

To understand the mechanism whereby this novel form of biological control is achieved, a description of crown gall disease induction is required, with emphasis on features that affect sensitivity to agrocins.

2. *Ti* PLASMIDS AND CROWN GALL INDUCTION

Crown gall is caused by strains of *A. tumefaciens* that harbor a large (ca. 200 kb) tumor-inducing (Ti) plasmid (Zaenen et al., 1974). During infection the soil-dwelling pathogens invade plant cells and part of the Ti plasmid, the T-DNA (ca. 24 kb) is transferred from the bacterium to the plant cell, possibly by means of a circular DNA intermediate (Koukolíková-Nicola et al., 1985). The T-DNA integrates into the plant nuclear DNA (Chilton et al., 1980) and as a result the plant cells grow and divide rapidly to form an undifferentiated gall (Fig. 1). The T-DNA also directs the synthesis in the plant cells of unique compounds called opines (Petit et al., 1978). There are a number of types of opines, including nopaline, octopine, agropine, and the recently discovered succinamopine (Chilton et al., 1984). These form the basis of the classification of Ti plasmids, which are divided into groups depending on the type of opines whose synthesis their T-DNA directs in the transformed plant cell. In addition, the Ti plasmid carries genes enabling its host bacterium to catabolize the relevant opine. *Agrobacterium tumefaciens* has therefore developed a remarkable symbiotic strategy with the plant, termed genetic colonization, whereby the plant cell synthesizes compounds that can be utilized only by the inciting bacteria.

Tumor-inducing plasmid-carrying *A. tumefaciens* strains can be classified, apart from their opine type, according to the host plants on which they can incite tumors. Thus, for example, strains carrying pTiA6, an octopine Ti plasmid, can induce galls on a broad range of hosts, including common test plants such as tomatoes, sunflowers, *Datura* spp., and *Kalanchoë* spp. (De Cleene and De Ley, 1976). In contrast, strains carrying the octopine Ti plasmid pTiAg63 have a narrow host range restricted almost entirely to grapevines. That the host range could, at least in part, be due to the Ti plasmid was shown when Thomashow et al. (1980) transformed strain A136 (a broad host-range strain cured of its nopaline Ti plasmid pTiC58 while retaining its cryptic plasmid pAtC58), with the Ti plasmid pTiAg63, from the narrow host-range strain Ag63. The transformants showed the same narrow host range as strain Ag63, forming large active galls on

Fig. 1. Crown gall disease on a Persimmon tree (photograph courtesy of F. van Zyl).

grapevines. The reverse type of experiment showed that a wide host-range Ti plasmid, when conjugated into Ag63, rendered the transconjugants infective on a broad range of host plants. In addition, these authors noted that whereas it had previously been thought that all octopine Ti plasmids were very homogeneous, their data showed that considerable differences in restriction patterns existed between broad and narrow host-range octopine Ti plasmids.

In an attempt to pinpoint which regions of the Ti plasmid were responsible for host range, Buchholz and Thomashow (1984) constructed derivatives of narrow-range pTiAg63 that contained some of the oncogenes of the broad range pTiA6. The resultant plasmid now became broad host range. Contributing data led these authors to postulate that oncogene 4 was responsible for the expanded host range and proposed that in pTiAg63 sequences related to pTiA6 oncogene 4 are defective or function poorly in most plants. However, they state that it is clear from their data that factors other than the oncogene 4 sequences also contribute to the different host-range characteristics of pTiA6 and pTiAg63 and conclude that undefined host factors may also play a role.

In similar experiments, Hoekema et al. (1984) extended the host range of strain LBA649 (pTiAg57), which is usually limited to grapevines, by the introduction of the T-region from the wide host-range octopine plasmid pTiAch5. By inactivation of pTiAch5 oncogene 4, which codes for an enzyme responsible for cytokinin biosynthesis (*cyt*), these workers eliminated the broad host-range transformation effect and thus also identified this gene as being sufficient to extend the host range of LBA649. The conclusion was that limited host range is not caused by a lack of recognition of certain plants by a particular biotype, but is mainly due to the absence or inactivity of a *cyt* gene in the T-region of the narrow host-range Ti plasmid.

One problem regarding host range still remains however, namely that of the biotypes or biovars of *Agrobacteria.* At present, three are recognized, biotype 1 having a broad host range, biotype 2 being commonly found on stonefruit and biotype 3 whose hose range is largely limited to grapevines (Kerr and Panagopoulos, 1977, Panagopoulos et al., 1978). The biochemical characteristics on which biotypes are based are clearly not determined by the Ti plasmid, but the fact that a narrow host-range biotype 3 organism such as Ag63 can be converted to a broad host-range organism by incorporation into its Ti plasmid of the *cyt* gene of a broad host-range Ti plasmid implies that the biotype cannot determine the host range of pathogenic *Agrobacteria.*

Why biotype 3 strains carrying Ti plasmids, usually of the octopine type, defective in the *cyt* gene, should be found specifically infecting grapevines is an unsolved problem. It cannot be that these bacteria are incapable of colonizing other plants, because the addition of the *cyt* gene, which enables them to form galls on other plants, is unlikely to affect this property. Moreover, since biotypes 1 and 2 can be found infecting grapevines (Panagopoulos *et al.*, 1978; Süle, 1978), it cannot be a host-determined exclusion or selection. The particular selective pressures that have led to this specific host-bacterial interaction would be extremely interesting to unravel.

Clearly there is a need to study the ecology of the three biotypes, with or without Ti plasmids, in soil and on root and crown surfaces. Moreover, as both Ti plasmid type and biotype can affect sensitivity to agrocin (see Section 3), this information is vital to the effective implementation of biological control of crown gall. To this end the selective media for the three biotypes developed by Brisbane and Kerr (1983) are of particular importance.

3. SENSITIVITY TO AGROCIN 84

The correlation between sensitivity to agrocin 84 and pathogenicity, mentioned above, was proven when Engler et al. (1975) cured *A. tumefaciens* strain C58 of its nopaline Ti plasmid by growth at 37°C. All of the 10 cured isolates were fully resistant to agrocin 84. Engler et al. also noted that not all oncogenic, Ti plasmid-harboring *A. tumefaciens* strains are sensitive to agrocin 84. The reason for this became apparent when a physical and genetic map of pTiC58, the nopaline Ti plasmid of strain C58, was prepared (Holsters et al., 1980). Transposon-induced deletions in the 87 Md region of the plasmid abolished sensitivity to agrocin 84. Furthermore, DNA homology studies between pTiC58 and the octopine Ti plasmid pTiAch5 revealed that this region was not present on the latter plasmid (Engler et al., 1981). Thus sensitivity to agrocin 84 is encoded in a region that is unique to nopaline Ti plasmids such as pTiC58 and strains harboring an octopine Ti plasmid are naturally resistant to the agrocin.

At first sight this seems to be an extraordinary situation, whereby a plasmid carries potentially suicidal genes. However, Ellis and Murphy (1981) showed that the agrocin 84 permease does not exist solely to transfer this antibiotic molecule into *A. tumefaciens* cells. They reasoned that if this permease could permit uptake of other nontoxic compounds, these would interfere with the uptake of agrocin 84. They found four new opines, all sugar phosphodiesters, which behaved in this fashion and called them agrocinopines. They postulated that these compounds are the "legitimate" substrates for the permease and that agrocin 84 has merely "pirated" an existing uptake system.

Our understanding of the behavior of strain 84 was further clarified when its plasmids were investigated. It harbors three plasmids (Merlo and Nester, 1977): a large cryptic plasmid, a small 47.7 kb agrocin-encoding plasmid, pAgK84, (Slota and Farrand, 1982) and a medium-sized 200 kb plasmid, pAt84b. This last plasmid carries genes for nopaline catabolism, and thus strain 84, although not capable of inducing gall development, can compete with pathogens for nopaline and can, in addition, kill such pathogens by the production of agrocin, which is taken up by a permease designed for the uptake of compounds such as an agrocinopine.

Although on the basis of this account it would appear that all strains of *A. tumefaciens* harboring a nopaline Ti plasmid should be sensitive to agrocin 84, this is unfortunately not the case. Perry and Kado (1982) found that all biotype 3 strains tested, regardless of whether they possessed nopaline Ti plasmids, were resistant *in vivo* to agrocin 84.

4. AGROCIN 84 PRODUCTION AND STRAIN IMPROVEMENT

That plasmid pAgK84 encodes agrocin synthesis was proven when Cooksey and Moore (1981) cured strain 84 of this plasmid with concomitant loss of the ability to produce the agrocin. Moreover, Ellis et al. (1979) transferred the plasmid by mobilization to other *Agrobacterium* strains and isolated transconjugants that could now produce the agrocin. Slota and Farrand (1982) constructed a physical map of pAgK84 and showed that the biosynthesis of the agrocin does not require the presence of other plasmids. Although the functional map has not yet been completed, approximately 20 kb are involved in agrocin biosynthesis (see Kerr and Tate, 1984).

Besides coding for nopaline catabolism, pAt84b carries genes for the mobilization of pAgK84 (Ellis et al., 1979). Although these transfer genes are normally repressed, they are inducible by nopaline or agrocinopine A (Ellis et al., 1982). Moreover, pAt84b is self-transmissible at a very low rate (Slota and Farrand, 1982). Therefore the possibility of a breakdown in biological control potentially exists whereby pAgK84 could be mobilized to a pathogenic strain that retains its pathogenicity, produces agrocin, and is immune to it. This strain will not be subject to biological control and may represent a threat. However, this would occur mainly in situations where a high ratio of pathogen to strain 84 exists, resulting in inadequate control, gall formation, and the subsequent depression of the mobilization functions of pAt84b by the nopaline synthesized in the gall (Ellis and Kerr, 1979; Panagopoulos et al., 1979). To overcome these problems and prevent the transfer of pAgK84 to pathogens, a strain carrying a derivative of pAgK84 deficient in transfer and mobilization functions, but still capable of synthesizing agrocin 84, is being developed (Kerr and Tate, 1984).

Another way in which biological control could break down is by the transfer of a Ti plasmid from a pathogen to strain 84, as it has been shown that pTiC58 carries genes for self-transmission which, although normally repressed, are inducible by nopaline. It is possible that under field conditions where strain 84 has not been applied in sufficient numbers to prevent gall development, such galls that produce nopaline could facilitate Ti plasmid transfer to strain 84. This could be prevented by transferring the agrocin-encoding plasmid to a bacterium unlikely to receive or maintain a Ti plasmid. To this end Hendson and Thomson (1986) used as an agrocin plasmid recipient *Rhizobium meliloti* which, though a rhizosphere-inhabiting bacterium closely related to *Agrobacterium*, has been shown by Hooykaas et al. (1977) not to act as a recipient for pTi. They were encouraged in this approach by a statement by Kerr (1974): "Most workers now realize that if biological control is to be effective, the controlling organism must be adapted to the site where it is to operate," who commented further on the suitability of rhizosphere bacteria in general to the control of root diseases such

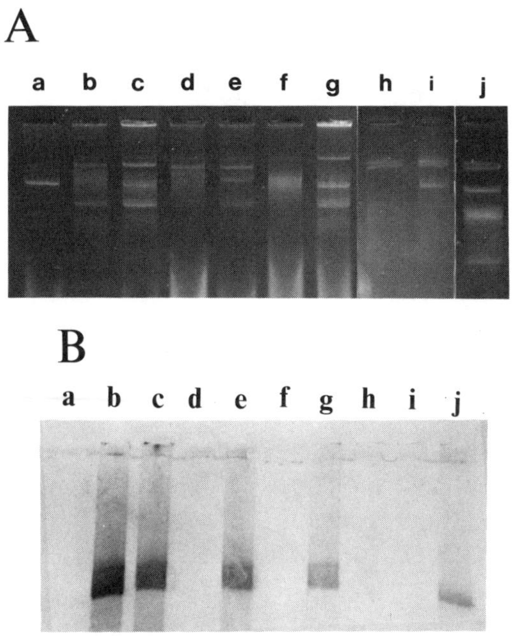

Fig. 2. Southern blot hybridization between ^{32}P-labeled pAg396 DNA and DNA extracted from (*a*) *E. coli* J53 (RP4), (*b*) *A. tumefaciens* 396, (*c*) 396 (RP4), (*d*) *A. tumefaciens* C58, (*e*) C58 (RP4) (pAg396), (*f*) C58Cl, (*g*) C58Cl (RP4) (pAg396), (*h*) *R. meliloti* RF10, (*i*) *R. meliloti* RF10 (RP4), and (*j*) RF10 (RP4) (pAg396). (*A*) Agarose gel electrophoresis (0.7%). (*B*) Autoradiography after transfer and hybridization same strains but different gel from *A*.

as crown gall. As the donor strain Hendson and Thomson chose *A. tumefaciens* 396, an oncogenic strain that harbors an agrocin with the same host range as 84 (Engler et al., 1975), but which grows faster than strain 84 (Hendson et al., 1983). They succeeded in mobilizing pAg396 with RP4 to *Rhizobium meliloti* RF10 (Fig. 2). This agrocin plasmid was stably maintained for at least 36 generations and produced yields of agrocin comparable to the parental strain 396 (Fig. 3). *Rhizobium meliloti* RF10 (pAg396) inhibited gall development by strain C58 on *Datura* plants in glasshouse trials (Table 1), but it remains to be seen whether it will be effective when applied in the normal way to seedlings before planting.

5. STRUCTURE AND MODE OF ACTION OF AGROCIN 84

Agrocin 84 is an N^6,5′-disubstituted adenine nucleotide with a fraudulent nucleoside sugar (Tate et al., 1979). As noted by Kerr and Tate (1984), the lower (or alpha) face of the β-D-3′-deoxyarabinofuranosyl group is structurally equivalent to a dideoxynucleoside, an ideal DNA chain terminator such as is used in DNA sequencing. However little is known of the mechanism of action of agrocin 84 in

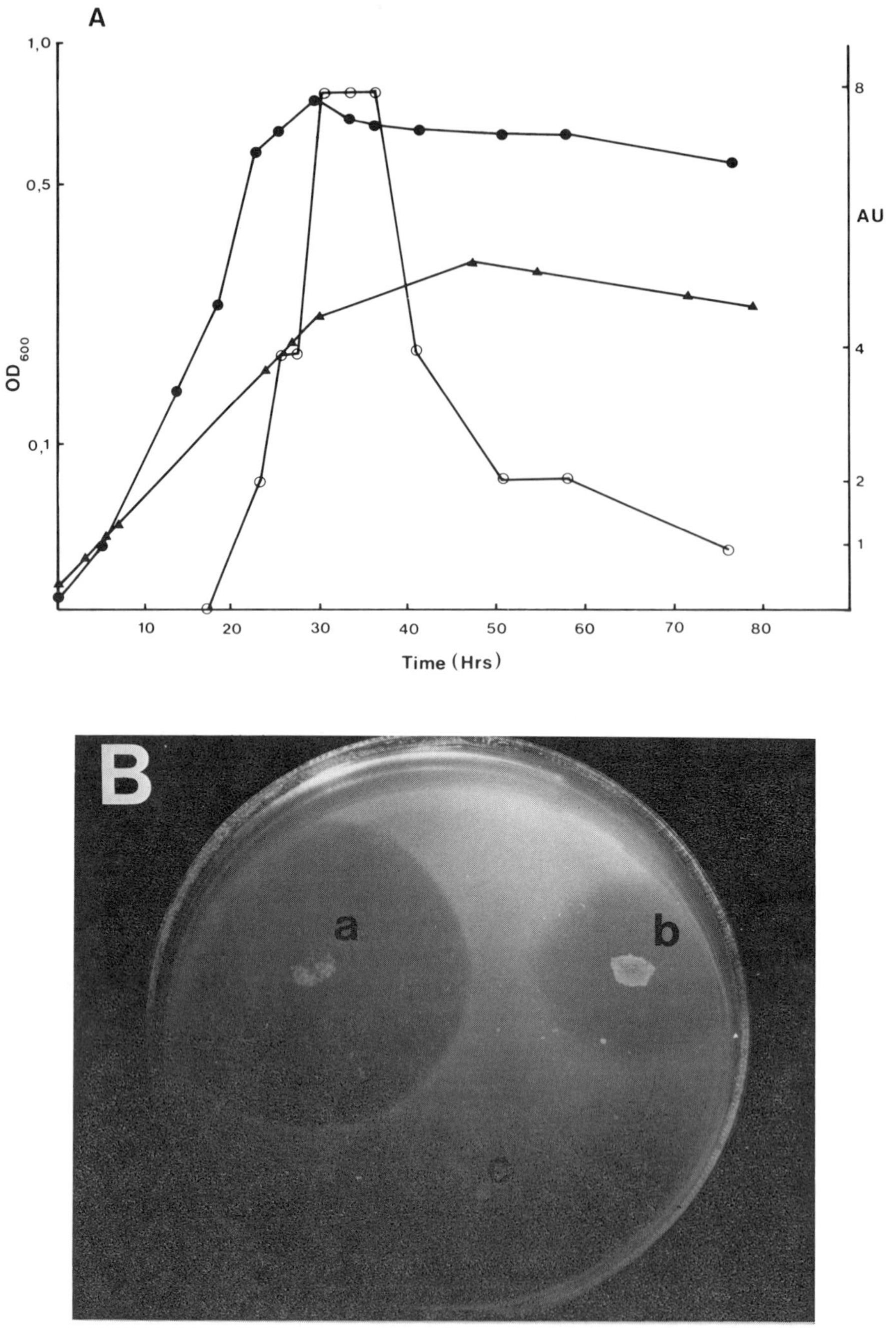

Fig. 3. (*A*) Semilogarithmic plot of growth (●) and production (○) of agrocin by *Rhizobium meliloti* RF10 (RP4) (pAg396) in minimal medium containing 1.06 × 10^{-2} *M* sodium glutamate and 1.11 × 10^{-3} *M* glucose compared with semilogarithmic plot of growth of RF10 (▲). No agrocin was produced by RF10. Cultures were assayed for agrocin production using the well-plate method and agrocin activity (AU) was calculated as the reciprocal of the highest dilution to give a detectable zone of inhibition. Each point is the average of four assays. (*B*) Zones of inhibition of growth of strain C58 by (*a*) K84, (*b*) RF10 (RP4) (pAg396), and (*c*) RF10.

TABLE 1. Biological Control of Crown Gall by *Rhizobium meliloti* RF10 (pAg396)

	Gall Index[a] in the Presence of Strain				
Plant Number	*C58*	*RF10*	*RF10 (pAg396)*	*C58 + RF10[b]*	*C58 + RF10 (pAg396)[b]*
1	3	0	0	1	1
2	4	0	0	2	1
3	3	0	0	3	1
4	3	0	0	3	1
5	4	0	0	3	1
	17	0	0	12	5

[a]Gall index: 1 = gall size 0–5 mm; 2 = gall size 5–10 mm; 3 = gall size 10–25 mm; 4 = gall size > 25 mm.

[b]Bacteria were coinoculated onto *Datura* seedlings at a ratio of 1 C58:50 *Rhizobium meliloti*.

inhibiting cell growth beyond the fact that it affects DNA synthesis (Das et al., 1978). The specific uptake system ($K_m = 5.9 \times 10^{-8}$ *M*; Murphy and Roberts, 1979) which is encoded by nopaline Ti plasmids, requires the N^6-D-glucofuranosyloxyphosphoramidate substituent of the agrocin, while the 5′-phosphoramidate substituent is essential for toxicity (Murphy et al., 1981). It appears that the N^6-substituent is "recognized" by a plasmid encoded periplasmic binding protein during uptake by sensitive cells. Data suggests that the 5′-oxyphosphoryl group is an absolute requirement for toxicity, which may reflect the absence of a kinase capable of phosphorylating the fraudulent nucleoside prior to its incorporation into a growing DNA chain (Tate et al., 1979).

6. AGROCINS WITH DIFFERENT HOST RANGES

The host range of agrocin 84 is very specific, being limited to strains carrying a nopaline-type Ti plasmid. Moreover, it is not effective against biotype 3 strains with host ranges limited to grapevines. In an attempt to overcome the first problem, Hendson et al. (1983) screened *Agrobacterium* strains for agrocins with a broader host range than agrocin 84. Working on the assumption that nonpathogenic agrocin-producing bacteria may well be capable of colonizing a gall, as had been shown in the case of strain 84, they screened strains isolated from crown galls on a variety of plants. Many agrocin-producing strains were isolated, but most had very limited host ranges. However, one strain D286, a nonpathogenic biotype 1 strain isolated from a gall on a Eucalyptus tree produced an agrocin with a broader host range *in vitro* than agrocin 84. *In vitro* results can, however, often be misleading, as various culture media can inhibit agrocin production, and a pathogenic strain may be insensitive *in vitro*, but be controlled *in vivo* (see Moore and Warren, 1979). Moreover many agrocins that are active *in vitro* are not *in vivo* as, for example, was shown by Kerr and Panagopoulos (1977). With this in mind, strain D286 was coinoculated onto tobacco and *Da-*

tura plants with a variety of pathogenic strains. Although only about 30% of the strains were sensitive *in vitro*, approximately 98% were controlled *in vivo* (van Zyl, personal communication).

This finding encouraged Hendson et al. (1983) to investigate strain D286 further and they found that sensitivity to agrocin D286 mapped in the region from 11 to 18 Md on the nopaline Ti plasmid pTiC58, an area that falls within one of the regions homologous to both pTiC58 and the octopine plasmids pTiB6S3 and pTiAch5. This may well explain the broad host range of agrocin D286. Although the structure of agrocin D286 has not yet been determined, it resembles agrocin 84 (Roberts et al., 1977) in having the characteristic adenine N^6-phosphoramidate UV absorption maximum at 264 nm with a shoulder at 270 nm. The agrocin was shown to inhibit DNA, RNA, and protein synthesis of sensitive strains.

Plasmid screens showed that D286 carried a large 300 Md plasmid, which may correspond to similar sized cryptic plasmids seen in many strains of *Agrobacterium*, and another slightly smaller than the 132 Md pTiC58 (Fig. 4 and Hendson et al., 1983). Whether the latter plasmid encodes agrocin production has not yet been determined, despite attempts to either cure it or mobilize it to other strains.

Recently field trials have been undertaken in which seedlings of two different cultivars of chrysanthemums were dipped into suspensions of strains D286 and 84 before planting in soil naturally infested with crown gall. Of the 140 undipped plants, 82 and 76% of the two varieties became galled. Strain 84 decreased this to 50 and 45%, respectively, while D286 decreased gall incidence to 15 and 12% (Van Zyl, personal communication). Thus although strain 84 did afford some control, unlike the findings of Bazzi (1983), where this strain did not prevent gall formation on chrysanthemums, strain D286 gave better protection.

Although strain D286 may well prove useful in biological control of strains carrying Ti plasmids other than the nopaline type, it is unfortunately not inhibi-

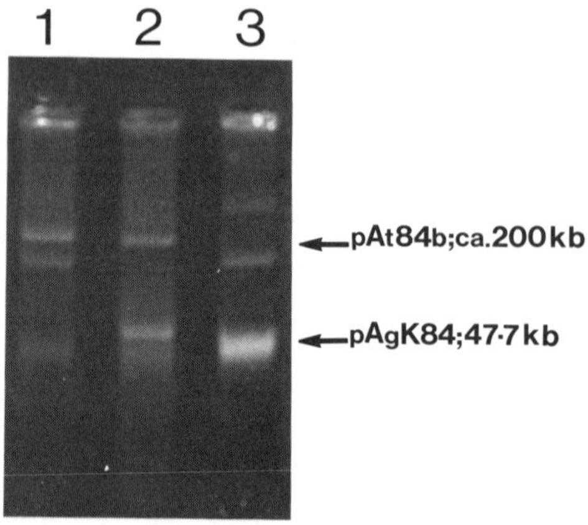

Fig. 4. Plasmid profiles of agrocin-producing *A. tumefaciens* strains. Lane 1, J73; lane 2, 84; lane 3, D286. Arrows refer to plasmids of strain 84.

tive to biotype 3 strains on grapevines, regardless of the catabolic nature of their Ti plasmids. Most, but not all, biotype 3 strains carry octopine Ti plasmids and some can catabolize both octopine and nopaline (Panagopoulos et al., 1978; our unpublished results). Perry and Kado (1982) found that all biotype 3 strains tested, including those that can utilize nopaline as sole carbon source, were resistant *in vivo* to agrocin 84. As crown gall disease of grapevines is a worldwide problem, Webster et al. (1986) screened isolates from crown galls for agrocins active against grapevine isolates. They have shown that strain J73, a biotype 2 isolate from a *Prunus salicinia* gall, inhibits grapevine pathogens both *in vitro* and *in vivo* (Table 2). *In vivo* tests on grapevines were possible because although J73 is a pathogenic strain carrying a Ti plasmid (Fig. 4), it is a biotype 2 strain and does not form galls on grapevines. However, before field trials can be embarked upon it will have to be cured of its Ti plasmid. It is interesting that although agrocin J73 is a small protease-resistant molecule with features similar to agrocin 84 (unpublished results), sensitivity to it cannot be entirely chromosomally encoded as strains either harboring, or cured of, their Ti plasmids are sensitive to the agrocin.

Recently Staphorst et al. (1985) reported greenhouse trials where gall formation on grapevines was decreased by the simultaneous application to wounds of nonpathogenic biotype 3 agrocin-producers and a variety of biotype 3 pathogens. The agrocin-producers were isolated from gall tissue of grapevines. These strains look promising but further trials will be necessary to assess their potential.

Although strain 84 was isolated in and is apparently indigenous to Australia, oncogenic strains have been isolated in other countries that produce agrocins with similar, if not identical, specificities. These include the biotype 1 strain Bo542 of European origin (Sciaky et al., 1978), which harbors the agrocin-encoding plasmid pAgBo542 (Slota and Farrand, 1982) and strain 396 mentioned above. The appearance in widely diverse geographical areas of *Agrobacterium* strains producing nucleotide agrocins capable of biological control of crown gall allows one to assume that more will be isolated in the future.

7. IS AGROCIN PRODUCTION ALL THAT IS REQUIRED FOR BIOLOGICAL CONTROL?

Although it has repeatedly been stated that strain 84 mediates biological control via the production of an agrocin, the possibility of a contributary role played by competition with a pathogen for the binding of "infection" sites on the plant, proposed by Lippincott and Lippincott (1969), has still not been completely ruled out.

A report by Cooksey and Moore at the 1980 Annual Meeting of the American Phytopathological Society stated that a derivative of 84 cured of its agrocin plasmid reduced galling on wounded tomato stems by *A. tumefaciens* K24 to 35% as long as it was inoculated into wounds 24 hr before the pathogen. Moreover,

TABLE 2. Sensitivity of Grapevine Isolates of *A. tumefaciens* to Strain J73 *In Vitro* and *In Vivo*

In Vitro[a]

Biotype	*Countries of Origin*	*Number of Strains Tested*	*Number of Sensitive Strains*
1	Greece, Hungary, United States, South Africa	11	10
2	South Africa	2	1
3	South Africa, United States, Iran	14	11

In Vivo[b]

Pathogenic Biotype 3 Strains[c]	*Plant Number*	*Gall Index*[d] *in the Presence of* *No. Controlling Organism*	*J73*
D/1	1	3	1
	2	4	2
	3	2	2
	4	3	2
		12	7
F1/4	1	2	1
	2	3	2
	3	2	0
	4	3	0
		10	3

[a]Strain J73 was plated on AB medium (McCardell and Pootjes, 1976), supplemented with 0.5% casamino acids and 0.5% sodium glutamate, grown at 26°C for 3 days, and overlayed with the grapevine isolates.
[b]Performed by F. van Zyl, Department of Plant Protection, Stellenbosch, South Africa.
[c]South African grapevine isolates.
[d]Gall index is the size of the gall (on an arbitrary scale from 0 to 4) obtained on individual grapevines.

other reports have shown that strain 84 can control agrocin 84-resistant strains (see Moore and Cooksey, 1981).

In a 1979 paper, Ellis and Kerr found that a C58Cl transconjugant carrying the agrocin plasmid from strain 84 was ineffective in biological control of the pathogenic strain 27 on tomatoes. They correlated this with slow growth of this transconjugant in the wounded tomato stem and concluded that a successful biological control agent must be able to produce agrocin and grow sufficiently well to produce enough of the antibiotic.

If as these data suggest competition between controlling and pathogenic bacteria for binding to sites on plant cells and efficient growth of the agrocin-producing strain at the wound site are contributing factors to biological control, the reason for the success of strain 84 may well be a combination of such factors. When considering biological control of strains resistant to agrocin 84, thought should be given to identifying strains of *Agrobacterium* that are efficient colonizers of the plant to be protected and introducing into them plasmids encoding broad host-range agrocins.

8. CONCLUSIONS

Biological control of crown gall disease of plants by the agrocin-producing biotype 2 strain 84 is practiced worldwide with considerable success. However, owing to the potential for breakdown of such control, strains are being developed, such as the strain carrying a derivative of pAgK84 deficient in transfer and mobilization functions, and *R. meliloti* (pAg396), which may circumvent such problems. As strain 84 acts neither on *A. tumefaciens* strains carrying Ti plasmids other than the nopaline type, nor on biotype 3 strains that infect grapevines, agrocin-producers active against such nonprotectable strains are being sought. To this end strains such as D286, which is active against nopaline, octopine and agropine Ti plasmids, and J73 active against grapevine pathogens, are being investigated. That a variety of agrocin-producing strains capable of biologically controlling crown gall disease have been identified augurs well for the future success of controlling this ubiquitous plant disease.

ACKNOWLEDGMENTS

I wish to thank Mavis Hendson, Jocelyn Webster, Jill Rodel, and Christine Botha for critical reading and assistance with the preparation of the manuscript. I acknowledge the help with *in vivo* trials, and the disclosure of unpublished results, by B. W. Strydom, J. L. Staphorst, F. G. H. van Zyl, and Z. E. Groenewald of the Department of Plant Protection. Finally I wish to thank Jackie Field for her painstaking typing of this manuscript.

REFERENCES

Alconero, R. (1980). Crown gall *Agrobacterium tumefaciens* of peaches *Prunus persica* from Maryland, South Carolina and Tennessee USA and problems with biological control. *Plant Dis.*, **64:**835–838.

Bazzi, C. (1983). Biological control of crown gall in Italy. *International Workshop on Crown Gall.* Swiss Federal Research Station for Fruit-Growing, Viticulture and Horticulture, pp. 1–21.

Brisbane, P. G. and A. Kerr (1983). Selective media for three biovars of *Agrobacterium. J. Appl. Bacteriol.*, **54:**425–431.

Buchholz, W. G. and M. F. Thomashow (1984). Host range encoded by the *Agrobacterium tumefaciens* tumor-inducing plasmid pTiAg63 can be expanded by modification of its T-DNA oncogene complement. *J. Bacteriol.*, **160:**327–332.

Chilton, M-D., R. K. Saiki, N. Yadav, M. P. Gordon, and F. Quetier (1980). T-DNA from *Agrobacterium* Ti-plasmid is in the nuclear fraction of crown gall tumor cells. *Pro. Natl. Acad. Sci. USA*, **77:**4060–4064.

Chilton, W. S., J. Tempé, H. Matzke, and M-D Chilton (1984). Succinamopine: A new crown gall opine. *J. Bacteriol.*, **157:**357–362.

Cooksey, D. A. and L. Moore (1981). An agrocin mutant of *Agrobacterium radiobacter* K84 and biological control of crown gall. *Phytopathology*, **71:**104.

Das, P. K., M. Basu, and G. C. Chatterjee (1978). Studies on the mode of action of agrocin 84. *J. Antibiot.*, **31:**490–492.

De Cleene, M. and J. De Ley (1976). The host range of crown gall. *Bot. Rev.*, **42:**389–411.

El-Fiki, F. and K. L. Giles (1981). *Agrobacterium tumefaciens* in agriculture and research. In *Biology of the Rhozobeaceae* (K. L. Giles and A. G. Atherley eds.). International Review of Cytology, Supplement 13, pp. 47–57.

Ellis, J. G. and A. Kerr (1979). Transfer of agrocin 84 production from strain 84 to pathogenic recipients: A comment on the previous paper. In *Soil-Borne Plant Pathogens* (B. Schippers and W. Gams eds.). Academic, London, pp. 579–583.

Ellis, J. G. and P. J. Murphy (1981). Four new opines from crown gall tumours—their detection and properties. *Mol. Gen. Genet.*, **181:**36–43.

Ellis, J. G., A. Kerr, M. Van Montagu, and J. Schell (1979). *Agrobacterium*: genetic studies on agrocin 84 production and the biological control of crown gall. *Physiol. Plant Pathol.*, **15:**311–319.

Ellis, J. G., A. Kerr, A. Petit, and J. Tempé (1982). Conjugal transfer of nopaline and agropine Ti-plasmids—the role of agrocinopines. *Mol. Gen. Genet.*, **186:**269–274.

Engler, G., M. Holsters, M. Van Montagu, J. Schell, J. P. Hernalsteens, and R. Schilperoort (1975). Agrocin 84 sensitivity: A plasmid determined property in *Agrobacterium tumefaciens. Mol. Gen. Genet.*, **138:**345–349.

Engler, G., A. Depicker, R. Maenhaut, R. Villarroel, M. Van Montagu, and J. Schell (1981). Physical mapping of DNA base sequence homologies between an actopine and a nopaline Ti plasmid of *Agrobacterium tumefaciens. J. Mol. Biol.*, **152:**183–208.

Hendson, M. and J. A. Thomson (1986). Expression of an agrocin-encoding plasmid of *Agrobacterium tumefaciens* in *Rhizobium meliloti. J. Appl. Bacteriol.* **60:** 147–154.

Hendson, M., L. Askjaer, J. A. Thomson and M. Van Montagu (1983). Broad-host-range agrocin of *Agrobacterium tumefaciens. Appl. Environ. Microbiol.*, **45:**1526–1532.

Hoekema, A., B. S. de Pater, A. J. Fellinger, P. J. J. Hooykaas, and R. A. Schilperoort (1984). The limited host range of an *Agrobacterium tumefaciens* strain extended by a cytokinin gene from a wide host range T-region. *EMBO J.*, **3:**3043–3047.

Holsters, M., B. Silva, F. Van Vliet, C. Genetello, M. De Block, P. Dhaese, A. Depicker, D. Inzé, G. Engler, R. Villarroel, M. Van Montagu, and J. Schell (1980). The functional organization of the nopaline *A. tumefaciens* plasmid pTiC58. *Plasmid*, **3:**212–230.

Hooykaas, P. J. J., P. M. Klapwijk, M. P. Nuti, R. A. Schilperoort, and A. Rörsch (1977). Transfer of the *Agrobacterium tumefaciens* Ti plasmid to avirulent Agrobacteria and to Rhizobium *ex planta. J. Gen. Microbiol.*, **98:**477–484.

Kennedy, B. W. and S. M. Alcorn (1980). Estimates of USA crop losses to prokaryotic plant pathogens. *Plant Dis.*, **64:**674–676.

Kerr, A. (1969). Crown gall of stone fruit. I. Isolation of *Agrobacterium tumefaciens* and related species. *Aust. J. Biol. Sci.*, **22:**111–116.

Kerr, A. (1974). Soil microbiological studies on *Agrobacterium radiobacter* and biological control of crown gall. *Soil Sci.*, **118:**168–172.

Kerr, A. and K. Htay (1974). Biological control of crown gall through bacteriocin production. *Physiol. Plant Pathol.*, **4:**37–44.

Kerr, A. and C. G. Panagopoulos (1977). Biotypes of *Agrobacterium tumefaciens* and their biological control. *Phytopathol. Zeitschr.*, **80:**172–179.

Kerr, A. and M. E. Tate (1984). Agrocins and the biological control of crown gall. *Microbiol. Sci.*, **1:**1–4.

Koukolíková-Nicola, Z., R. D. Shillito, B. Hohn, K. Wang, M. Van Montagu, and P. Zembryski (1985). Involvement of circular intermediates in the transfer of T-DNA from *Agrobacterium tumefaciens* to plant cells. *Nature*, **313:**191–195.

Lehoczyky, J. (1978). Root system of the grape vine as a reservoir of *Agrobacterium tumefaciens* cells. In *Proceedings of the 4th International Conference on Plant Pathogenic Bacteria.* Angers, France, pp. 239–244.

Lippincott, B. B. and J. A. Lippincott (1969). Bacterial attachment to a specific wound site as an essential stage in tumour initiation by *Agrobacterium tumefaciens. J. Bacteriol.*, **97:**620.

McCardell, B. A. and C. F. Pootjes (1976). Chemical nature of agrocin 84 and its effects on a virulent strain of *Agrobacterium tumefaciens. Antimicrob. Agents Chemother.*, **10:**498–502.

Merlo, P. J. and E. W. Nester (1977). Plasmids in avirulent strains of *Agrobacterium. J. Bacteriol.*, **129:**76–80.

Moore, L. W. and D. A. Cooksey (1981). Biology of *Agrobacterium tumefaciens:* Plant interactions. In *Biology of the Rhizobeaceae* (K. L. Giles and A. G. Atherly eds.). International Review of Cytology, Supplement 13. Academic, New York, pp. 15–45.

Moore, L. W. and G. Warren (1979). *Agrobacterium radiobacter* strain 84 and biological control of crown gall. *Annu. Rev. Phytopathol.*, **17:**163–179.

Murphy, P. J. and W. P. Roberts (1979). A basis for agrocin 84-sensitivity in *Agrobacterium radiobacter. J. Gen. Microbiol.*, **114:**207–213.

Murphy, P. J., M. E. Tate, and A. Kerr (1981). Substituents at N^6 and $C5'$ control selective uptake and toxicity of the adenine-nucleotide bacteriocin, agrocin 84 in agrobacteria. *Eur. J. Biochem.*, **115:**539–543.

New, P. B. and A. Kerr (1972). Biological control of crown gall: field measurements and glasshouse experiments. *J. Appl. Bacteriol.*, **35:**279–287.

Panagopoulos, C. G., P. G. Psallidas, A. S. Alivizatos (1978). Studies on biotype 3 of *Agrobacterium radiobacter* var. *tumefaciens. Proceedings of the 4th International Conference of Plant Pathogenic Bacteria.* Angers, France, pp. 221–228.

Panagopoulos, C. G., P. G. Psallidas, and A. S. Alivizatos (1979). Evidence of a breakdown in the effectiveness of biological control of crown gall. In *Soil-Borne Plant Pathogens* (B. Schippers and W. Gams eds.). Academic, London, pp. 569–578.

Panagopoulos, C. G., P. G. Psallidas, and D. C. Stylianidis (1983). Current work on the effectiveness of biocontrol of crown gall. *International Workshop on Crown Gall.* Swiss Federal Research Station for Fruit-Growing, Viticulture and Horiticulture, pp. 67–78.

Perry, K. L. and C. I. Kado (1982). Characteristics of Ti plasmids from broad host range and ecologically specific biotype 2 and 3 strains of *Agrobacterium tumefaciens. J. Bacteriol.*, **151:**343–350.

Petit, A., Y. Dessaux, and J. Tempé (1978). The biological significance of opines: 1. A study of opine catabolism by *Agrobacterium tumefaciens. Proceedings of the 4th International Conference on Plant Pathogenic Bacteria.* Angers, France, pp. 143–152.

Roberts, W. P. and A. Kerr (1974). Crown gall induction: serological reactions, isozyme patterns and sensitivity to mitomycin C and to bacteriocin, of pathogenic and non-pathogenic strains of *Agrobacterium radiobacter. Physiol. Plant Pathol.*, **4:**81–91.

Roberts, W. P., M. E. Tate, and A. Kerr (1977). Agrocin 84 is a 6-*N*-phosphoramidate of an adenine nucleotide analogue. *Nature*, **265:**379–380.

Schroth, M. N., A. R. Weinhold, A. H. McCain, D. C. Hilderbrand, and N. Ross (1971). Biology and control of *Agrobacterium tumefaciens. Hilgardia*, **40:**537–552.

Sciaky, D., A. L. Montoya, and M-D Chilton (1978). Fingerprints of *Agrobacterium* Ti plasmids. *Plasmid*, **1**:238–253.

Slota, J. E. and S. K. Farrand (1982). Genetic isolation and physical characterization of pAgK84—the plasmid responsible for agrocin 84 production. *Plasmid*, **8**:175–186.

Staphorst, J. L., F. G. H. van Zyl, B. W. Strydom, and Z. E. Groenewold (1985). Agrocin producing pathogenic and nonpathogenic biotype 3 strains of *Agrobacterium tumefaciens* active against biotype 3 pathogens. *Curr. Microbiol.* **12:** 45–52.

Stonier, T. (1960). *Agrobacterium tumefaciens* Conn. II. Production of an antibiotic substance. *J. Bacteriol.*, **79**:889–898.

Süle, S. (1978). Pathogenicity and agrocin 84 resistance of *Agrobacterium tumefaciens. Proceedings of the 4th International Conference on Plant Pathogenic Bacteria.* Angers, France, pp. 255–257.

Tate, M. E., P. J. Murphy, W. P. Roberts, and A. Kerr (1979). Adenine N^6 substituent of agrocin 84 determines its bacteriocin-like specificity. *Nature*, **280**:697–699.

Thomashow, M. F., C. G. Panagopoulos, M. P. Gordon, and E. W. Nester (1980). Host range of *Agrobacterium tumefaciens* is determined by the Ti plasmid. *Nature*, **283**:794–796.

Webster, J., M. Dos Santos, and J. A. Thomson (1986). An agrocin-producing *Agrobacterium tumefaciens* active against grapevine isolates. Appl. Env. Microbiol. (in press).

Zaenen, I., N. Van Larebeke, H. Teuchy, M. Van Montagu, and J. Schell (1974). Supercoiled circular DNA in crown-gall; including *Agrobacterium* strains. *J. Mol. Biol.*, **86**:109–127.

11

BIOCHEMICAL RESPONSES OF PLANTS TO PATHOGENS

KLAUS HAHLBROCK
DIERK SCHEEL
Max Planck Institut für Züchtungsforschung, Cologne, West Germany

1. INTRODUCTION

The rapidly developing fields of genetic engineering and modern breeding technology of higher plants will soon offer new tools for introducing valuable traits into crop plants. Developing plant resistance to pathogens is one major goal of plant breeding programs. The prerequisites for the application of gene transfer techniques in this area are the isolation and characterization of the responsible genes. This has so far not been achieved for any pathosystem.

In general, plants respond in two different ways to pathogens. There is either no obvious interaction, or an interaction occurs and is, in the extreme cases, either incompatible (the plant is resistant) or compatible (the plant is susceptible). The interaction of a potential pathogen with a nonhost plant usually results in a typical resistance response of the plant (basic incompatibility, nonhost or species resistance), whereas the majority of host plants is susceptible (basic or species compatibility). However, certain cultivars of susceptible species are resistant to certain races of the pathogen (cultivar resistance) (Bailey, 1983). It is by now accepted that operationally defined gene-for-gene relationships exist for pathosystems with cultivar resistance (Bailey, 1983; Ellingboe, 1984). Under normal conditions, plants are resistant to most of the microorganisms in their environment. Only a relatively small number of true host–pathogen combinations have evolved. The biochemical events occurring in interactions of host and nonhost plants with potential pathogens are similar, but their intensities and patterns of appearance depend on environmental and physiological conditions as well as the specific tissue challenged by the pathogen.

Figure 1 summarizes the biochemical reactions frequently observed after the inoculation of various host or nonhost plants with several potential pathogens. Not all of the reactions have so far been studied in any one plant, but the sum of all available data suggests that most, or all of them, are part of a typical resistance response of plants (Bell, 1981; Sequeira, 1983). Variations might occur more at the level of relative amounts, location, and timing of induction than in terms of an all-or-none response. Furthermore, it is important to realize that none of these reactions is absolutely specific for plant–pathogen interactions. Most of them can be triggered by endogenous as well as abiotic signals, for example, during certain stages of development or by treatment with toxic chemicals. It is possible that the reactions shown in Fig. 1 and further discussed below are not only typical for responses to pathogens, but also represent a more general defense response, either in the same or in modified combinations, to certain forms of environmental hazard. It is unlikely that genes encoding enzymes that are involved in these defense reactions are identical to the operationally defined genes for cultivar resistance (Ellingboe, 1982).

In spite of the obvious similarities of the defense responses of different plants, specific recognition processes are probably involved in pathosystems with cultivar resistance, resulting either in the rapid activation or in the suppression of the defense response. Although even less is known at present about the mechanisms of recognition than about the subsequent biochemical reactions of infected plants, it is possible that genes conferring resistance on otherwise susceptible plants might be involved at all regulatory levels in plant–pathogen interactions, including the initial steps of recognition. This should be borne in mind, even if more discussion is devoted in the following pages to the more frequently studied late steps in disease resistance.

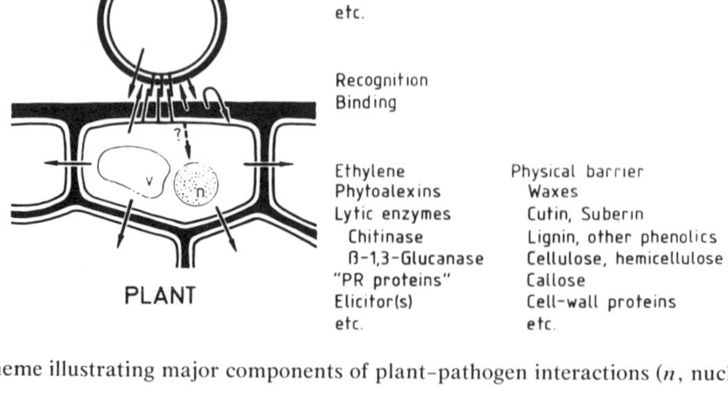

Fig. 1. Scheme illustrating major components of plant–pathogen interactions (*n*, nucleus; *v*, vacuole).

2. STRUCTURAL BARRIER

Before pathogens can establish an infection, they have to overcome the barriers protecting all plants against their environment. While viruses and bacteria enter plants either through wounds or with the help of herbivores, pathogenic fungi grow through wounds or natural openings, such as stomata, or actively penetrate the surface layers of plants. These structural barriers include the cell wall, which is usually impregnated by lignin, and the cuticle, which is attached to the pectinaceous layer of the cell wall and covers all aerial parts of plants. Its major constituent is cutin, an insoluble hydroxyfatty acid polymer, which is embedded in soluble waxes. Underground parts of plants and wounds of aerial plant parts are protected by suberin, a mixed polymer of phenolics and lipid-derived material, which is tightly attached to the cell wall carbohydrates.

2.1. Waxes

The most common components of plant surface waxes are C_{21}–C_{37} *n*-alkanes, C_{29} and C_{31} being the most frequently occurring chain lengths. Often one or a few alkanes predominate. The minor constituents are monoesters composed of fatty acids and long-chain alcohols with chain lengths from C_{32}–C_{64}, fatty acids, aldehydes, alcohols, diesters involving fatty alcohols, hydroxyfatty acids, *p*-hydroxy and *p*-methoxycinnamic acids, polyesters of ω-hydroxyfatty acids and diols, and terpenes (Kolattukudy, 1980).

There is no single role of wax components in plant defense against pathogens. The hydrophobic character of certain plant surface waxes has been discussed as an indirect defense mechanism against fungal attack (Martin, 1964; Kubo and Matsumoto, 1984), since moisture, which is essential for spore germination, cannot remain on these surfaces. Lipophilic products, such as furanocoumarins, have been identified as constituents of leaf surface waxes from carrot (Städler and Buser, 1984) and parsley (Knogge and Hahlbrock, unpublished) and may directly inhibit the growth of potential pathogens. However, wax components stimulate fungal development (Martin, 1964) or induce synthesis and excretion of degradative enzymes like cutinase by pathogenic fungi (Kolattukudy, 1980). Resistance-related changes in surface wax composition in response to pathogen attack have not been observed.

2.2. Cutin

The predominating constituents of cutin are hydroxylated C_{16} and C_{18} fatty acids. 10,16-Dihydroxypalmitic acid and its positional isomers are the main components of the C_{16} family. Products derived from these compounds by oxidation or reduction have been described as rare or minor components. Among the C_{18} monomers, stearic, oleic, linoleic, 18-hydroxyoleic, 18-hydroxy-9,10-epoxystearic acid and their Δ^{12} unsaturated analogues occur in major amounts and their Δ^{12}–Δ^{15} unsaturated analogues as well as derivatives with further hydroxy and epoxy groups in trace amounts (Kolattukudy, 1980). Structurally, cutin is a polyester of these hydroxyfatty acids; ester linkages involving primary alcohols are most important for its structure. *p*-Coumaric acid, small amounts of ferulic acid, and in certain species other secondary plant products, such as flavonoids, are covalently attached to cutin.

Several fungal plant pathogens excrete cutinase, an enzyme catalyzing the degradation of cutin. Its synthesis is induced by cutin. Specific inhibition of this enzyme protects the plant from fungal infection. The virulence of different strains of *Fusarium solani* f. sp. *pisi* on pea correlates with the level of cutinase generated by the germinating spores. In this context cutin has been designated as the main barrier of plants against invasion by pathogenic microorganisms (Kolattukudy, 1984). On the other hand, it has been shown, at least for pathosystems with cultivar resistance, that the penetration of the cuticle occurs freely, irrespective of fungal virulence or the degree of susceptibility of the host (Martin, 1964). In those systems the cuticle does not play a major role in the protection of plants against pathogens.

2.3. Suberin

Underground parts of plants, aerially formed periderms, including wound periderms, and barks are covered by suberin. Suberin is an insoluble mixed polymer of aliphatic and aromatic constituents, tightly attached to the cell wall. A suberin-enriched fraction from potato tuber, for example, has an approximate composition of 50% carbohydrates, 5–10% other aliphatic components, and 40–45% phenolic components. The monomeric composition of suberin is only poorly understood. Its major aliphatic components are C_{14}–C_{32} ω-hydroxyfatty acids and the corresponding dicarboxylic acids. Fatty acids and fatty alcohols with 20–32 carbon atoms, are further constituents occurring in substantial amounts. The phenolic composition of suberin is still obscure. *p*-Coumaric acid, ferulic acid, vanillin, *p*-hydroxybenzaldehyde, and very small amounts of syringaldehyde have been described as minor components (Kolattukudy, 1981). In a hypothetic structural model of suberin suggested by Kolattukudy (1984), a lignin-like phenolic matrix is attached to the cell-wall carbohydrates (Fig. 2). The aliphatic components are esterified to the phenolics. Long-chain fatty acids and alcohols are in terminal positions providing the hydrophobic domain for the interaction with the waxes that always accompany suberin.

Fungal pathogens are able to penetrate suberin after enzymatic depolymerization. Culture fluid from *Fusarium solani* f. sp. *pisi* releases radiolabel from the aliphatic part as well as from the phenolic matrix of radioactively labeled suberin (Kolattukudy, 1984). Recently two extracellular esterases have been purified from *F. solani* f. sp. *pisi* grown on suberin. Both esterases depolymerize the aliphatic components of suberin (Fernando et al., 1984) and are very similar, if not identical, to cutinases purified from the culture fluid of the same fungus after growth on cutin.

Some confusion exists about the participation of induced suberization in plant defense reactions because of analytical uncertainties regarding the distinction of suberin and lignin (Kolattukudy, 1980; Friend, 1985).

2.4. Lignin

During secondary thickening, plant cell walls and intercellular spaces are impregnated with the phenolic polymer lignin. Lignification occurs by enzymatic dehydrogenation and subsequent radical condensation of coumaryl, coniferyl, and sinapyl alcohols. Lignins from different plant species differ in the relative proportions of these building blocks (Freudenberg, 1965; Nimz, 1974). Lignin is covalently linked to hemicellulose and cellulose by ester and ether bonds (Eriksson et al., 1980) and to cell-wall proteins, probably hydroxyproline-rich glycoproteins (Whitmore, 1978). The precursors originate from the general phenylpropanoid pathway according to Fig. 2. Many of the enzymes involved, as well as their regulation, have been studied (Grisebach, 1981), providing a valuable basis for examining their participation in defense reactions of plants to pathogens. As illustrated in Fig. 2, this evaluation is complicated because products of the phenylpropanoid pathway are precursors of many secondary plant products, including coumarins, flavonoids, isoflavonoids, pterocarpans, stilbenoids, dihydrophenanthrenes, lignin, suberin, and other phenols, all of which can also be involved in disease resistance (Dixon et al., 1983).

Preformed lignin is thought to play a role in the limitation of fungal growth in certain tissues, for example, in wheat and bean (Ride, 1983). Local lignin deposition, sometimes occurring throughout the protoplasm, is frequently observed as an early reaction after fungal attack in incompatible and nonhost/pathogen interactions (Vance et al., 1980; Ride, 1983; Hammerschmidt, 1984; Hammerschmidt et al., 1984; Friend, 1985). Chemical analysis indicates a different composition of the induced material as compared with lignin from healthy tissue (Ride, 1975; Asada et al., 1979). Rapid accumulation of lignin precursors (Keen and Littlefield, 1979) and local increase in the activities of phenylalanine ammonia-lyase (PAL) and peroxidases (Coffey and Cassidy, 1984), apparently due to increased *de novo* synthesis of the enzymes (Asada et al., 1979), accompany lignin deposition. Treatment of the plant tissue with the PAL inhibitor α-aminooxyacetic acid (AOA) prior to inoculation inhibits deposition of the phenolic polymer and promotes fungal invasion in wheat (Tiburzy, 1984) and potato (Friend, 1985). In potato tuber slices esterification of wall components with *p*-coumaric and ferulic acids leads to further structural and chemical modifications of the cell wall (Friend, 1985).

Increased local lignification, accompanied by increased enzyme activities of PAL and peroxidases, has also been proposed to be an important mechanism reducing the spread of viruses in plant tissue (Vance et al., 1980). Several enzymes involved in lignin biosynthesis and their regulation have been studied in hypersensitively reacting tobacco leaves infected by tobacco mosaic virus (Le-

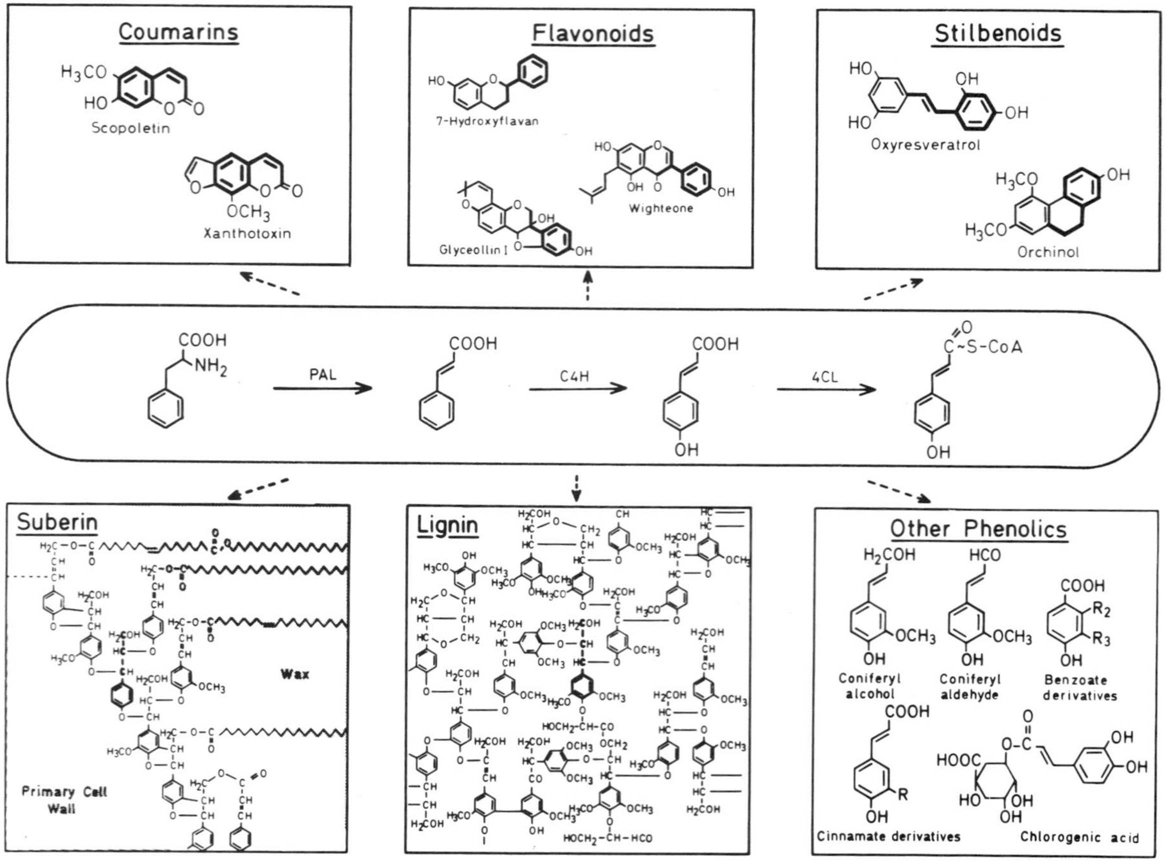

Fig. 2. Scheme illustrating the common biosynthetic origin of various defense-related phenolic materials in higher plants (Nimz, 1974; Bailey and Mansfield, 1982; Kolattukudy, 1984). Thick lines indicate some of the phenylpropanoid units derived from phenylalanine (C4H = cinnamic acid 4-hydroxylase; 4CL = 4-coumarate : CoA ligase; PAL = phenylalanine ammonia-lyase).

grand, 1983). Activities of PAL, cinnamic acid 4-hydroxylase, CoA ligases, *O*-methyltransferases, and peroxidases increase drastically in areas around necrotic lesions, with the increases spreading radially in advance of necrosis (Legrand et al., 1976). Application of AOA or α-aminooxy-β-phenylpropionic acid (AOPP) to these leaves increases the size of lesions and virus-containing areas (Massala et al., 1980).

In most cases where the deposition of polymeric phenolic material after pathogen attack is studied, insufficient chemical analysis of the newly synthesized material and structural similarities between suberin and lignin prevent a clear-cut distinction between these two polymers (Vance et al., 1980; Kolattukudy, 1981; Friend, 1985). While Kolattukudy (1981) believes that most phenolic polymers deposited in response to injury by mechanical wounding or pathogen attack represent suberin, most other authors designate this material as lignin or lignin-like. Furthermore, studies with potato tuber slices as host tissue result in a combination of wound, senescence, and resistance reactions whose relative contributions might vary with the age of the slices.

2.5. Polysaccharides of the Primary Cell Wall

The primary cell wall of plant cells consists of approximately 90% polysaccharides and 10% glycoproteins. In addition, methyl ethers and esters as well as acetyl and feruloyl esters are found (Dey and Brinson, 1984; McNeil et al., 1984). Depending on their solubility, the wall polysaccharides are separated into three classes, pectic polysaccharides, hemicelluloses, and cellulose.

Cellulose, the structural polysaccharide of cell walls, is composed of linear chains of β-(1→4)-linked D-glucosyl residues. The chains aggregate to microfibrils, which are stabilized by hydrogen bonds. In addition, gramineae contain linear chains of β-(1→3)- and β-(1→4)-linked glucosyl residues (Delmer, 1983; Dey and Brinson, 1984).

Xyloglucan is the major constituent of hemicelluloses in dicotyledonous plants. It consists of a backbone of β-(1→4)-linked D-glucosyl residues that are linked to cellulose by hydrogen bonds. Single xylosyl residues are α-(1→6)-linked to the backbone. Hemicellulose of monocotyledonous plants is mainly composed of xylan with a backbone of β-(1→4)-linked xylosyl residues and L-arabinofuranosyl groups as the most common side chains (Dey and Brinson, 1984).

The pectic fraction accounting for about 35% of the primary wall of dicotyledonous plants contains mainly galactosyluronic acid, rhamnose, arabinose, and galactose residues and is characterized by its ability to form gels. The structure and composition of pectic polymers of monocotyledonous plants are still obscure. The major components of dicotyledonous pectins are homogalacturonans, rhamnogalacturonan I and II. Homogalacturonans are insoluble unbranched polymers of α-(1→4)-linked galactosyluronic acid residues. Rhamnogalacturonan I, apparently the backbone of the pectic polymers, is composed of α-(1→4)-linked D-galactosyluronic acid and 2-linked L-rhamnosyl residues, half of the latter being branched with glucosyl substituents at the O-4 position. Rhamnogalacturonan II is the most complex plant polysaccharide, with a structure very different from that of rhamnogalacturonan I. This highly branched molecule with about 60 glycosyl residues contains large proportions of galactosyluronic acid and rhamnosyl residues, but also very rare components. Further constituents of pectic polysaccharides are pure arabinans, galactans, and arabinogalactans (Dey and Brinson, 1984).

Primary cell walls of suspension-cultured spinach cells contain ferulic and *p*-coumaric acids that are specifically attached to galactopyranose and arabinopyranose residues (Fry, 1983). The pectic fraction is estimated to contain one feruloyl residue per 60 sugar residues. Wall-bound peroxidases are believed to catalyze the cross-linking of pectic molecules via diferulate bridges and thereby to cause wall rigidification, which could also play a role in pathogen defense.

Callose, an unbranched β-(1→3)-glucan, is a normal constituent of sieve plates, pollen, pistils, and plasmodesmata of higher plants. Plant cells are able to synthesize callose rapidly in response to stress, such as wounding or pathogen attack (Aist, 1976; Bell, 1981). The formation of callose-containing papillae is frequently involved in the defense of plants against fungi and viruses (Bell, 1981). Rapid accumulation of large amounts of callose in papillae is correlated with the resistance of barley *ml-o* mutants to powdery mildew (Skou et al., 1984; Stolzenburg et al., 1984a), but the question as to whether callose finally causes resistance, has not been definitely answered (Stolzenburg et al., 1984b). In suspension cultures of soybean, callose formation together with phytoalexin accumulation, is elicited by chitosan, a D-glucosamine polymer from some fungal cell walls (Köhle et al., 1984).

Apart from forming a general barrier against attempted infections, cell-wall polysaccharides may play a role as a source of so-called "endogenous" elicitors. Homogalacturonan fragments eliciting phytoalexin biosynthesis in soybean cotyledons are liberated from soybean cell walls or citrus pectin either after partial acid hydrolysis or after treatment with endo-α-1,4-polygalacturonic acid lyase (McNeil et al., 1984). The predominant oligosaccharide of the elicitor fraction prepared by acid hydrolysis is an α-(1→4)-linked D-dodecagalacturonide, whereas enzymatic hydrolysis results in the release of deca-α-(1→4)-D-galacturonide with an unsaturated galactosyluronic acid residue at the nonreducing terminal. Pure endo-α-1,4-polygalacturonic acid lyase from *Rhizopus stolonifer*, a fungal pathogen of castor bean, releases oligogalacturonide elicitors from host-plant cell walls (West et al., 1984). Castor bean seedlings treated with this elicitor fraction synthesize the phytoalexin casbene and exhibit increased activity of casbene synthase. The most active component of the oligogalacturonide fraction is trideca-α-(1→4)-D-galacturonide. Suspension-cultured carrot cells accumulate the phytoalexin 6-methoxymellein after infection with the fungus *Chaetomium globosum* (Kurosaki and Nishi, 1983) as well as after treatment with hydrolysates obtained by pectinolytic or proteolytic digestion of carrot cells or after direct addition of pectinase, pronase, or trypsin to the culture fluid (Kurosaki and Nishi, 1984). An endogenous elicitor from melon cell walls stimulates the synthesis of ethylene and hydroxyproline-rich glycoprotein in melon petioles (Roby et al., 1985). Other fragments produced from plant cell walls by partial acid hydrolysis which are toxic to plant cells are perhaps involved in hypersensitive cell death (McNeil et al., 1984).

2.6. Glycoproteins of the Primary Cell Wall

Proteins associated with plant primary cell walls are mostly glycoproteins. They are either covalently bound or soluble to variable degrees. Some of them act as enzymes, whereas others appear to be structural components of the wall with additional functions in cell-cell recognition, cell extension, and pathogen defense.

At least three classes of glycoproteins containing large amounts of hydroxyproline are secreted by plant cells: extensin, arabinogalactan proteins, and lectins (Lamport, 1980). Extensin is secreted as soluble hydroxyproline-rich glycoprotein (HRGP) and slowly cross-linked in the cell wall via isodityrosine bridges, finally forming an insoluble wall component (Cooper and Varner, 1984). Approximately two-thirds of the soluble form consist of carbohydrates and one-third of protein. The apparent molecular weight of extensin from carrot is about 86,000 (Stuart and Varner, 1980), and a M-36,000 protein is released upon deglycosylation (van Holst and Varner, 1984). Peroxidase-catalyzed cross-linking appears to be a specific reaction since dityrosine bridges are rapidly formed *in vitro* (Cooper and Varner, 1984). The amino-acid sequence, derived from a cDNA sequence, is characterized by several repetitive units, including SerProProProPro, HisLysProPro, TyrTyrProPro, TyrLysProPro, and the TyrLysTyrLys site of an intramolecular isodityrosine bridge (Cooper et al., 1984). Extensin has an unusual amino-acid composition of approximately 41 mol % hydroxyproline, 12 mol % serine, 10 mol % lysine, 8 mol % tyrosine, and 0–11 mol % histidine (McNeil et al., 1984). The lack of acidic amino acids and large amounts of lysine and histidine result in a basic isoelectric point in the range of 10–12. The protein may be modified to soluble extensin by prolyl hydroxylase and hydroxyproline arabinosyltransferase. Both enzymes seem to be localized in the golgi apparatus (Sadava and Chrispeels, 1971; Owens and Northcote, 1981). Tri- and tetra-arabinoside side chains linked to most of the hydroxyproline residues represent 96% of the total carbohydrate portion. The remaining 4% are serine-linked single galactosyl residues.

In contrast to the insoluble extensin, arabinogalactan proteins are soluble HRGPs (McNeil et al., 1984) and are found extracellularly as well as in the cytoplasm. The most common amino acids of the acidic protein moiety are hydroxyproline, serine, alanine, and glycine. Arabinoside and arabinogalactan chains with high degrees of polymerization as well as the branched carbohydrate core of β-D-galactopyranosyl residues are linked to hydroxyproline. Arabinosides and glycosyluronic acids are further sugars attached to the core. Proline hydroxylation and arabinosyl transfer to hydroxyproline appear to occur in the same manner as described for extensin (Fincher et al., 1983).

Changes in hydroxyproline or HRGP content of the cell wall occur in several plant–pathogen interactions (Touzé and Esquerré-Tugayé, 1982), but it is not known whether these changes are causally related to resistance. In muskmelon (*Cucumis melo*) cell walls, the amount of HRGP increases tenfold after infection with the fungus *Colletotrichum lagenarium* (Esquerré-Tugayé and Lamport, 1979), concomitant with a drastic increase in ethylene production by the host (Toppan et al., 1982). Elicitors prepared either from mycelium of this fungus or from melon cell walls cause similar reactions (Roby et al., 1985). Treatment of the seedlings with ethylene prior to infection increases both, HRGP content of the wall and pathogen resistance, whereas inhibition of HRGP synthesis by 2-*trans*-hydroxyproline lowers the amount of wall-associated HRGP and enhances the colonization of the host by the pathogen (Esquerré-Tugayé et al., 1979). Specific inhibition of ethylene biosynthesis with L-canaline or aminoethoxyvinylglycine in petioles from infected or elicitor treated tissue also inhibits deposition in the cell wall of [^{14}C]hydroxyproline originating from [^{14}C]proline (Toppan et al., 1982; Roby et al., 1985). The ethylene precursor 1-aminocyclopropane-1-carboxylic acid stimulates HRGP synthesis in untreated tissue and restores ethylene and HRGP production in petioles treated with both elicitor and aminoethoxyvinylglycine (Roby et al., 1985). An elicitor from the soybean pathogen *Phytophthora megasperma* f. sp. *glycinea* stimulates the synthesis of ethylene and HRGP in soybean hypocotyls (Roby et al., 1985). Enhancement of hydroxyproline content of the wall is also observed in the incompatible but not in the compatible interaction of cucumber (*Cucumis sativus* L.) with *Cladosporium cucumerinum* (Hammerschmidt et al., 1984). In contrast, hydroxyproline concentrations in cell walls of wheat (*Triticum aestivum* L. em Thell) increase after inoculation of susceptible lines with *Erysiphe graminis* (Clarke et al., 1981).

The accumulation of HRGP and of its mRNAs in infected tissue and in elicitor-treated cell cultures has been studied in the pathosystem bean (*Phaseolus vulgaris*)/*Colletotrichum lindemuthianum* (Bolwell, 1984; Bolwell et al., 1985a; Showalter et al., 1985). Transient activity increases of prolyl hydroxylase and protein arabinosyltransferase show similar time courses. The arabinosyltransferase is associated with the Golgi complex, whereas prolyl hydroxylase was purified to homogeneity from the endoplasmic reticulum (Bolwell et al., 1985b). In this system HRGP content of the cell wall seems to increase even before phytoalexin accumulation (Robbins et al., 1985), whereas the putative extensin mRNA accumulates less rapidly (Showalter et al., 1985).

Lectins of the solanaceae are the third class of secreted HRGPs. They occur extra- and intracellularly (Sequeira, 1978; Lamport, 1980; Fincher, et al., 1983). Their protein content is about 50–60%; hydroxyproline, serine, glycine, and cysteine are the predominant amino acids. Arabinoside residues are linked to hydroxyproline, and galactosyl residues to serine. The HRGPs from potato agglutinating avirulent, but not virulent, strains of the bacterial wilt pathogen *Pseudomonas solanacearum* have been named agglutinins to distinguish them from the structurally similar lectins, because they show only low hemagglutinating activity (Leach et al., 1982a). These agglutinins are soluble components of the cell wall (Leach et al., 1982b) which contain about 61% carbohydrates (Leach et al., 1982a). The most common amino acids of their basic protein moieties are hydroxyproline, lysine, serine, and proline. The carbohydrate portion consists mainly of arabinose and small amounts of galactose, glucose, and glucosamine. The entire molecule has an apparent molecular weight of 91,000, which is reduced to 56,000 upon deglycosylation. It is unclear whether lectins play a role in plant–pathogen interactions.

Plant cell walls contain a number of enzymes which, as far as their structures have been elucidated, are glycoproteins (McNeil et al., 1984). Some of them, in particular peroxidases and glycosyl hydrolases, are believed to play a role in plant defense. Specific peroxidases may be involved in lignin biosynthesis (Asada et al., 1979) and in the formation of the isodityrosine bridges of extensin (Cooper and Varner, 1984) and the diferulate bridges of wall-bound ferulic acid residues (Fry, 1983). Though activity increases of peroxidases after pathogen attack have been demonstrated (Legrand et al., 1976; Asada et al., 1979; Vance et al., 1980; Coffey and Cassidy, 1984), no specific involvement of any isoenzyme in defense-related processes has been shown with certainty. Glycosyl hydrolases as common components of plant cell walls (McNeil et al., 1984) may be responsible for the release of elicitors from fungal or plant cell walls or for the inactivation of elicitors.

3. PHYTOALEXINS

In addition to the structural barriers discussed above, plants contain a large diversity of other defense materials. Some of them are normal constituents of plant cells (Friend, 1981; Mansfield, 1983; Stoessl, 1983), whereas others, including phytoalexins, are often newly synthesized after pathogen attack as one important part of the multicomponent defense response. Several recent reviews deal extensively with phytoalexins (Bailey and Mansfield, 1982; Dixon et al., 1983; Grisebach, 1983; Darvill and Albersheim, 1984; Friend and Rathmell, 1984; Kuć and Rush, 1985). We will only summarize and briefly discuss some important aspects.

By definition, "phytoalexins are low molecular weight antimicrobial compounds that are both synthesized by and accumulate in plants after their exposure to microorganisms" (Paxton, 1982). Compounds released from immediate precursors during the infection process are not included in this definition. Evidence for an important role of phytoalexins in resistance is not required by this definition and is still an unsolved problem for most plant–pathogen interactions. Direct evidence has been obtained only in one system. A genetical approach has been applied by VanEtten's group (VanEtten et al., 1980; Tegtmeier and

VanEtten, 1982; Kistler and VanEtten, 1984) to show that pisatin plays a decisive role in resistance of pea to *Nectria haematococca*, because only pathogen strains that are able to detoxify the phytoalexin invade the plant.

Using immunohistochemical methods Grisebach and co-workers have demonstrated that tissue distribution and timing of induction of phytoalexin synthesis correlate with cultivar resistance/susceptibility of soybean to *Phytophthora megasperma* f. sp. *glycinea* (Hahn et al., 1985). Staskawicz et al. (1984) were able to identify an avirulence gene of *Pseudomonas syringae* pv. *glycinea*, race 6, by transposon mutagenesis. This gene confers race-specific avirulence to different *P. syringae* strains. In all incompatible interactions of the transformed strains phytoalexin accumulation was comparable to that induced by race 6 in the respective soybean cultivars.

Phytoalexins have not been observed in all infected plant tissues, but it is possible that insufficient isolation or detection methods were used in some cases. Generally phytoalexins of the same chemical class are synthesized within one plant family. Altogether, a large structural diversity of phytoalexins exists within the plant kingdom, as indicated by only a few examples in Fig. 3. The total amount of phytoalexins, the timing of induction, and the relative amounts of the individual components vary with the species or race of the infecting microorganism, the plant tissue, and the physiological and environmental conditions.

Phytoalexin production, together with other defense reactions, can be induced by various elicitors of microbial or plant origin and by a number of other

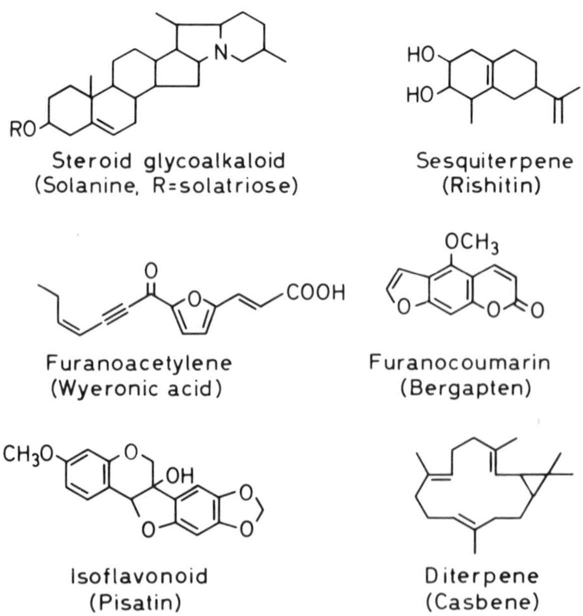

Fig. 3. Selected phytoalexins of different biosynthetic origin.

chemical and physical stimuli. As further discussed below (Section 6), the existence of specific elicitors for phytoalexins is doubtful.

The accumulation of phytoalexins upon infection of resistant hosts is often associated with rapid cell death, which is usually confined to small local lesions in the infected tissue (hypersensitive response). One much discussed, but unanswered question, is whether there is a causal relationship between the two events. It is conceivable that phytoalexins, which are toxic to plant cells as well as to microorganisms, are one, or perhaps the, causal agent of cell death in host tissue. However, it is also possible that phytoalexin accumulation and cell death occur independently of each other, or that cell death and the concomitant release of endogenous elicitors (see Section 2.5) trigger phytoalexin synthesis in the cells directly surrounding the necrotic tissue. Arguments in favor of this third possibility have been discussed by Bailey (1982) and Mansfield (1984), but all indications are circumstantial. Direct evidence for a spatial separation and consecutive occurrence of cell death and phytoalexin synthesis is lacking. A conclusive answer to this question would require that both the timing and the location of the two events relative to each other would be determined, and this has been impossible for three reasons. First, the precise localization of phytoalexin synthesis in individual cells of infected tissue has not been determined; second, accurate methods for the measurement of very small amounts of phytoalexins during the early stages of induction have just become available (Moesta et al., 1982; Hahn et al., 1985); and third, cell death is a dynamic process and therefore its moment of occurrence is difficult to measure precisely with sufficient accuracy.

It should be noted that it has not been possible to give a clear-cut biochemical definition of the hypersensitive response, although this response is the most obvious, and probably the most important, event in disease resistance of plants. Neither browning of the tissue, whose biochemical mechanisms are not understood in any detail, nor staining for viability of the cells would be sufficiently accurate measures for such a highly complex and dynamic phenomenon as the hypersensitive response, which might well be the visible result of the sum of all of the plant's defense reactions.

4. LYTIC ENZYMES

In addition to the inhibition of growth or even the killing of microorganisms with toxic low-molecular-weight substances, plants are capable of excreting enzymes to degrade fungal and bacterial cell walls. Two such enzymes have so far been studied, chitinase and β-1,3-glucanase. Both of them have endoglycolytic activities directed against cell-wall components of many microorganisms (Fig. 4). Various plant systems have been reported to respond to infections, or to treatments with elicitor or ethylene (see below), with increased rates of production of these two enzymes (Boller et al., 1983; Mauch et al., 1984; Chappell et al., 1984; and references cited therein).

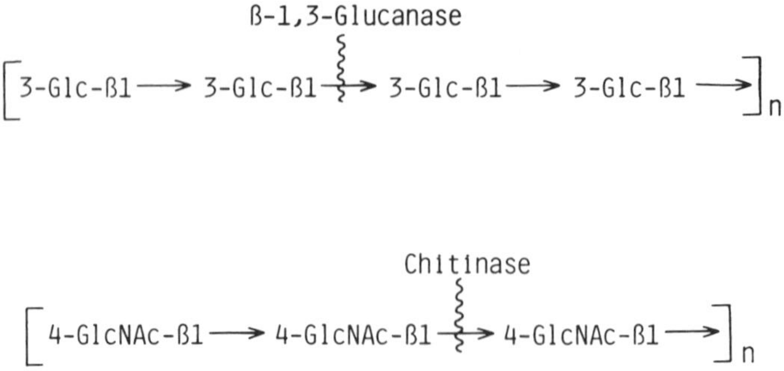

Fig. 4. Modes of action of β-1,3-glucanase and chitinase (Glc = glucose; GlcNAc = *N*-acetylglucosamine).

Chitinase from higher plants has the interesting property of also acting as a lysozyme (Boller et al., 1983). Thus the same enzyme is probably involved in the degradation of both fungal and bacterial cell walls, despite the different chemical nature of the respective substrates. Since substrates of plant origin which could be attacked by chitinase are not known, it is likely that a major function of this enzyme is to act in the defense against pathogens.

An unequivocal assessment of the function(s) of β-1,3-glucanase is not yet possible. Possible substrates, β-1,3-linked glucose oligomers or polymers, occur as cell-wall components in microorganisms as well as in plants, for example, in the form of callose. The concomitant activity increases of chitinase and β-1,3-glucanase in many plant–pathogen interactions suggest that both enzymes take part in the plants' defense responses. It cannot be excluded, however, that β-1,3-glucanase has an additional role in the plants' own metabolism.

5. PATHOGENESIS-RELATED PROTEINS

The term "pathogenesis-related protein" (PR protein) has been applied to several characteristic proteins of unknown function whose synthesis is induced upon infection of plants with fungi, bacteria, viruses, or viroids. They have rather low molecular weights (approximately 14,000–30,000), and some of the most prominent ones are quite acidic (pI values around 4) (Van Loon, 1985). However, less acidic, neutral, and even very basic PR proteins (pI values > 10) have also been isolated (e.g., from tomato) (Camacho-Henriquez and Sänger, 1982a,b). Each plant expresses a characteristic set of PR proteins, which are usually detected by comparison of protein extracts from uninfected and infected tissues on one-dimensional or two-dimensional gels. Preferential solubility of some of the PR proteins at low pH and resistance to various proteinase treatments has aided their isolation and characterization from some of the most thoroughly studied systems—tobacco, tomato, and bean (for further details see Camacho-Henriquez and Sänger, 1982a,b; Van Loon, 1985; Matsuoka and Ohashi, 1984).

Unfortunately, the nomenclature used in the literature by different authors is confusing and is mostly based on migration properties on gels, elution behavior from chromatography columns, or molecular weight. As far as comparisons within and between species have been attempted, it appears that immunological cross-reactivity between species is poor (but has been observed) and occurs within species only for some of the PR proteins (Van Loon, 1985; Matsuoka and Ohashi, 1984). Two recently identified cDNA clones were found to be homologous to mRNAs from parsley, which coded for one and three distinct PR proteins, respectively, as analyzed by hybrid-select translation and comparison with authentic PR proteins from elicitor-treated parsley cell cultures (Somssich et al., 1986). Taken together, the restricted cross-reactivities of PR proteins with antisera and of PR-protein mRNAs with cDNAs suggest that at least some of them are closely related but structurally distinct products encoded by small gene families.

The synthesis of PR proteins is induced not only by infection with pathogens or treatment with pathogen-derived elicitors, but also by other causes of pathogenesis, such as exposure of plant tissues or cultured plant cells to various inorganic or organic chemicals (Van Loon, 1983). Furthermore, PR proteins may be formed under normal conditions during certain stages of development, such as flowering or leaf senescence. Their localization within cells or intercellular spaces, or both, has not been fully established, although massive accumulations in intercellular spaces have been reported (for references see Van Loon, 1985). Thus, although PR proteins are clearly different from heat-shock proteins, the two protein classes are analogous in many respects. Both of them fall into subclasses encoded by gene families, their synthesis is inducible by specific environmental stresses, including certain chemicals, and they are synthesized in the absence of apparent exogenous treatments during specific stages of development. Members of each class seem to occur in different intra- and intercellular locations and, perhaps partly for this reason, have been difficult to correlate with specific biochemical functions.

6. SIGNAL MOLECULES

The schematic view of a host–pathogen interaction given in Fig. 1 includes the action of elicitors as signal molecules. Elicitors are commonly defined as compounds inducing phytoalexin synthesis, but since they have frequently been found to induce other defense reactions as well (e.g., Kombrink and Hahlbrock, 1986), and since elicitors of other defense reactions usually also induce phytoalexin synthesis, the term elicitor will here be applied in all cases where a typical defense reaction is induced.

We have discussed above (Section 2.5) the possibility that endogenous elicitors, generated from the plant's own cell wall by microbial enzymes or by activation of plant enzymes in the course of plant–pathogen interactions, may have an important role in the induction of defense reponses in plants. Recent observations indicating that such endogenous elicitors act synergistically with pathogen-derived elicitors support this hypothesis (see below).

Many different types of elicitor from exogenous sources, biotic as well as abiotic, induce phytoalexin synthesis in plant tissue. The chemical nature of these elicitors seems to be extremely variable. They range from simple inorganic substances, such as copper or mercury salts, to complex carbohydrates, lipids, and proteins (Darvill and Albersheim, 1984). Since by the usual definition (Paxton, 1982), all molecules inducing phytoalexin synthesis in plant cells are called elicitors, even enzymes triggering this response have been included in this class of compounds. The best-studied example is endopolygalacturonic acid lyase (Lee and West, 1981), which was subsequently shown to act through the release of endogenous elicitor-active oligogalacturonides from plant cell walls (Davis et al., 1984; Jin and West, 1984). A similar effect has recently been reported for formate, acetate, and propionate buffers (Davis et al., 1986), whose action as abiotic (exogenous) elicitors might thus be an indirect one, that is, to induce the formation of the more directly acting endogenous elicitor(s). Both examples demonstrate the imprecise definition of elicitors, resulting from the lack of detailed knowledge of their mechanisms of action.

An interesting hypothesis postulates the synergistic action of microbial (exogenous) and plant-derived (endogenous) elicitors (Davis et al., 1986). According to this hypothesis, relatively low concentrations of the two combined signal molecules trigger a large defense response, as compared with the effects of each component alone. Such synergistic actions would greatly increase the variation potential of the responding plant cell, especially if even more than two different signal molecules would act synergistically in various numbers and combinations. Alternatively, if the ability of a given cell to respond to a given signal is rather fixed and largely determined by its differentiation state and environmental conditions, different cells might respond to different signals.

Although many elicitor-active molecules, including enzymatically released and chemically defined fragments of microbial (Sharp et al., 1984) and plant cell walls (Davis et al., 1984) have been described, only one defined elicitor with race-cultivar specificity has so far been isolated from its natural site of action. Intercellular fluids of tomato leaves inoculated with different races of the fungal pathogen *Cladosporium fulvum* contains race-specific elicitors produced by the fungus (de Wit et al., 1985). One of them has been partially purified and shown to be a peptide with an apparent molecular weight of about 5500. It seems to be a product of a specific avirulence gene and induces necrosis in tomato cultivars carrying the corresponding resistance gene. The peptide may thus be regarded as a resistance-inducing toxin (de Wit et al., 1985), as opposed to the more frequently described host-specific toxins inducing susceptibility (Daly and Knoche, 1982). The question arises again in this context as to which criterion, or criteria, should be used to best define a naturally occurring elicitor.

Other signal molecules in plant–pathogen interactions, especially those transmitting the elicitor signal intracellularly, have not been identified. The site and mode of action and the ultimate fate of the elicitor have not been determined and might vary with its chemical nature as well as with the biological system. Ethylene is often induced very rapidly and to a large extent upon infection of plant tissue, but its role is doubtful (Mauch et al., 1984). The induction of chitinase (Boller et al., 1983) and of HRGP (see Section 2.6) in ethylene-treated plants has been reported. Various metabolic control substances, such as Ca^{2+} and inositol phosphates, might be involved in signal transmission. However, the establishment of a specific role for such common cellular components requires that specific and secondary effects are clearly distinguished.

7. REGULATORY MECHANISMS

The multicomponent nature of defense responses in plants and their variability in magnitude, timing, and relative composition with environmental conditions, genetic constitution, and differentiation state of a given cell make it difficult to establish whether one particular reaction is regulated by the same mechanism under all circumstances. Moreover, the same signal might, directly or indirectly, trigger different mechanisms of activation of biochemical reactions, either in the same or in neighboring cells of infected tissue. It is, furthermore, important to note that the infection site may consist of different cell types, possibly with different contributions to the overall resistance response. Individual cells at infection sites have not been studied with regard to the regulatory mechanisms of the resistance response. To circumvent the technical difficulties of such studies, the large complexity of plant–pathogen interactions has so far often been reduced to treatment of relatively homogenous populations of cultured plant cells with pathogen-derived elicitors.

Two such combinations have so far been used to demonstrate that induction occurs at the level of gene expression. Run-off transcription experiments with isolated nuclei from *Petroselinum crispum* cells that had been treated with an elicitor preparation from *Phytophthora megasperma* f. sp. *glycinea* revealed a rapid activation of genes encoding two enzymes of phenylpropanoid metabolism, phenylalanine ammonia-lyase and 4-coumarate:CoA ligase (Chappell and Hahlbrock, 1984), several PR proteins (Somssich et al., 1986), and many other unidentified proteins (Bollmann et al., unpublished results). Using a different technique, which is based on the isolation of newly synthesized 4-thiouridine-labeled RNA by organomercurial affinity chromatography, Cramer et al. (1985) have also demonstrated phenylalanine ammonia-lyase gene activation, together with the activation of genes encoding chalcone synthase, chalcone isomerase, and a large number of unidentified proteins, upon treatment of cultured *Phaseolus vulgaris* cells with elicitor from *Colletotrichum lindemuthianum*. These results are in line with observations that mRNAs encoding phytoalexin-biosynthetic enzymes in whole-plant tissues of *Phaseolus vulgaris* (Bell et al., 1984; Ryder et al., 1984; Cramer et al., 1985) and *Glycine max* (Schmelzer et al., 1984) are rapidly induced upon incompatible interactions with host-specific pathogens. Thus there is strong, though indirect, evidence that regulation of these and probably several other pathways, for example, HRGP synthesis (Section 2.6), is exerted through gene activation in infected tissue.

It is likely that not all pathogen-induced changes in metabolic activities of plants are controlled at the level of transcription. Possible examples are elicitor-induced increases in β-1,3-glucan synthase activity in *Glycine max* cells (Köhle et al., 1985) and in the activity of 1-aminocyclopropane-1-carboxylic acid synthase, an enzyme of ethylene biosynthesis, in *Petroselium crispum* cells (Chappell et al., 1984). These increases are much more rapid than those of the enzyme activities mentioned above. In the case of β-1,3-glucan synthase in *Glycine max* cells, convincing data have been presented indicating that a preformed enzyme is activated by an allosteric mechanism (Köhle et al., 1985).

8. OPEN QUESTIONS

This brief summary of the most frequently observed biochemical defense reactions of plants to pathogens intends to emphasize how far we still are from being able to deduce a universal picture from the available data. Many thorough and very detailed studies have been made of certain aspects of the cytology, physiology, or biochemistry of a rather large diversity of systems. However, an overall view is still lacking, even for the most extensively investigated systems. This is certainly due not only to the very high complexity of the interaction of two very different and highly differentiated organisms, but also to the further complication by the large dependence of all interactions on genotype, cell type, and environmental conditions. Other reasons are the abovementioned technical difficulties associated with the measurement of biochemical reactions occurring in only one or a small number of cells.

These difficulties are expected to be overcome very soon by the development of new or improved methods of measuring the concentrations of various types of defense-related compounds *in situ* (in defined single cells or intercellular spaces). Interest will range from signal molecules to the immediate (RNA), subsequent (enzymes, other proteins), and final products of gene activation (phytoalexins, wall-bound phenolics, other cell-wall components, etc.) or metabolites resulting from other activation mechanisms.

The availability of such techniques will then undoubtedly prompt comprehensive studies of a sufficient number of systems to distinguish common from species-specific reactions, and possibly direct from indirect effects. The results will, perhaps, allow general conclusions to be drawn as to whether, or how, an average defense response can be defined biochemically. Of the many defense reactions so far described, only phytoalexin accumulation has been studied in a large number of species, but many more careful investigations of individual cells and small sections of infection sites like those conducted by Moesta et al. (1982) and Hahn et al. (1985) will be necessary to enable a better assessment of the specific role of phytoalexins in disease resistance.

Some of the major questions to be addressed by investigations involving *in situ* measurements are:

1. Is resistance to pathogens invariably associated with induced phytoalexin accumulation, or with the induction or activation of some or all of the other frequently observed defense reactions?
2. Is the timing of the response the same for all activated genes?
3. Is there a difference in timing, quality, or quantity of the response between individual cells of infection sites?
4. What is the difference in timing, quality, and quantity of the response between resistant and susceptible cultivars of the same host to a given race of a pathogen?
5. Is there a difference in timing, quality, and quantity of the response between host and nonhost resistance of a given plant cell?
6. Is there a difference in timing, quality, and quantity of the response of the same resistant host plant to different pathogens, or of the same nonhost plant to different nonpathogens?

These are only a few of the immediately arising questions regarding the biochemistry of plants' responses to pathogens. There are many more, not only in biochemistry, but also in related fields, such as physiology, cytology, and genetics. Even a brief discussion of these additional questions regarding related fields would be beyond the scope of this chapter. However, in view of the following conclusions, three particularly important points should be emphasized. First, physiological and environmental conditions can have large effects on the resistance response to a pathogen. Second, none of the many operationally defined "resistance genes" has been identified with regard to its biochemical function. Circumstantial evidence (Ellingboe, 1982) suggests that "resistance genes" or their products have direct or indirect regulatory functions in the expression of the resistance response, rather than coding for one of the enzymes involved in the defense-related pathways. And third, the relative importance of individual defense materials in disease resistance might vary from system to system. So far, only two systems have been studied in sufficient detail to ascribe a decisive role to one particular component of the defense response: phytoalexins in the resistance of *Pisum sativum* cultivars to avirulent, phytoalexin demethylase-deficient races of the fungus *Nectria haematococca* (Kistler and VanEtten, 1984), and cutin in the resistance of the same plant to avirulent, cutinase-deficient races of the fungus *Fusarium solani* f. sp. *pisi* (Kolattukudy and Soliday, 1984).

9. CONCLUSIONS REGARDING NONCONVENTIONAL APPROACHES TO DISEASE CONTROL

In the context of nonconventional approaches to disease control, it is worth noting that previous (conventional) approaches had little connection with biochemical research. It is, however, foreseeable that this situation will change drastically. So far, disease control has mainly been a matter of plant breeding and agricultural practices, which more and more include the application of synthetic fungicides and other agrochemicals in rapidly increasing amounts. There is an urgent need for improvement, complementation, and partly replacement, of these approaches by new methods. The extensive monocultivation of specially selected cultivars of crop species enables various pests, not least new virulent pathogen races, to develop rapidly, and the massive use of agrochemicals meets with increasingly large ecological problems.

In all likelihood, biochemistry will contribute much to new (nonconventional) approaches. At least three areas of biochemical research might soon lead to practical applications. One area would be to study enzymes with specific roles in the virulence of a pathogen to inhibit their activities with selectively acting chemicals. The inhibition of cutinase by organophosphorous and other synthetic or natural compounds to protect *Pisum sativum* plants from infection by otherwise virulent races of *Fusarium solani* f. sp. *pisi* (Köller et al., 1982) might serve as one example.

A second area of intensified biochemical research will probably be concerned with synergistically acting effectors ("sensitizers") of naturally occurring defense reactions. Although a few such cases have been observed, it will be necessary to study these phenomena in sufficient detail to fully exploit their potentials and design specific effector molecules with minimized danger of potential side effects and of toxicological and ecological problems. This topic is discussed more thoroughly in the chapter by Baker. A large variety of reactions involved in plant–pathogen interactions, including the early events of recognition, binding, and elicitation, might be specifically affected by exogenously applied substances (Rathmell, 1984).

Finally and probably most importantly, the new methods of recombinant DNA technology will very likely have a large impact in plant disease control. The ultimate goal is to transfer single, biochemically defined genes, or combinations of defined genes, to plants that are susceptible to a certain pathogen, thus conferring at least cultivar resistance, and possibly even nonhost resistance, on the transformed plants. This will require the isolation, structural identification, and elucidation of the mechanisms of action of both the operationally defined genes for cultivar resistance ("resistance genes") and the unidentified genes responsible for nonhost resistance. Of course, it is possible that approaches not involving detailed molecular studies of the genes themselves and their modes of action, such as plant transformation using "shot-gun cloning" procedures or genes operationally identified and isolated by "tagging" with transposable elements, will soon be successful. On a long-term basis, however, these approaches will probably not be applicable to the most desirable case of nonhost resistance and will generally be limited in applicability if a basic understanding of the molecular mechanisms is lacking.

Hence further investigations of the biochemical reactions outlined in this chapter, extension of these studies to unidentified reactions of the defense response to pathogens, and, with priority, studies of the regulatory mechanisms controlling disease resistance to identify and characterize the responsible genes will provide the basis for the expected impact of new approaches in plant disease control. As in animal and human medicine, the better the molecular mechanisms of symptom development and mode of actions of effectors are understood, the more successfully a cure, or even the eradication of a disease, can usually be effected.

REFERENCES

Aist, J. R. (1976). Papillae and related wound plugs of plant cells. *Ann. Rev. Phytopathol.*, **14:**145-163.

Asada, Y., T. Ohguchi, and I. Matsumoto (1979). Induction of lignification in response to fungal infection. In *Recognition and Specificity in Plant Host-Parasite Interactions* (J. M. Daly and I. Uritani, eds.). Tokyo University Press, Tokyo, pp. 99-112.

Bailey, J. A. (1982). Mechanisms of phytoalexin accumulation. In *Phytoalexins* (J. A. Bailey and J. W. Mansfield, eds.). Blackie, Glasgow, pp. 289-318.

Bailey, J. A. (1983). Biological perspectives of host-pathogen interactions. In *The Dynamics of Host Defence* (J. A. Bailey and B. J. Deverall, eds.). Academic, New York, pp. 1-32.

Bailey, J. A. and J. W. Mansfield (eds.) (1982). *Phytoalexins*. Blackie, Glasgow.

Bell, A. A. (1981). Biochemical mechanisms of disease resistance. *Ann. Rev. Plant Physiol.*, **32:**21-81.

Bell, J. N., R. A. Dixon, J. A. Bailey, P. M. Rowell, and C. J. Lamb (1984). Differential induction of chalcone synthase mRNA activity at the onset of phytoalexin accumulation in compatible and incompatible plant-pathogen interactions. *Proc. Natl. Acad. Sci. USA*, **81:**3384-3388.

Boller, T., A. Gehri, F. Mauch, and U. Vögeli (1983). Chitinase in bean leaves: Induction by ethylene, purification, properties, and possible function. *Planta*, **157:**22-31.

Bolwell, G. P. (1984). Differential patterns of arabinosylation by membranes of suspension-cultured cells of *Phaseolus vulgaria* (French bean) after subculture or elicitation. *Biochem. J.*, **222:**427-435.

Bolwell, G. P., M. P. Robbins, and R. A. Dixon (1985a). Metabolic changes in elicitor-treated bean cells. Enzymic responses associated with rapid changes in cell wall components. *Eur. J. Biochem.*, **148:**571-578.

Bolwell, G. P., M. P. Robbins, and R. A. Dixon (1985b). Elicitor-induced prolyl hydroxylase from French bean (*Phaseolus vulgaris*). Localization, purification and properties. *Biochem. J.*, **299:**693-699.

Camacho-Henriquez, A. and H. L. Sänger (1982a). Gelectrophoretic analysis of phenol-extractable leaf proteins from different viroid/host combinations. *Arch. Virol.*, **74:**167-180.

Camacho-Henriquez, A. and H. L. Sänger (1982b). Analysis of acid-extractable tomato leaf proteins after infection with a viroid, two viruses and a fungus and partial purification of the "pathogenesis-related" protein p14. *Arch. Virol.*, **74:**181-196.

Chappell, J. and K. Hahlbrock (1984). Transcription of plant defence genes in response to UV light or fungal elicitor. *Nature*, **311:**76–78.

Chappell, J., K. Hahlbrock, and T. Boller (1984). Rapid induction of ethylene biosynthesis in cultured parsley cells by fungal elicitor and relationship to the induction of phenylalanine ammonia-lyase. *Planta*, **161:**475–480.

Clarke, J. A., N. Lisker, D. T. A. Lamport, and A. H. Ellingboe (1981). Hydroxyproline enhancement as a primary event in the successful development of *Erysiphe graminis* in wheat. *Plant Physiol.*, **67:**188–189.

Coffey, M. D. and D. S. M. Cassidy (1984). Peroxidase activity and induced lignification in rusted flax interactions varying in their degree of incompatibility. *Can. J. Bot.*, **62:**134–141.

Cooper, J. B. and J. E. Varner (1984). Cross-linking of soluble extensin in isolated cell walls. *Plant Physiol.*, **76:**414–417.

Cooper, J. B., J. A. Chen, and J. E. Varner (1984). The glycoprotein component of plant cell walls. In *Structure, Function, and Biosynthesis of Plant Cell Walls* (W. M. Dugger and S. Bartnicki-Garcia, eds.). American Society of Plant Physiologists, Rockville, pp. 75–88.

Cramer, C. L., T. B. Ryder, J. N. Bell, and C. J. Lamb (1985). Rapid switching of plant gene expression induced by fungal elicitor. *Science*, **227:**1240–1243.

Daly, J. M. and H. W. Knoche (1982). The chemistry and biology of pathotoxins exhibiting host-selectivity. *Adv. Plant Pathol.*, **1:**83–138.

Darvill, A. G. and P. Albersheim (1984). Phytoalexins and their elicitors—A defense against microbial infection in plants. *Ann. Rev. Plant Physiol.*, **35:**243–275.

Davis, K. R., G. D. Lyon, A. G. Darvill, and P. Albersheim (1984). Host-pathogen interactions. XXV. Endopolygalacturonic acid lyase from *Erwinia carotovora* elicits phytoalexin accumulation by releasing plant cell wall fragments. *Plant Physiol.*, **74:**52–60.

Davis, K. R., A. G. Darvill, and P. Albersheim (1986). Host-pathogen interactions. XXXI. Several biotic and abiotic elicitors act synergistically in the induction of phytoalexin accumulation in soybean. *Plant Mol. Biol.*, **6:**23–32.

De Wit, P. J. G. M., A. E. Hofman, G. C. M. Velthuis, and J. A. Kuć (1985). Isolation and characterization of an elicitor of necrosis isolated from intercellular fluids of compatible interactions of *Cladosporium fulvum* (syn. *Fulvia fulva*) and tomato. *Plant Physiol.*, **77:**642–647.

Delmer, D. P. (1983). Biosynthesis of cellullose. *Adv. Carbohydr. Chem. Biochem.*, **41:**105–153.

Dey, P. M. and K. Brinson (1984). Plant cell-walls. *Adv. Carbohydr. Chem. Biochem.*, **42:**265–382.

Dixon, R. A., P. M. Dey, and C. J. Lamb (1983). Phytoalexins: Enzymology and molecular biology. *Adv. Enzymol.*, **55:**1–136.

Ellingboe, A. H. (1982). Genetical aspects of active defence. In *Active Defense Mechanisms in Plants* (R. K. S. Wood, ed.). Plenum, New York, pp. 179–192.

Ellingboe, A. H. (1984). Genetics of host-parasite relations: an assay. *Adv. Plant Pathol.*, **2:**131–151.

Eriksson, Ö., D. A. I. Goring, and B. O. Lindgren (1980). Structural studies on the chemical bonds between lignins and carbohydrates in spruce wood. *Wood Sci. Technol.*, **14:**267–279.

Esquerré-Tugayé M.-T. and D. T. A. Lamport (1979). Cell surfaces in plant-microorganism interactions. I. A structural investigation of cell wall hydroxyproline-rich glycoproteins which accumulate in fungus-infected plants. *Plant Physiol.*, **64:**314–319.

Esquerré-Tugayé, M.-T., C. Lafitte, D. Mazau, A. Toppan, and A. Touzé (1979). Cell surfaces in plant-microorganism interaction. II. Evidence for the accumulation of hydroxyproline-rich glycoproteins in the cell wall of diseased plants as a defense mechanism. *Plant Physiol.*, **64:**320–326.

Fernando, G., W. Zimmermann, and P. E. Kolattukudy (1984). Suberin-grown *Fusarium solani* f. sp. *pisi* generates a cutinase-like esterase which depolymerizes the aliphatic components of suberin. *Physiol. Plant Pathol.*, **24:**143–155.

Fincher, G. B., B. A. Stone, and A. E. Clarke (1983). Arabinogalactan-proteins: Structure, biosynthesis, and function. *Ann. Rev. Plant Physiol.*, **34:**47–70.

Freudenberg, K. (1965). Lignin: its constitution and formation from *p*-hydroxycinnamyl alcohols. *Science*, **148:**595–600.

Friend, J. (1981). Plant phenolics, lignification and plant disease. *Progr. Phytochem.*, **7:**197–261.

Friend, J. (1985). Phenolic substances and plant disease. In *The Biochemistry of Plant Phenolics. Annual Proceedings of the Phytochemical Society of Europe*, Vol. 25 (C.F. van Sumere and P. J. Lee, eds.). Oxford University Press, Oxford, pp. 367–392.

Friend, J. and W. G. Rathmell (1984). Phytoalexins. In *Antimicrobials and Agriculture* (M. Woodbine, ed.). Butterworths, London, pp. 95–115.

Fry, S. C. (1983). Feruloylated pectins from the primary cell wall: Their structure and possible functions, *Planta*, **157:**111–123.

Grisebach, H. (1981). Lignins. In *The Biochemistry of Plants*, Vol. 7 (E. E. Conn, ed.). Academic, New York, pp. 457–478.

Grisebach, H. (1983). Phytoalexins. In *Secondary Metabolism and Differentiation in Fungi* (I. L. Bennet and A. Ciegler, eds.). Marcel Dekker, New York, pp. 377–428.

Hahn, M. G., A. Bonhoff, and H. Grisebach (1985). Quantitative localization of the phytoalexin glyceollin I in relation to fungal hyphae in soybean roots infected with *Phytophthora megasperma* f. sp. *glycinea*. *Plant Physiol.*, **77:**591–601.

Hammerschmidt, R. (1984). Rapid deposition of lignin in potato tuber tissue as a response to fungi non-pathogenic on potato. *Physiol. Plant Pathol.*, **24:**33–42.

Hammerschmidt, R., D. T. A. Lamport, and E. P. Muldoon (1984). Cell wall hydroxyproline enhancement and lignin deposition as an early event in the resistance of cucumber to *Cladosporium cucumerinum*. *Physiol. Plant Pathol.*, **24:**43–47.

Jin, D. F. and C. A. West (1984). Characteristics of galacturonic acid oligomers as elicitors of casbene synthetase activity in castor bean seedlings. *Plant Physiol.*, **74:**989–992.

Keen, N. T. and L. J. Littlefield (1979). The possible association of phytoalexins with resistance gene expression in flax to *Melampsora lini*. *Physiol. Plant Pathol.*, **14:**265–280.

Kistler, H. C. and H. D. VanEtten (1984). Regulation of pisatin demethylation in *Nectria haematococca* and its influence on pisatin tolerance and virulence. *J. Gen. Microbiol.*, **130:**2605–2613.

Köhle, H., D. H. Young, and H. Kauss (1984). Physiological changes in suspension-cultured soybean cells elicited by treatment with chitosan. *Plant Sci. Lett.*, **33:**221–230.

Köhle, H., W. Jeblick, F. Poten, W. Blaschek, and H. Kauss (1985). Chitosan-elicited callose synthesis in soybean cells as a Ca^{2+}-dependent process. *Plant Physiol.*, **77:**544–551.

Kolattukudy, P. E. (1980). Cutin, suberin, and waxes. In *The Biochemistry of Plants*, Vol. 4 (P. K. Stumpf, ed.). Academic, New York, pp. 571–645.

Kolattukudy, P. E. (1981). Structure, biosynthesis, and biodegradation of cutin and suberin. *Ann. Rev. Plant Physiol.*, **32:**539–567.

Kolattukudy, P. E. (1984). Fungal penetration of defensive barriers of plants. In *Structure, Function, and Biosynthesis of Plant Cell Walls* (W. M. Dugger and S. Bartnicki-Garcia, eds.). American Society of Plant Physiologists, Rockville, pp. 302–343.

Kolattukudy, P. E. and C. L. Soliday (1984). Biochemistry of barrier layers erected in response to stress. *J. Cell Biochem., Suppl.*, **8B:**228.

Köller, W., C. R. Allan, and P. E. Kolattukudy (1982). Protection of *Pisum sativum* from *Fusarium solani* f. sp. *pisi* by inhibition of cutinase with organophosphorus pesticides. *Phytopathology*, **72:**1425–1430.

Kombrink E. and K. Hahlbrock (1986). Responses of cultured parsley cells to elicitors from phytopathogenic fungi. Timing and dose dependency of elicitor-induced reactions. Plant Physiol., **81:**(in press).

Kubo, I. and A. Matsumoto (1984). Secreted oleanolic acid on the cuticle *Olea europaea* (Oleaceae); a chemical barrier to fungal attack. *Experientia*, **40:**937–938.

Kuć, J. and J. S. Rush (1985). Phytoalexins. *Arch. Biochem. Biophys.*, **236:**455–472.

Kurosaki, F. and A. Nishi (1983). Isolation and antimicrobial activity of the phytoalexin 6-methoxymellein from cultured carrot cells. *Phytochemistry*, **22:**669–672.

Kurosaki, F. and A. Nishi (1984). Elicitation of phytoalexin production in cultured carrot cells. *Physiol. Plant Pathol.*, **24:**169–176.

Lamport, D. T. A. (1980). Structure and function of plant glycoproteins. In *The Biochemistry of Plants*, Vol. 3 (J. Preiss ed.). Academic, New York, pp. 501–541.

Leach, J. E., M. A. Cantrell, and L. Sequeira (1982a). Hydroxyproline-rich bacterial agglutinin from potato. Extraction, purification, and characterization. *Plant Physiol.*, **70:**1353–1358.

Leach, J. E., M. A. Cantrell, and L. Sequeira (1982b). A hydroxyproline-rich bacterial agglutinin from potato: its localization by immunofluorescence. *Physiol. Plant Pathol.*, **21:**319–325.

Lee, S.-C. and C. A. West (1981). Polygalacturonase from *Rhizopus stolonifer*, an elicitor of casbene synthetase activity in castor bean (*Ricinus communis* L.) seedlings. *Plant Physiol.*, **67:**633–639.

Legrand, M. (1983). Phenylpropanoid metabolism and its regulation in disease. In *Biochemical Plant Pathology* (J. A. Callow, ed.). Wiley, New York, pp. 367–384.

Legrand, M., B. Fritig, and L. Hirth (1976). Enzymes of the phenylpropanoid pathway and the necrotic reaction of hypersensitive tobacco to tobacco mosaic virus. *Phytochemistry*, **15:**1353–1359.

Mansfield, J. W. (1983). Antimicrobial compounds. In *Biochemical Plant Pathology* (J. A. Callow, ed.). Wiley, New York, pp. 237–265.

Mansfield, J. W. (1984). Plant cell death during infection by fungi. In *Cell Ageing and Cell Death* (I. Davies and D. C. Sigee, eds.). Cambridge University Press, Cambridge, pp. 323–345.

Martin, J. T. (1964). Role of cuticle in the defense against plant disease. *Ann. Rev. Phytopathol.*, **2:**81–100.

Massala, R., M. Legrand, and B. Fritig (1980). Effect of α-aminooxyacetate, a competitive inhibitor of phenylalanine ammonia-lyase, on the hypersensitive resistance of tobacco to tobacco mosaic virus. *Physiol. Plant Pathol.*, **16:**213–226.

Matsuoka, M. and Y. Ohashi (1984). Biochemical and serological studies of pathogenesis-related proteins of *Nicotiana* species. *J. Gen. Virol.*, **65:**2209–2215.

Mauch, F., L. A. Hadwiger, and T. Boller (1984). Ethylene: symptom, not signal for the induction of chitinase and β-1,3-glucanase in pea pods by pathogens and elicitors. *Plant Physiol.*, **76:**607–611.

McNeil, M., A. G. Darvill, S. C. Fry, and P. Albersheim (1984). Structure and function of the primary cell walls of plants. *Ann. Rev. Biochem.*, **53:**625–663.

Moesta, P., U. Seydel, B. Lindner, and H. Grisebach (1982). Detection of glyceollin on the cellular level in infected soybean by laser microprobe mass analysis. *Z. Naturforsch.*, **37c:**748–751.

Nimz, H. (1974). Das Lignin der Buche—Entwurf eines Konstitutionsschemas. *Angew. Chem.*, **86:**336–344.

Owens, R. J. and D. H. Northcote (1981). The location of arabinosyl : hydroxyproline transferase in the membrane system of potato tissue culture cells. *Biochem. J.*, **195:**661–667.

Paxton, J. D. (1982). Phytoalexins. In *Active Defense Mechanisms in Plants* (R. K. S. Wood, ed.). Plenum, New York, pp. 344–346.

Rathmell, W. G. (1984). The discovery of new methods of chemical disease control: current developments, future prospects and the role of biochemical and physiological research. *Adv. Plant Pathol.*, **2:**259–288.

Ride, J. P. (1975). Lignification in wounded wheat leaves in response to fungi and its possible role in resistance. *Physiol. Plant Pathol.*, **5:**125–134.

Ride, J. P. (1983). Cell walls and other structural barriers in defence. In *Biochemical Plant Pathology* (J. A. Callow, ed.). Wiley, New York, pp. 215–236.

Roby, D., A. Toppan, and M.-T. Esquerré-Tugayé (1985). Cell surfaces in plant-microorganism interactions. V. Elicitors of fungal and of plant origin trigger the synthesis of ethylene and of cell wall hydroxyproline-rich glycoprotein in plants. *Plant Physiol.*, **77:**700–704.

Ryder, T. B., C. L. Cramer, J. N. Bell, M. P. Robbins, R. A. Dixon, and C. J. Lamb (1984). Elicitor rapidly induces chalcone synthase mRNA in *Phaseolus vulgaris* cells at the onset of the phytoalexin defense response. *Proc. Natl. Acad. Sci. USA*, **81:**5724–5728.

Sadava, D. and M. J. Chrispeels (1971). Intracellular site of proline hydroxylation in plant cells. *Biochemistry.*, **10:**4290–4294.

Schmelzer, E., H. Börner, H. Grisebach, J. Ebel, and K. Hahlbrock (1984). Phytoalexin synthesis in soybean (*Glycine max*). Similar time courses of mRNA induction in hypocotyls infected with a fungal pathogen and in cell cultures treated with fungal elicitor. *FEBS Lett.*, **172:**59–63.

Sequeira, L. (1978). Lectins and their role in host-pathogen specificity. *Ann. Rev. Phytopathol.*, **16:**453–481.

Sequeira, L. (1983). Mechanisms of induced resistance in plants. *Ann. Rev. Microbiol.*, **37:**51–79.

Sharp, J. K., B. Valent, and P. Albersheim (1984). Purification and partial characterization of a β-glucan fragment that elicits phytoalexin accumulation in soybean. *J. Biol. Chem.*, **259:**11312–11320.

Showalter, A. M., J. N. Bell, C. L. Cramer, J. A. Bailey, J. E. Varner, and C. J. Lamb (1985). Accumulation of hydroxyproline-rich glycoprotein mRNAs in response to fungal elicitor and infection. *Proc. Natl. Acad. Sci. USA*, **82:**6551–6555.

Skou, J. P., J. H. Jørgensen, and U. Lilholt (1984). Comparative studies on callose formation in powdery mildew compatible and incompatible barley. *Phytopath. Z.*, **109:**147–168.

Somssich, I. E., E. Schmelzer, J. Bollman, and K. Hahlbrock (1986). Rapid activation by fungal elicitor of genes encoding 'pathogenesis-related' proteins in cultured parsley cells. Proc. Natl. Acad. Sci. USA, **83:**2427–2430.

Städler, E. and H.-R. Buser (1984). Defense chemicals in leaf surface wax synergistically stimulate oviposition by a phytophagous insect. *Experientia*, **40:**1157–1159.

Staskawicz, B. J., D. Dahlbeck, and N. T. Keen (1984). Cloned avirulence gene of *Pseudomonas syringae* pv. *glycinea* determines race-specific incompatibility on *Glycine max* (L.) Merr. *Proc. Natl. Acad. Sci. USA*, **81:**6024–6028.

Stoessel, A. (1983). Secondary plant metabolites in preinfectional and postinfectional resistance. In *The Dynamics of Host Defence*. (J. A. Bailey and B. J. Deverall, eds.). Academic, New York, pp. 72–122.

Stolzenburg, M. C., J. R. Aist, and H. W. Israel (1984a). The role of papillae in resistance to powdery mildew conditioned by the *ml-o* gene in barley. I. Correlative evidence. *Physiol. Plant Pathol.*, **25:**337–346.

Stolzenburg, M. C., J. R. Aist, and H. W. Israel (1984b). The role of papillae in resistance to powdery mildew conditioned by the *ml-o* gene in barley. II. Experimental evidence. *Physiol. Plant Pathol.*, **25:**347–361.

Stuart, D. A. and J. E. Varner (1980). Purification and characterization of a salt-extractable hydroxyproline-rich glycoprotein from aerated carrot disks. *Plant Physiol.*, **66:**787–792.

Tegtmeier, K. J. and H. D. VanEtten (1982). The role of pisatin tolerance and degradation in the virulence of *Nectria haematococca* on peas: A genetic analysis. *Phytopathology*, **72:**608–612.

Tiburzy, R. (1984). Untersuchungen zur Bedeutung der hypersensitiven Reaktion befallener Weizenzellen als Resistenzfaktor gegen *Puccinia graminis* f. sp. *tritici*. *Dissertation*. Technische Hochschule, Aachen.

Toppan, A., D. Roby, and M.-T. Esquerré-Tugayé (1982). Cell surfaces in plant–microorganism interactions. III. *In vivo* effect of ethylene on hydroxyproline-rich glycoprotein accumulation in the cell wall of diseased plants. *Plant Physiol.*, **70:**82–86.

Touzé, A. and M.-T. Esquerré-Tugayé (1982). Defence mechanisms of plants against varietal non-specific pathogens. In *Active Defense Mechanisms in Plants* (R. K. S. Wood, ed.). Plenum, New York, pp. 103–117.

Van Holst, G.-J. and J. E. Varner (1984). Reinforced polyproline II conformation in a hydroxyproline-rich cell wall glycoprotein from carrot root. *Plant Physiol.*, **74:**247–251.

Van Loon, L. C. (1983). The induction of pathogenesis-related proteins by pathogens and specific chemicals. *Neth. J. Plant Path.*, **89:**265–273.

Van Loon, L. C. (1985). Pathogenesis-related proteins. *Plant Mol. Biol.*, **4:**111–116.

Vance, C. P., T. K. Kirk and R. T. Sherwood (1980). Lignification as a mechanism of disease resistance. *Ann. Rev. Phytopathology*, **18:**259–288.

VanEtten, H. D., P. S. Matthews, K. J. Tegtmeier, M. F. Dietert, and J. I. Stein (1980). The association of pisatin tolerance and demethylation with virulence on pea in *Nectria haematococca*. *Physiol. Plant Pathol.*, **16:**257–268.

West, C. A., R. J. Bruce, and D. F. Jin (1984). Pectic fragments of plant cell walls as mediators of stress responses. In *Structure, Function, and Biosynthesis of Plant Cell Walls* (W. M. Dugger and S. Bartnicki-Garcia, eds.). American Society of Plant Physiologists, Rockville, pp. 359–379.

Whitmore, F. W. (1978). Lignin-protein complex catalyzed by peroxidase. *Plant Sci. Lett.*, **13:**241–245.

12

PLANT IMMUNIZATION AND ITS APPLICABILITY FOR DISEASE CONTROL

JOSEPH KUĆ

Department of Plant Pathology, University of Kentucky, Lexington, Kentucky

If medical doctors controlled disease in humans as plant pathologists control disease in plants, we would be repeatedly coated with substances toxic to infectious agents, sprayed with or doused in poisonous substances that are absorbed through the skin and circulate through the body, or subjected to breeding experiments to develop "resistant human cultivars" that would be resistant to a disease, or at best several diseases, of importance at the moment. Hopefully, disease-resistant human cultivars would still function productively in society, although repeated backcrosses might be necessary to "reintroduce" genes for productivity and quality "lost" when genes for resistance were introduced. Major efforts would be exerted to search for "wild human species" or "near human species" to incorporate genes for resistance into the "less wild human species." The importance of sanitation for the control of infectious disease would be recognized by the medical doctors, but they would not have, as they lack at present, a chemical agent to administer for the control of diseases caused by viruses. As in current medical practice, antibiotics would be used to avoid infections in high-risk situations as well as to cure infected individuals. With respect to the use of antibiotics, the plant pathologist has been far more conservative than his counterpart in the animal field. He has not used antibiotics on the scale used in the animal feed industry. The motivation for the conservatism is probably due to the relative ineffectiveness and high cost of antibiotics for field application and because those working with animals were using antibiotics as feed additives when regulations concerning their use were less restrictive.

Forgotten, however, would be the basic tenet of preventive medicine and disease control, that with few exceptions, animals have the genetic information for resistance mechanisms to all infectious agents in their environment and resistance is determined by the speed and magnitude of gene expression and the activity of resultant gene products. Thus the strategy would be to rely upon the ability of the pesticide to kill or inhibit development of an infectious agent, without the understanding, implicit in medical practice, that even antibiotics do not replace the immune system, but merely allow the immune system to regain its advantage over the pathogen. Except for relatively infrequent genetic errors of metabolism, the immune system is and logically would have to be the ultimate and major mechanism for disease resistance and hence survival. Humans have been "surviving" since their creation, long before the discovery of antibiotics and surgical and diagnostic technology. Fortunately, most of us concede that more people survive longer and better in our present day than in the time of our cave or tree-dwelling ancestors.

Of course, the modern medical doctor does not face the same problems as the plant pathologists and other plant scientists, who are seeking to feed an ever-increasing human population. Mass monoculture, with demands for increased uniformity, yield, quality, and profit from the sale of products have not been major motives for progress in the medical profession. It, therefore, appears a paradox that recent human wisdom has devised a scheme to tax all of us in the United States to reward people for not producing food in a world plagued by hunger or alternatively to purchase surplus food and store and ultimately destroy it rather than use it to feed people. If all of this appears to be science fiction, perhaps science fiction is reality.

Some defense of the plant pathologist, however, is in order. Integrated pest management, with its warning systems and advisories of when to apply pesticides, has restricted their excessive and indiscriminate use, and undoubtedly the "value" of human life in nonwar situations has influenced the development of disease-control strategy in medicine. In addition, fungi are a major cause of disease in plants, whereas they produce few diseases in animals relative to those caused by other infectious agents. Most fungi penetrate directly through the cuticle of plants or through natural openings such as stomata. Relatively few are wound pathogens. Thus the plant pathologist, with some justification, is still guided in disease control strategy by the dictum "coat the plant surface and keep out the pathogen." The development of systemic fungicides, the antibiotics of plant disease control, is a marked departure from this strategy, but one that still ignores the immune system inherent in all plants. The assumption that has been the basis for plant disease control strategy is that resistant plants have genes for resistant mechanisms and susceptible plants do not; ergo, the only method to protect susceptible plants is to use chemicals that per se inhibit the development of pathogens. Because of this misconception, the plant scientist is only beginning to appreciate the importance of expressing genes for disease resistance present in plants and the regulation of activity of gene products. Until very recently, the plant scientist has ignored immunization—the major technology for preventive medicine in animals.

Questions I consider in this chapter include: Are biochemical and economic definitions of resistance and susceptibility synonymous? Are susceptibility and resistance to disease absolute or a continuum? Do plants have an immune system? Can plants be immunized against disease? Can immunization be used to control disease of crop plants in the field? What are the advantages and disadvantages of immunization? Where are we today and what are possible directions for the future?

1. ARE BIOCHEMICAL AND ECONOMIC DEFINITIONS OF RESISTANCE AND SUSCEPTIBILITY SYNONYMOUS?

When plant pathologists, agronomists, or farmers consider disease resistance in plants, they consider the economic loss as the determinant of resistance (for simplicity I include tolerance and escape under the umbrella of resistances) or susceptibility.

Clearly the persistent use of resistant cultivars is predicated on their performance. They will yield a profit under conditions where disease problems exist. They may not, however, produce the highest yields or highest quality crops where disease is not a problem.

A restricted lesion on foliage or fruit that results in economic loss would, therefore, be a symptom of susceptibility, even though the restriction of the lesion may be evidence for the presence of a biochemical mechanism for resistance. A plant with a resistance mechanism that would permit reproduction and hence survival of the species might still be considered susceptible economically. It is clear that survival of a species is not in itself sufficient to meet the demands of modern agriculture, but regulation of gene expression and activity of gene products may raise the level or change the timing of the mechanism to make the biochemically resistant plant also economically resistant.

A logical question at this point would be are humans resistant or susceptible to disease? They have survived the selective pressure of pathogens and they can be immunized against many diseases that periodically devastated mankind. The economic definition of resistance does not preclude the existence of a potential for effective mechanisms for resistance in susceptible plants, just as susceptibility to smallpox, typhoid, tetanus, or polio does not mean the lack of a potential for an effective disease resistance in man. This logically leads us to the second consideration.

2. ARE SUSCEPTIBILITY AND RESISTANCE TO DISEASE ABSOLUTE OR A CONTINUUM?

Observations of disease in the field indicate that disease resistance and susceptibility are not absolute. They are subjective evaluations within certain parameters of growth and environment which include light quality and intensity, temperature, age of plant, tissue differentiation, pathogen pressure, moisture, and level of fertilization.

The first three or four expanding leaves of apple shoots are often susceptible to apple scab, caused by *Venturia inaequalis*, and all others are resistant (Biehn et al., 1966; Grijseels et al., 1964; Nusbaum and Keitt, 1938). From a practical viewpoint, the reaction of leaves at the growing point is determining resistance or susceptibility. Nevertheless, all apple trees have resistance to apple scab caused by all races of *V. inaequalis* if we consider the older leaves. The resistance is even evident in apple varieties that the grower would consider completely susceptible or lacking in genes for resistance. Many similar situations exist. Cells of etiolated green bean hypocotyls close to the growing point are susceptible to anthracnose caused by *Colletotrichum lindemuthianum*. Resistance increases as the root zone is approached, and tissue near and in the root zone, as well as the roots themselves, are highly resistant (Elliston et al., 1971, 1976a–c; Kuć, 1982a; Kuć and Caruso, 1977). This is even true of bean cultivars which breeders or growers might consider as lacking in genes for resistance to *C. lindemuthianum*.

Cucumber seedlings of cultivars susceptible to scab, caused by *Cladosporium cucumerinum*, are highly susceptible at the growing point and for a distance of approximately 0.5 cm below the cotyledon. Tissue beneath this zone is highly resistant (Kuć, 1962). Tissue sections taken from the susceptible zone become macerated within 1 hr when placed in culture filtrates of the fungus or solutions containing a combination of proteolytic, pectinolytic, and cellulolytic enzymes, whereas sections taken from the resistant zone remain intact 24 hr after treatment. Temperature also has a profound effect on resistance. All cucumber cultivars are resistant to scab at temperatures of 22–24°C, and some are at 20–22°C. Cultivar resistance is determined at 17–20°C, which is the optimum temperature for scab. The pathogen grows and sporulates well *in vitro* at 17–24°C, and spores germinate and the pathogen penetrates the foliage at these temperatures.

The fungi *Physalospora obtusa*, *Botryosphaeria ribis*, and *Glomerella cingulata* cause disease on ripening apple fruit late in the season (Wallace et al., 1962; Kuć et al., 1967). The transition from immunity to susceptibility occurs within 1–2 wk in the field and the period of change varies from year to year depending on climatic conditions. The fungi are rapidly contained when they are introduced into wounds made in green apple fruit in June. However, when introduced into ripening fruit late in July, or perhaps early in August, the fungi spread through the tissue and a rapidly expanding area of rot is evident within 2–3 days. Fungitoxic compounds were not detected in resistant apples and resistance was apparently due to the inability of the fungi to degrade cell walls, possibly because of the presence of pectin–protein–polyvalent cation complexes.

Many pathogens exist as distinct races for which there are differential cultivars of a host, that is, cultivars resistant to some but susceptible to other races of the pathogen. Often cultivars resistant to all or none of the races are also found. The presence of differential cultivars would suggest either highly race-specific disease resistance mechanisms and/or differences in the speed and magnitude with which resistance mechanisms are expressed by different races on different cultivars. Evidence for the former has not been reported, whereas data with phytoalexins, lignification, and other inducible resistant mechanisms support the latter (Kuć, 1982a,b, 1983; Kuć and Tuzun, 1983; Kuć and Preisig, 1984; Kuć and Rush, 1985). Resistance is apparent in cucumber cultivars considered to lack genes for resistance mechanisms at low inoculum pressure (Dean and Kuć, 1983, 1986c).

The concept that resistance and susceptibility are not absolute is further strongly supported by the reports of Lazarovits, Stossel, and Ward working with the interaction of soybean and the fungus *Phytophthora megasperma* f. sp. *glycinea* (Lazarovits et al., 1981; Stossel, et al., 1980, 1981; Ward and Buzzell, 1983; Ward et al., 1981). They clearly established that the nature of the resistance gene or genes (Rps) in soybean, the race of the fungus, other genes in the host, temperature, light, wounding, and cell maturity individually or in combination can be the overriding factors that control the outcome of the reaction.

An additional but pertinent observation is that leaves are generally immune to root pathogens and roots are generally immune to leaf pathogens, even when the pathogens are introduced directly into tissues. The picture is made even more complex by the observation that some pathogens penetrate a host directly through the cuticle, and others penetrate through stomata or wounds.

3. DO PLANTS HAVE AN IMMUNE SYSTEM?

Clearly plants have an immune system, though its components and regulation differ from that of the immune system found in animals. Nevertheless, the ultimate contribution of the immune system in plants and animals is to facilitate the survival of the species, the *sine qua non* of biology (Dawkins, 1976). Plants, animals, and infectious agents have survived the selection pressure of evolution. Though infectious disease has devastated plant and animal populations, I am unaware of a single instance where it has been unequivocally established that infectious disease alone has resulted in the loss of a species.

The immune system in plants includes numerous mechanisms we know of and probably many of which we are still unaware. One mechanism for disease resistance in plants is the rapid accumulation of low molecular antimicrobial substances (phytoalexins) at and immediately around sites of infection. The accumulation of phytoalexins has been demonstrated in at least 17 plant families (Coxon, 1982; Ingham, 1982; Kuć, 1982a; Stossl, 1982, 1983). Similarities are evident among phytoalexins from plants within a family. Plants from the Leguminosae, Solanaceae, Compositae, and Convolvulaceae produce predominantly isoflavonoids, carbocyclic sesquiterpenoids, polyacetylenes, and furanosesquiterpenoids, respectively. Of the 102 phytoalexins reported in the Leguminosae, 84 are isoflavonoid derivatives, 8 are furanoacetylenes, 4 are stilbenes, 3 are benzofurans, 2 are chromones, and 1 is a flavanone (Ingham, 1982). Similarly, in the Solanaceae, 34 phytoalexins reported are terpenoid derivatives, 6 are phenylpropanoid phenols, and 3 are polyacetylenes (Kuć,

1982a). Though some phytoalexins are distributed among many families (e.g., caffeic acid derivatives accumulate in potato, carrot, and sweet potato), plants in the Leguminosae have not been reported to produce sesquiterpenoid phytoalexins and those in the Solanaceae have not been reported to produce isoflavonoids.

In all cases reported to date, phytoalexin accumulation as a factor for disease resistance is not determined by the presence or absence of genetic information for the requisite biosynthetic pathways (Kuć, 1972, 1982a,b; Kuć and Caruso, 1977; Kuć and Rush, 1985). Thus phytoalexins as factors in disease resistance would be compatible with the hypothesis that all plants contain the genetic potential for resistance mechanisms. This observation is consistent with the ability of dogs, cats, and man to produce antibodies, and all three, though susceptible to diseases, can be immunized for life against some of them.

Phytoalexins are degraded by host and microbial enzymes (Smith, 1982; VanEtten et al., 1982). Thus accumulation is a function of both rate of synthesis and degradation. They are not notably active as antibiotic agents (ED_{50} ca. 1-5 $\times$ 10^{-4} M) though the quantity that accumulates at sites of infection is sufficient to markedly inhibit the growth of some fungi and bacteria *in vitro*. Phytoalexins have a broad spectrum of activity against the pathogens and nonpathogens of a host, and a broad range of pathogens, nonpathogens, and physiological stresses elicit accumulation of the same phytoalexin. A case for the involvement of phytoalexins in disease resistance depends upon the speed and magnitude with which they are produced and not selective toxicity per se or selectivity of elicitation (Kuć, 1972; Kuć and Preisig, 1984; Kuć and Rush, 1985; Smith, 1982). This is markedly different from the rather high level of specificity associated with the induction of antibodies and their interaction with antigens in animals, though the efficacy of the antibody-antigen interaction in the disease resistance of animals is also highly dependent upon the speed and magnitude of antibody production. Unlike protein antibodies in animals, the phytoalexins are low molecular weight, generally lipophilic substances that are products of secondary metabolism and they are not translocated. A signal that conditions plants to accumulate phytoalexins rapidly when infected, however, is translocated (Kuć, 1982a,b; Kuć and Caruso, 1977; Kuć and Tuzun, 1983; Kuć, 1984; Kuć and Preisig, 1984; Kuć and Rush, 1985). Thus phytoalexins accumulate rapidly after infection or stress in plants "sensitized" to respond by prior infection. The energy and precursors required for their synthesis are conserved until when and where they are needed. Some phytoalexins are phytotoxic at concentrations that inhibit the development of microorganisms. Attempts to protect plants against disease by eliciting and maintaining high levels of phytoalexins systemically in foliage proved unsuccessful. The frequent application of fungus-derived glucan elicitors to the foliage of green-bean and soybean plants caused severe necrotization and stunting of growth.

Though protecting plants by directly eliciting phytoalexin accumulation has not been successful, sensitizing plants to rapidly mobilize defense mechanisms and accumulate phytoalexins after infection has proven successful in laboratory and field experiments. The sensitization has been accomplished by limited infection with fungi, bacteria, or viruses; products from immunized plants; or synthetic chemicals (Caruso and Kuć, 1977a; Farih et al., 1981; Kuć, 1982a,b; Kuć and Caruso, 1977; Kuć and Tuzun, 1983; Lancake, 1981; Sutton, 1979).

Though infection with fungi, bacteria, and viruses elicits phytoalexin accumulation, other stress-inducing factors, such as mercuric or cupric chloride, ethylene, UV radiation, metabolic inhibitors, cell-wall components of fungi and bacteria and their derivatives, microbial metabolites, and insect and nematode damage, are also active elicitors (Bailey, 1982, 1983; Bostock et al., 1981, 1982; Keen, 1982a,b; Kuć et al., 1984; Ossowski et al., 1984; Sharp et al., 1984a–c; Yoshikawa and Masago, 1982). This is especially true for plants in the Leguminoseae. This rather low level of specificity as related to elicitation neither supports nor detracts from the possible contribution of phytoalexins in disease resistance, and race specificity in elicitation of phytoalexin has been reported for a bacterial and fungal disease in soybean (Keen, 1982a,b; Yoshikawa and Masago, 1982).

In some susceptible reactions between fungi and plants, phytoalexins accumulate to higher levels than in resistant interactions. This is often evident when biotrophic growth terminates, sporulation occurs, and lesions develop (Kuć, 1972; Kuć and Caruso, 1977). In this case phytoalexin accumulation probably occurred too late to contain the fungus and restrict disease development.

Two distinct aspects of phytoalexin elicitation are apparent. The first is necrotization or damage to plant cells by chemical or physical agents, which releases a nonphytotoxic substance. The second is migration of the nonphytotoxic substance to healthy cells where it conditions enhanced phytoalexin synthesis (Bailey, 1982, 1983; Kuć, 1982a,b; Kuć and Caruso, 1977). The phytoalexins are then transported out of the producer cells and accumulate in the necrotized or damaged tissues. The actual elicitor of phytoalexin synthesis and accumulation may not be the fungal glucan, toxic chemical, or UV radiation, but rather the substance released by the damaged tissues. Thus a major emphasis should be placed on why many fungal cell-wall glucans are highly phytotoxic, rather than on how they elicit phytoalexin accumulation. Recent studies support the hypothesis that infection and physiological stress cause the release of plant cell-wall fragments containing galacturonic acid residues, and these fragments elicit the accumulation of phytoalexins (Bruce and West, 1982; Davis et al., 1982; Hahn et al., 1981; Keen, 1982a,b; Nothnagel et al., 1983; West et al., 1985; Walker-Simmons et al., 1983; Ryan et al., 1985). It is possible that phytotoxic elicitors from microbial cell walls damage plant cells, bringing plant hydrolases in contact with plant cell walls, and cause the release of galacturonic acid-containing oligosaccharides and polysaccharides. These fragments may be the actual elicitors of phytoalexin accumulation. Fungal or bacterial hydrolases active on plant cell walls could also act as the plant enzymes. Alternatively, the microbial cell walls and products from plant cell walls may directly elicit phytoalexin accumulation. It is not unlikely that endogenous elicitors are present in all plants reported to accumulate phytoalexins (Bailey, 1982, 1983). The release of the endogenous elicitors may be analagous to the release of alarm signals that activate resistance mechanisms. Thus we are presented with a paradox, all plants have the potential for a resistance response and all infectious agents have the ability to elicit a resistance response, yet susceptibility occurs in nature. Though most reports to date of resistance mechanisms emphasize the accumulation of phytoalexins as a result of elicitor action, other defense mechanisms may be activated as well.

Mechanical wounding or short-term injury generally are not adequate for appreciable accumulation of phytoalexins (Kuć, 1972, 1982a,b). It appears that a low-level persistent stress, perhaps caused by the continued presence of the infectious agent or elicitor, is required for the accumulation of high levels of phytoalexins. This is consistent with the observation that, when the infectious agent is contained, or in the case of fungi and bacteria perhaps killed, phytoalexin levels drop to levels that approach those in uninfected plants. Similarly, the hydrolysis of glucan elicitors and dilution or sequestering of inorganic toxicants would, with time, remove the source of stress and phytoalexin levels would drop, unless the elicitors or toxicants caused the release of additional elicitors of host origin.

It is generally conceded, however, that phytoalexin accumulation is only one component of the plant's mechanisms for disease resistance, and evidence is lacking that phytoalexins have a role in plant resistance to diseases caused by viruses (Kuć, 1983; Kuć and Preisig, 1984; Kuć and Rush, 1985; Stossl, 1982, 1983). Other components of the disease-resistance complex include the presence of preformed antimicrobial agents and physical barriers, lignification, papillae, suberization, and the formation of callose, agglutinins, and enzyme inhibitors (Aist, 1983; Bailey, 1982, 1983; Coxon, 1982; Deverall, 1982; Friend, 1976, Ingham, 1982; Keen, 1982a,b; Kuć, 1972, 1982a,b, 1983, 1984; Stossl, 1982, 1983; Vance et al., 1980).

Disease resistance in plants likely depends upon the activation of coordinated, multicomponent defense mechanisms. A single mechanism or compound with a single metabolic site of action would be unsatisfactory, as illustrated by the rapid development of resistant strains of pathogens to single-site-dependent systemic fungicides and antibiotics. The mechanisms for resistance probably include those that exclude pathogens, those that are general and nonspecific responses to wounding or metabolic insult, and those directed, with various degrees of specificity, against particular groups of pathogens. Undoubtedly, each different mechanism contributes to the overall outcome of the interaction. This layering of mechanisms, an apparent duplication of effort under some circumstances, provides depth and hence stability to resistance.

4. CAN PLANTS BE IMMUNIZED AGAINST DISEASE?

4.1. Use of Infectious Agents

Reports by Kuć and his colleagues indicate that cucumbers, watermelons, and muskmelons can be systemically immunized against diseases caused by viruses, bacteria, or fungi by restricted infection with viruses, bacteria, or fungi (Caruso and Kuć, 1977a,b, 1979; Gessler and Kuć, 1982; Hammerschmidt and Kuć, 1982; Hammerschmidt et al., 1976, 1982; Jenns and Kuć, 1977, 1979, 1980; Jenns et al., 1979; Kuć and Richmond, 1977; Kuć et al., 1975; Richmond et al., 1979). In addition, green bean has been systemically immunized against anthracnose, caused by *Colletotrichum lindemuthianum* (Elliston et al., 1971, 1976a–c, 1977; Sutton, 1979); tobacco against blue mold caused by *Peronospora tabacina* (Cohen and Kuć, 1981; Cruickshank and Mandryk, 1960; Kuć and Tuzun, 1983; Tuzun and Kuć, 1983, 1985a), bacterial wilt caused by *Pseudomonas solanaciarum* (Sequeira, 1983), and tobacco mosaic virus (Ross, 1966). Immunization protected cucumber and tobacco plants for an entire season, and restricted infection with *C. lagenarium* or tobacco necrosis virus (TNV) protected cucumber against at least 12 unrelated diseases. Immunization was effective against diseases caused by obligate and facultative fungi, local lesion and systemic viruses, fungi and bacteria that cause wilts, and those that cause restricted and nonrestricted lesions on foliage and fruit (Kuć, 1982a, 1983, 1984).

Reports indicate that penetration from appressoria of *C. lagenarium* is markedly reduced on immunized cucumber; rapid lignification after challenge is a mechanism for restricting the development of *C. lagenarium* and *Cladosporium cucumerinum* in immunized cucumber (Kuć, 1983; Hammerschmidt and Kuć, 1982; Hammerschmidt et al., 1982; Richmond et al., 1979); and rapid accumulation of isoflavonoid phytoalexins after challenge is a mechanism for restricting development of *C. lindemuthianum* in immunized green beans (Kuć, 1984). It is unlikely, however, that a single mechanism for disease resistance would be effective against the broad spectrum of pathogens against which immunization is effective.

The immunizing signal is graft transmissible in cucumber (Jenns and Kuć, 1979; Dean and Kuć, 1986a,b) and tobacco (Kuć and Tuzun, 1983; Tuzun and Kuć, 1983, 1985a), and immunization increases the growth of tobacco (Kuć and Tuzun, 1983; Tuzun and Kuć, 1983, 1985a). It has been clearly established in cucumber that the inducer (infected) leaf is the source of the signal, and that protected but uninfected leaves do not produce the signal (Dean and Kuć, 1986a,b). It appears that in tobacco, immunization is an irreversible process and immunized buds grafted on nonimmunized rootstocks developed into fully grown immunized plants (Tuzun and Kuć, 1985a). The irreversibility of the phenomenon in tobacco would suggest that gene expression can be either permanently turned on (or off), which leads to disease resistance. The irreversibility of the phenomenon is supported further by the report of Roberts (1983) that acquired resistance to tobacco mosaic virus is transmitted to the progeny of hypersensitive tobacco. The genetic and regulatory implications of these observations are profound, as are the possibilities they present for the economic control of plant disease in the field.

Immunization has also been demonstrated in other diverse plant–pathogen interactions, including pear and apple-*Erwinia amylovora* (Goodman, 1967, 1980; McIntyre et al., 1973), peach-*Cytospora cincta* (Brown and Helton, 1971), coffee-*Hemileia vastatrix* (Moraes et al., 1976; Beretta et al., 1977), tomato-*Fusarium oxysporum* (Davis, 1968), cotton-*Tetranychus uritaceae*, a mite (Karban and Carey, 1984), bean-*Uromyces phaseoli* and sunflower-*Puccinia helianthi* (Yarwood, 1956), tobacco-*Phytophthora parasitica* var *nicotiana* (McIntyre and Miller, 1977), tobacco-diverse pathogens and an aphid (McIntyre et al., 1981).

It appears that immunization involves sensitization of plants to respond rapidly to infection and that the response is multicomponent. One component of the response in green bean is the rapid accumulation of phaseollin, whereas in cucumber it is evidenced by rapid lignification. Phaseollin accumulation and lignification occur at the sites of infection. It is possible, therefore, to separate two components of the immunization process: the chemical agents, including phytoalexins, which accumulate around the site of infection and contribute to the inhibition of fungal and bacterial development, and the signal that sensitizes cells removed from the site of an inducing inoculation to rapidly accumulate phytoalexins when challenged. As with phytoalexin accumulation, immunization appears to be due to a low-level and persistent stress. In cucurbits the stress elicits resistance against diseases caused by fungi, bacteria, and viruses (Kuć, 1985; Stermer and Hammerschmidt, 1985). Unlike phytoalexin accumulation, the response has a systemic sensitization component before the response that restricts development of the infectious agent.

4.2. Use of Chemical Agents

Reports are also available of the effectiveness of chemical compounds (inactive as antimicrobial agents *in vitro*) as elicitors of resistance in susceptible plant hosts. The D- and DL-isomers of phenylalanine markedly increased the resistance of seven apple varieties to apple scab caused by *Venturia inaequalis* (Kuć et al., 1959). A solution of the amino acid was infused into growing shoots through leaf midribs for a 24-hr period, and leaves above the points of infusion were challenged with the fungus. Solutions of D-alanine, DL-α-amino-*n*-butyric acid, and D- and DL-leucine also elicited resistance, whereas all L-isomers were inactive. In subsequent studies, α-aminoisobutyric acid (AIB) was reported to markedly increase the resistance of apple foliage to scab (MacLennan et al., 1963). The acid was not fungitoxic at a concentration as high as 0.4 *M in vitro* and AIB was stable in cultures of the fungus and in apple foliage.

Recently Garas and Kuć (1984) extracted an inducer of immunization from cucumber leaves infected with *C. lagenarium* or *P. lachrymans*. Infiltration of active extracts into cotyledons and the first true leaf of cucumber induced immunity in the foliage discussed above to disease caused by *C. lagenarium*. A lag of 4–5 days was required between infiltration of the inducer and expression of immunity.

Injection of cell-free sonicates of virulent and avirulent isolates of *Erwinia amylovora*, the incitant of fire blight, or the nonpathogen *Erwinia herbicula* into

"Bartlett" pear seedlings protected them against fire blight (McIntyre et al., 1973). Sonicates were not inhibitory to the pathogen and protection was lost when nucleic acids were precipitated from sonicates (McIntyre et al., 1975). Deoxyribonuclease, but not ribonuclease, destroyed the activity of nucleic acids isolated from sonicates. Activity was associated with the deoxyribonucleic acid (DNA) fraction from linear log sucrose gradients. The DNA prepared by the Marmur technique protected more effectively than DNA prepared from sonicates.

Fire blight symptoms were apparent on all control seedlings, from which the radicle was excised, 4 days after they were dipped into a suspension of *E. amylovora*, and a few days later they were entirely destroyed. In contrast, 12 days after challenging seedlings treated with DNA, about 50% had fire-blight symptoms and only 20–30% of those with symptoms were entirely destroyed. The remaining seedlings with symptoms showed necrosis only at the hypocotyl base and this was often dry in appearance, as compared to the normal water-soaked appearance of infected tissues. The DNA-protected seedlings, from which the radicle had been excised, were often observed to form roots. Those that did were transplanted into vermiculite and soil, at high relative humidity, for 1 wk, and then challenged by injecting 10^2 cells at 0.5 cm below the cotyledons. Control seedlings showed fire-blight symptoms within 2 days of challenge, whereas seedlings protected with DNA showed no symptoms for at least 1 wk. After 1 wk, restricted symptoms developed on seedlings treated with DNA. This indicated the lasting nature of protection with DNA and that protection occurs even at sites removed from the point of initial DNA application. It is unlikely that protection by DNA is dependent upon the specific source or nucleotide sequence of the molecules, since DNA from various bacterial sources was active and resistance to some fungal, bacterial, and viral diseases in plants has been elicited by anionic substances such as polyacrylic and salicylic acids (Mills and Wood, 1984).

Other chemical agents that elicit disease resistance but are not antimicrobial *in vitro* include: 2,2-dichloro-3,3-dimethyl-cyclopropane carboxylic acid, 3-allyloxy-1,2-benzisothiazole-1,1-dioxide, and perhaps aluminum tris-*O*-ethyl phosphonate, though phosphorous acid may directly be an antimicrobial agent for the latter compound (Lancake, 1981; Sekizawa and Mase, 1980; Farih et al., 1981). At least in part, all of the chemicals cited in this section appear to act by sensitizing the plant to rapidly activate resistance mechanisms when challenged by the pathogen. Some of the resistance-related mechanisms activated by the chemicals after challenge (e.g., phytoalexin accumulation, lignification, and enhanced peroxidase activity) are also common to plants systemically protected by restricted infection. The data suggest that a metabolic perturbation, resulting in a low level of rather persistent stress, is responsible for the sensitization. The data for plant immunization using infectious agents or chemicals strongly support the concept that disease resistance in plants is determined by the speed and magnitude with which genes for resistance mechanisms are expressed and the activity of gene products rather than the presence or absence of such genes.

5. CAN IMMUNIZATION BE USED TO CONTROL DISEASE OF CROP PLANTS IN THE FIELD?

Temperature, cultivar, locus of inoculation, and prior-inoculation (inoculation 24 hr earlier with three nonpathogenic fungi) affected the severity of fusarium wilt caused by *Fusarium oxysporium* f. *niveum* in watermelons (Shimotsuma et al., 1972). Prior inoculation with *Helminthosporium carbonum* was found to be more effective in cross protection than were *Fusarium oxysporium* f. *lycopersici* or *Verticillium alboatrum*. Temperatures of 27°C resulted in more fusarium infection than at 20°C and obscured the expression of both genetic resistance and resistance induced by the prior-inoculations. Induced resistance increased plant survival and plant growth for three cultivars in field trials, but possibly produced deleterious effects in a resistant cultivar.

Protection against anthracnose in cucumbers, watermelons, and muskmelons under field conditions was demonstrated by Caruso and Kuć (1977b). They reported that plants receiving a restricted infection with *C. lagenarium* prior to transplanting to the field had fewer and smaller lesions following challenge with high inoculum concentrations of the pathogen in the field than plants that were not infected prior to transplant. The survival of protected watermelon plants was 98%, as compared to 32% for controls.

Sanudo (1982) reported protection of green beans (*Phaseolus vulgaris*) against anthracnose in the field by using a cultivar nonpathogenic race of the fungus.

More recently immunization in the field has been demonstrated to be as effective in protecting tobacco against blue mold as metalaxyl, the most effective chemical control agent against the disease. Immunization also increased the marketable yield of tobacco (Tuzun et al., 1984). Sporangial suspensions of *P. tabacina* (Kentucky isolate 82) were injected into stem tissue, external to the cambium, of plants growing at three field locations during 1983–1984: Kentucky 14 burley tobacco in Lawrence Co., Kentucky; MS 14XL8 burley hybrid in Owen Co., Kentucky; and PR 5-65 cigar tobacco planted in Gurabo, Puerto Rico. Metalaxyl was sprayed on the foliage and added to the soil of comparable plants at the time of stem injection and untreated plants served as controls. Treatments were arranged into a randomized complete block design with three or more replications. Plants receiving the stem injection had significantly greater height and fresh weight than plants treated with metalaxyl or control plants. Marketable yield was increased by up to 25% over the controls at the two Kentucky locations. Stem-injected plants were protected against blue mold as well as or better than metalaxyl-treated plants in Puerto Rico.

The major problem encountered with the use of immunization for the control of plant disease in the field to date is not its effectiveness but rather the economics of the technology and the delivery of the immunizing agent. Injecting inoculum into stems or restricted foliar inoculation with avirulent or virulent pathogens is clearly not applicable to modern agriculture, and the preparation and cost of inoculum and its delivery makes the process noncompetitive with current fungicides. Nevertheless, the problem is logistical and not conceptual. I will discuss approaches to bridging the "logistical gap" in the final section of this chapter.

6. WHAT ARE THE ADVANTAGES AND DISADVANTAGES OF IMMUNIZATION?

Induced systemic resistance has numerous positive features, some of which may offer advantages over established methods of disease control by the use of pesticides or breeding.

1. Immunization is effective against viral, bacterial, and fungal diseases.
2. It is likely that immunization is dependent upon the activation of several different mechanisms and, therefore, is probably stable. Systemic fungicides with a single metabolic site of action are generally unstable with the development of new races of pathogens.
3. Immunization is systemic.
4. Immunization is persistent. In some cases it persists for the life of annual plants.
5. Since immunization utilizes mechanisms for resistance present in plants, it may be considered natural and as safe for man and the environment as disease-resistant plants.
6. Immunization in cucurbits and tobacco is graft transmissible from rootstock to scion. This could be significant since many plants are propagated by grafting.
7. Immunization in tobacco appears nonreversible. Buds from immunized plants grafted onto susceptible rootstocks developed into fully grown immunized tobacco plants.
8. Immunization systemically sensitizes the plant to respond, but the major expenditure of energy and the expression of mechanisms for resistance generally are localized and occur only in the presence of pathogens, that is, when and where needed.
9. The ability to immunize susceptible plants implies that the genetic potential for resistance is present in all plants.
10. Recent results from our laboratory indicate that plants can be immunized by chemicals extracted from immunized plants. This suggests the possibility of immunization by field spraying or seed treatment.

Immunization at present, however, is not without problems.

1. The natural chemical signals for systemic immunization have not been characterized.

2. Immunization is not economically competitive with our present technology in modern agriculture, although it appears as effective as available systemic pesticides.

3. Immunization has not received sufficient field testing to determine its stability and persistence under high natural-pathogen pressure.

4. People have difficulty in accepting the reality that plants can be systemically immunized against disease.

7. WHERE WE ARE TODAY AND POSSIBLE DIRECTIONS FOR THE FUTURE

It is apparent that plants have a high degree of genetic instability and the stability of the genome can be disrupted by stress (Marx, 1984). Different stresses may induce changes in the repetitive DNA of plants. It is possible that such changes can result in changes in the resistance of plants to diseases caused by fungi, bacteria, and viruses (Kuć, 1985). Tuzun and Kuć (1985a) reported that buds from tobacco plants immunized against blue mold when grafted onto nonimmunized rootstocks, developed into fully grown, immunized plants. In recent experiments they report further that immunization to blue mold can be transmitted from tobacco plants to tobacco plants via callus (Tuzun and Kuć, 1985b). The persistence of immunization in tobacco to blue mold suggests a permanent change in the genome and/or its expression as related to disease resistance. Plants grown from seed of immunized plants and plants derived from callus are currently being tested for resistance to blue mold (Tuzun and Kuć, unpublished data). The data support the possibility of using induced resistance to produce long-term effects on the genome or the expression of the genome of plants as related to disease resistance. This might be accomplished via tissue culture or seed.

The limiting factor for the field application of induced resistance at this time is economic feasibility rather than the effectiveness or duration of the phenomenon. A major effort in my laboratory is being devoted to the characterization of the translocatable "immunity signal," which is nonphytotoxic and sensitizes the plant to react rapidly when challenged by the pathogen. In addition, we are investigating factors that give rise to the production of the translocatable signal. The signal might be used as a spray or seed treatment to immunize plants. Synthetic chemicals that mimic the natural signal might also be economic and highly effective. As discussed earlier in this chapter, a precedent for the use of such chemicals is established.

I believe the phenomenon of plant immunization has opened an entire new area of molecular biology as related to the role of stress and infection on the genome and its expression. It has also provided the possibility for a new method of disease control based on the expression of latent disease resistance mechanisms present in existing high-yielding, high-quality crops.

ACKNOWLEDGMENTS

The author's research reported in this chapter was supported in part by grants 78-59-2211-0-1063 and 7B30-0-185 from the U.S. Department of Agriculture as well as grants from the Ciba-Geigy Corporation, R. J. Reynolds Corporation, The Rockefeller Foundation, and the Graduate School of the University of Kentucky.

REFERENCES

Aist, J. (1983). Structural responses as resistance mechanisms. In *The dynamics of Host Defense* (J. Bailey and B. Deverall, eds.). Academic, New York, pp. 33–70.

Bailey, J. (1982). Mechanisms of phytoalexin accumulation. In *Phytoalexins* (J. Bailey and J. Mansfield, eds.). Blackie and Sons, Glasgow, pp. 289–318.

Bailey, J. (1983). Biological perspectives of host-pathogen interactions. In *The dynamics of Host Defense* (J. Bailey and B. Deverall, eds.). Academic, New York, pp. 1–32.

Beretta, M., E. Martins, and W. Moraes (1977). Induced protection to *Hemileia vastatrix* at a distance from the site of the inducing action in coffee plants. *Summa Phytopathol.*, **3:**66–70.

Biehn, W., E. Williams, and J. Kuć (1966). Resistance of mature leaves of *Malus atrosanguinea* 804 to *Venturia inaequalis* and *Helminthosporium carbonum. Phytopathology*, **56:**588-589.

Bostock, R., J. Kuć, and R. Laine (1981). Eicosapentaenoic and arachidonic acids from *Phytophthora infestans* elicit fungitoxic sesquiterpenes in potato. *Science*, **212:**67–69.

Bostock, R., R. Laine, and J. Kuć (1982). Factors affecting the elicitation of sesquiterpenoid phytoalexin accumulation by eicosapentaenoic and arachidonic acids in potato. *Plant Physiol.*, **70:**1417–1424.

Brown, J. and A. Helton (1971). Induced resistance to *Cytospora* in *Prunus persica. Phytopathology*, **61:**685–687.

Bruce, R. and C. West (1982). Elicitation of casbene synthetase activity in castor bean, the role of pectic fragments of the plant cell wall in elicitation by a fungal endopolygalacturonase. *Plant Physiol.*, **69:**1181–1188.

Caruso, F. and J. Kuć (1977a). Protection of watermelon and muskmelon against *Colletotrichum lagenarium* by *Colletotrichum lagenarium. Phytopathology*, **67:**1285–1289.

Caruso, F. and J. Kuć (1977b). Field protection of cucumber, watermelon and muskmelon against *Colletotrichum lagenarium* by *Colletotrichum lagenarium. Phytopathology*, **67:**1290–1292.

Caruso, F. and J. Kuć (1979). Induced resistance of cucumber to anthracnose and angular leaf spot by *Pseudomonas lachrymans* and *Colletotrichum lagenarium. Physiol. Plant Pathol.*, **14:**191–201.

Cohen, Y. and J. Kuć (1981). Evaluation of systemic resistance to blue mold induced in tobacco leaves by prior stem inoculation with *Peronospora hyoscyami tabacina. Phytopathology*, **71:**783–787.

Coxon, D. (1982). Phytoalexins from other families. In *Phytoalexins* (J. Bailey and J. Mansfield, eds.). Blackie and Sons, Glasgow, pp. 106–132.

Cruickshank, I. and M. Mandryk (1960). The effect of stem infestation of tobacco with *Peronospora tabacina* Adam on foliage reaction to blue mold. *J. Aust. Inst. Agric. Sci.*, **26:**369–372.

Davis, D. (1968). Partial control of Fusarium wilt in tomato by formae of *Fusarium oxysporum. Phytopathology*, **58:**121–122.

Davis, K., G. Lyon, A. Darvill, and P. Albersheim (1982). A polygalacturonic acid lyase isolated from *Erwinia carotovora* is an elicitor of phytoalexins in soybeans. *Plant Physiol.*, (Suppl.) **69:**142.

Dawkins, R. (1976). *The Selfish Gene*, Oxford University Press, New York, 244 pp.

Dean, R. and J. Kuć (1983). The level of induced systemic resistance in cucumber cultivars is revealed by increasing concentrations of challenge inoculum. *Phytopathology*, **78:**821.

Dean, R. and J. Kuć (1986a). Induced systemic protection in cucumbers: the source of the signal. *Physiol and Mol. Plant Pathol.* **28:**227–233.

Dean, R. and J. Kuć (1986b). Induced systemic protection in cucumbers: time of production and movement of the "signal." *Phytopathology* (In Press).

Dean, R. and J. Kuć (1986c). Induced systemic protection in cucumber: effects of inoculum density on symptom development of *Colletotrichum lagenarium* in protected and unprotected plants. *Phytopathology* **76:**186–189.

Deverall, B. (1982). Introduction. In *Phytoalexins* (J. Bailey and J. Mansfield, eds.). Blackie and Sons, Glasgow, pp. 1–20.

Elliston, J., J. Kuć, and E. Williams (1971). Induced resistance to anthracnose at a distance from the site of the inducing interaction. *Phytopathology*, **61:**1110–1112.

Elliston, J., J. Kuć, and E. Williams (1976a). Protection of bean against anthracnose by *Colletotrichum* species nonpathogenic on bean. *Phytopathol. Z.*, **86:**117–126.

Elliston, J., J. Kuć, and E. Williams (1976b). A comparative study of the development of compatible, incompatible and induced incompatible interactions between *Colletotrichum* species and *Phaseolus vulgaris. Phytopathol. Z.*, **87:**289–303.

Elliston, J., J. Kuć, and E. Williams (1976c). Effect of heat treatment on the resistance of *Phaseolus vulgaris* to *Colletotrichum lindemuthianum* and *Colletotrichum lagenarium. Phytopathol. Z.*, **88:**43–52.

Elliston, J., J. Kuć, E. Williams, and J. Rahe (1977). Relation of phytoalexin accumulation to local and systemic protection of bean against anthracnose. *Phytopathol. Z.*, **88:**114–130.

Farih, A., P. Tsao, and J. Menge (1981). Fungitoxic activity of efosite aluminum on growth sporulation and germination of *Phytophthora parasitica* and *P. citrophthora. Phytopathology*, **71:**934–936.

Friend, J. (1976). Lignification in infected tissue. In *Biochemical Aspects of Plant–Parasite Relationships* (J. Friend and D. Threlfall, eds.). Academic, New York, pp. 291–303.

Garas, N. and J. Kuć (1984). The extraction, assay and general properties of a resistance-inducing factor from cucumber leaves induced with *Pseudomonas lachrymans. Phytopathology*, **74:**873.

Gessler, C. and J. Kuć (1982). Induction of resistance to Fusarium wilt in cucumber by root and foliar pathogens. *Phytopathology*, **72:**1439–1441.

Goodman, R. (1967). The protection of apple tissue against *Erwinia amylovora* infection by avirulent strains and three other bacterial species *Phytopathology*, **57:**22–24.

Goodman, R. (1980). Defenses triggered by previous invaders: bacteria. In *Plant Diseases*, Vol. V, (J. Horsfall and E. Cowling, eds.). Academic, New York, pp. 305–317.

Grijseels, A., Williams, E. and J. Kuć (1964). Hypersensitive response in selections of *Malus* to fungi nonpathogenic to apple. *Phytopathology*, **54:**1152–1154.

Hahn, M., A. Darvill, and P. Albersheim (1981). Host-pathogen interactions XIX: The endogenous elicitor, a fragment of a plant cell wall polysaccharide that elicits phytoalexin accumulation in soybeans. *Plant Physiol.*, **68:**1161–1169.

Hammerschmidt, R. and J. Kuć (1982). Lignification as a mechanism for induced systemic resistance in cucumber. *Physiol. Plant Pathol.*, **20:**61–71.

Hammerschmidt, R., E. Nuckles, and J. Kuć (1982). Association of enhanced peroxidase activity with induced systemic resistance of cucumber to *Colletotrichum lagenarium. Physiol. Plant Pathol.*, **20:**73–82.

Ingham, J. (1982). Phytoalexins from the leguminosae. In *Phytoalexins* (J. Bailey and J. Mansfield, eds.). Blackie and Sons, Glasgow, pp. 21–80.

Jenns, A. and J. Kuć (1977). Localized infection with tobacco necrosis virus protects cucumber against *Colletotrichum lagenarium. Physiol. Plant Pathol.*, **11**:207–212.

Jenns, A. and J. Kuć (1979). Graft transmission of systemic resistance of cucumber to anthracnose induced by *Colletotrichum lagenarium* and tobacco necrosis virus. *Phytopathology*, **69**:753–756.

Jenns, A. and J. Kuć (1980). Characteristics of anthracnose resistance induced by localized infection with tobacco necrosis virus. *Physiol. Plant Pathol.*, **17**:81–91.

Jenns, A., F. Caruso, and J. Kuć (1979). Nonspecific resistance to pathogens induced systemically by local infection of cucumber with tobacco necrosis virus, *Colletotrichum lagenarium.* (Pass) Ell et Halst or *Pseudomonas lachrymans* (Sm et Bryan) Carsner. *Phytopathol. Mediterr.*, **18**:129–134.

Karban, R. and J. Carey (1984). Induced resistance of cotton seedlings to mites. *Science*, **225**:53–54.

Keen, N. (1982a). Specific recognition in gene for gene host-parasite systems. In *Advances in Plant Pathology*, Vol. I (D. Ingram and P. Williams, eds.). Academic, London, pp. 35–82.

Keen, N. (1982b). Phytoalexins—progress in regulation of their accumulation in gene-for-gene interactions. In *Plant Infection* (Y. Asada, W. Bushnell, S. Ouchi, and C. Vance, eds.). Japan Scientific Society Press, Tokyo and Springer-Verlag, Berlin, pp. 281–299.

Kuć, J. (1962). Production of extracellular enzymes by *Cladosporium cucumerinum. Phytopathology*, **52**:961–963.

Kuć, J. (1972). Phytoalexins. *Annu. Rev. Phytopathol.*, **10**:207–232.

Kuć, J. (1982a). Phytoalexins from the Solanaceae. In *Phytoalexins* (J. Bailey and J. Mansfield, eds.). Blackie and Sons, Glasgow, pp. 81–105.

Kuć, J. (1982b). Induced immunity to plant disease. *Bioscience*, **32**:854–860.

Kuć, J. (1983). Induced systemic resistance in plants to diseases caused by fungi and bacteria. In *The dynamics of Host Defense* (J. Bailey and B. Deverall, eds.). Academic, Sydney, pp. 191–221.

Kuć, J. (1984). Phytoalexins and disease resistance from a perspective of evolution and adaptation. In *Origins and Development of Adaptation.* Ciba Foundation Symposium 102, Pitman, London, pp. 100–118.

Kuć, J. (1985). Expression of latent genetic information for disease resistance in plants. In *Cellular and Molecular Biology of Plant Stress* (J. Key and T. Kosuge, eds.). A. Liss, New York, pp. 303–318.

Kuć, J. and F. Caruso (1977). Activated coordinated chemical defense against disease in plants. In *Host Resistance To Pests* (P. Hedin, ed.). American Chemical Society Press, Washington, DC, pp. 78–89.

Kuć, J. and C. Preisig (1984). Fungal regulation of disease resistance mechanisms in plants. *Mycologia*, **76**(5):767–784.

Kuć, J. and S. Richmond (1977). Aspects of the protection of cucumber against *Colletotrichum lagenarium* by *Colletotrichum lagenarium. Phytopathology*, **67**:533–536.

Kuć, J. and J. Rush (1985). Phytoalexins. *Arch. Biochem. Biophys.*, **236**:379–389.

Kuć, J. and S. Tuzun (1983). Immunization for disease resistance in tobacco. *Rec. Adv. Tobacco Sci.*, **9**:179–213.

Kuć, J., A. Barnes, A. Daftsios, and E. Williams (1959). The effect of amino acids on the susceptibility of apple varieties to scab. *Phytopathology*, **49**:313–315.

Kuć, J., E. Williams, M. Maconkin, J. Ginzel, A. Ross, and L. Freedman (1967). Factors in the resistance of apple to *Botryosphaeria ribis. Phytopathology*, **57**:38–42.

Kuć, J., G. Schockley, and K. Kearney (1975). Protection of cucumber against *Colletotrichum lagenarium* by *Colletotrichum lagenarium. Physiol. Plant Pathol.*, **7**:195–199.

Kuć, J., E. Tjamos, and R. Bostock (1984). Metabolic regulation of terpenoid accumulation and disease resistance in potato. In *Isopentenoids in Plants—Biochemistry and Function* (D. Ness, G. Fuller, and L. Tsai, eds.). Marcel Dekker, New York, pp. 103–126.

Lancake, P. (1981). Alternative chemical agents for controlling plant disease. *Phil. Trans. R. Soc. London*, **B295:**83–101.

Lazarovits, G., P. Stossel, and E. Ward (1981). Age-related changes in specificity and glyceollin production in the hypocotyl reaction of soybeans to *Phytophthora megasperma* var. *sojae*. *Phytopathology*, **71:**94–97.

MacLennan, D., J. Kuć, and E. Williams. (1963). Chemotherapy of the apple scab disease with butyric acid derivatives. *Phytopathology*, **53:**1261–1266.

Marx, J. (1984). Instability in plants and the ghost of Lamarck. *Science*, **224:**1415–1416.

McIntyre, J. and P. Miller (1977). Protection of tobacco against *Phytophthora parasitica* var. *nicotiana* by cultivar nonpathogenic races and *Pratylenchus penetrans*. *Phytopathology*, **68:**235–239.

McIntyre, J., J. Kuć, and E. Williams (1973). Protection of pear against fireblight by bacteria and bacterial sonicates. *Phytopathology*, **63:**872–877.

McIntyre, J., J. Kuć, and E. Williams (1975). Protection of Bartlett pear against fireblight with deoxyribonucleic acid from virulent and avirulent *Erwinia amylovora*. *Physiol. Plant Pathol.*, **7:**153–170.

McIntrye, J., J. Dodds, and J. Hare (1981). Effects of localized infections of *Nicotiana tabacum* by tobacco mosaic virus on systemic resistance against diverse pathogens and an insect. *Phytopathology*, **71:**297–301.

Mills, P. and R. Wood (1984). The effects of polyacrylic acid, acetylsalicylic acid, and salicylic acid on resistance of cucumber to *Colletotrichum lagenarium*. *Phytopathol. Z.*, **111:**209–216.

Moraes, W., E. Martins, M. Musumeci and M. Beretta (1976). Induced protection to *Hemileia vastatrix* in coffee plants. *Summa Phytopathol.*, **2:**39–43.

Nothnagel, E., M. McNeil, P. Albersheim, and A. Dell (1983). Host-pathogen interactions XXII. A galacturonic acid oligosaccharide from plant cell walls elicits phytoalexins. *Plant Physiol.*, **71:**916–926.

Nusbaum, C. and G. Keitt (1938). A cytological study of host-parasite relations with *Venturia inaequalis* on apple leaves. *J. Agric. Res.*, **56:**595–618.

Ossowski, P., A. Pilotti, P. Garegg, and B. Lindberg (1984). Synthesis of a glucoheptaose and a glucooctaose that elicit phytoalexin accumulation in soybean. *J. Biol. Chem.*, **259:**11337–11340.

Richmond, S., J. Kuć, and J. Elliston (1979). Penetration of cucumber leaves by *Colletotrichum lagenarium* is reduced in plants systemically protected by previous infection with the pathogen. *Physiol. Plant Pathol.*, **14:**329–338.

Roberts, D. (1983). Acquired resistance to tobacco mosaic virus transmitted to the progeny of hypersensitive tobacco. *Virology*, **121:**161–163.

Ross, A. (1966). Systemic effects of local lesion formation. In *Viruses of Plants* (A. Beemster and S. Dijkstra, eds.). North Holland Publishing Co., Amsterdam, pp. 127–150.

Ryan, C., P. Bishop, M. Walker-Simmons, W. Brown, and J. Graham (1985). Pectic fragments regulate the expression of proteinase inhibitor genes in plants. In *Cellular and Molecular Biology of Plant Stress* (J. Key and T. Kosuge, eds.). A. Liss, New York, pp. 319–334.

Sanudo, B. (1982). Algunas consideraciones sobre resistencia inducida a la antracnosis (*Colletotrichum lindemuthianum*) en el frijol Diacol Cindino. *Ascolfi Informa*, **8**(3):28–29.

Sekizawa, Y. and S. Mase (1980). Recent progress in studies of the non-fungicidal controlling agent, probenazole, with reference to induced resistance mechanisms of the rice plant. *Rev. Plant Prot. Res.*, **13:**114–121.

Sequeira, L. (1983). Mechanisms of induced resistance in plants. *Annu. Rev. Microbiol.*, **37**:51–79.

Sharp, J., M. McNeil, and P. Albersheim (1984a). The primary structures of one elicitor-active and seven elicitor-inactive Hexa (β-D-glucopyranosyl)-D-glucitols isolated from the mycelial walls of *Phytophthora megasperma* f. sp. *glycinea*. *J. Biol. Chem.*, **259**:11321–11336.

Sharp, J., B. Valent, and P. Albersheim (1984b). Purification and partial characterization of a β-glucan fragment that elicits phytoalexin accumulation in soybean. *J. Biol. Chem.*, **259**:11312–11320.

Sharp, J., P. Albersheim, P. Ossowski, A. Pilotti, P. Garegg, and B. Lindberg (1984c). Comparison of the structures and elicitor activities of a synthetic and a mycelial-wall derived hexa(β-D-glucopyranosyl)-D-glucitol. *J. Biol. Chem.*, **259**:11341–11345.

Shimotsuma, M., J. Kuć, and C. Jones (1972). The effects of prior inoculations with nonpathogenic fungi on Fusarium wilt of watermelon. *Hortscience*, **7**(1):72–73.

Smith, D. (1982). Toxicity of phytoalexins. In *Phytoalexins* (J. Bailey and J. Mansfield, eds.). Blackie and Sons, Glasgow, pp. 218–252.

Stermer, B. and R. Hammerschmidt (1985). The induction of disease resistance by heat shock. In *Cellular and Molecular Biology of Plant Stress* (J. Key and T. Kosuge, eds.). A. Liss, New York, pp. 291–302.

Stossel, P., G. Lazarovits, and E. Ward (1980). Penetration and growth of compatible and incompatible races of *Phtophthora megasperma* var. *sojae* in soybean hypocotyls differing in age. *Can. J. Bot.*, **58**:2594–2601.

Stossel, P., G. Lazarovits, and E. Ward (1981). Electron microscope study of race-specific and age-related resistant and susceptible reactions of soybeans to *Phytophthora megasperma* var. *sojae*. *Phytopathology*, **71**:617–623.

Stossl, A. (1982). Biosynthesis of phytoalexins. In *Phytoalexins* (J. Bailey and J. Mansfield, eds.). Blackie and Sons, London, pp. 133–180.

Stossl, A. (1983). Secondary plant metabolites in preinfectional and postinfectional resistance. In *The dynamics of Host Defense* (J. Bailey and B. Deverall, eds.). Academic, New York, pp. 71–122.

Sutton, D. (1979). Systemic cross protection in bean against *Colletotrichum lindemuthianum*. *Australas. Plant Pathol.*, **8**:4–5.

Tuzun, S. and J. Kuć (1983). A new technique which immunizes against blue mold (*Peronospora hyoscyami* f. sp. *tabacina*) and increases growth of tobacco. *Phytopathology*, **73**:823.

Tuzun, S. and J. Kuć (1985a). Movement of a factor in tobacco infected with *Peronospora tabacina* Adam which systemically protects against blue mold. *Physiol. Plant Pathol.* **26**:321–330.

Tuzun, S. and J. Kuć (1985b). Transfer of induced systemic resistance to blue mold *Peronospora tabacina* Adam via callus. *Phytopathology*, **75**:1304.

Tuzun, S., W. Nesmith, and J. Kuć (1984). The effect of stem infections with *Peronospora tabacina* and metalaxyl treatment on growth of tobacco and protection against blue mold in the field. *Phytopathology*, **74**:804 (Abstract).

Vance, C., T. Kirk, and R. Sherwood (1980). Lignification as a mechanism for disease resistance. *Annu. Rev. Phytopathol.*, **18**:259–288.

VanEtten, H., D. Matthews, and D. Smith (1982). Metabolism of phytoalexins. In *Phytoalexins* (J. Bailey and J. Mansfield, eds.). Blackie and Son, Glasgow, pp. 181–217.

Walker-Simmons, M., L. Hadwiger, and C. Ryan (1983). Chitosans and pectic polysaccharides both induce the accumulation of the antifungal phytoalexin pisatin in pea pods and antinutrient proteinase inhibitors in tomato leaves. *Biochem. Biophys. Res. Comm.*, **110**:194–199.

Wallace, J., J. Kuć, and H. Draudt (1962). Biochemical changes in the water insoluble material of maturing apple fruit and their possible relationship to disease resistance. *Phytopathology*, **52**:1023–1027.

Ward, E. and R. Buzzell (1983). Influence of light, temperature and wounding on the expression of soybean genes for resistance to *Phytophthora megasperma* f. sp. *glycinea*. *Physiol. Plant Pathol.*, **23:**401–409.

Ward, E., P. Stossel, and G. Lazarovits (1981). Similarities between age-related and race-specific resistance of soybean hypocotyls to *Phytophthora megasperma* var. *sojae*. *Phytopathology*, **71:**504–508.

West, C., P. Moesta, D. Jin, A. Lois, and K. Wickham (1985). The role of pectic fragments of the plant cell wall in the response to biological stress. In *Cellular and Molecular Biology of Plant Stress* (J. Key and T. Kosuge, eds.). A. Liss, New York, pp. 335–349.

Yarwood, C. (1956). Cross protection with two rust fungi. *Phytopathology*, **46:**540–544.

Yoshikawa, M. and H. Masago (1982). Biochemical mechanism of glyceollin accumulation in soybean. In *Plant Infection* (Y. Asada, W. Bushnell, S. Ouchi, and C. Vance, eds.). Japan Society Scientific Press, Tokyo and Springer-Verlag, Berlin, pp. 265–279.

13

BIOCHEMICAL CONTROL OF FOREST PATHOGENS INSIDE THE TREE

ALOYS HÜTTERMANN
ANNEGRET HAARS
Forstbotanisches Institut, Universität Göttingen, West Germany

The root and butt rot caused by the Basidiomycete *Heterobasidion annosum* (Fr.) Bref. [formerly named *Fomes annosus* (Fr.) Cooke] is one of the most important diseases of conifers of the northern hemisphere. The yearly damage it causes to the coniferous forests of the north temperate zone of the world amounts to probably more than one billion U. S. dollars (Dimitri, 1980). This damage occurs in virtually all types of forests: the virgin, nonmanaged forests of Siberia or Turkey; the regulated and strictly planned sustained-yield forests of central Europe; and the short-rotation plantations of conifers in the southern United States (Dimitri, 1980).

With regard to its host and its biology, *Heterobasidion annosum* is very different from all other systems described in this book; therefore, it might be appropriate to allot more space for the description of the system than otherwise might be considered appropriate.

1. BIOLOGY OF *HETEROBASIDION ANNOSUM*

Heterobasidion annosum belongs to the *Hymenomycetidae*, a subclass of the Basidiomycetes, Family Poriaceae. Among these Taxa, it has the exceptional feature that in addition to fructification via the fruiting bodies, it forms conidiophores of the genus *Oedocephalum* form *Spiniger* (Stalpers, 1974). Both forms of sporulation now can be obtained under sterile and controlled conditions in the laboratory. Conidiophores are formed within a few days of starvation (Flemming and Hüttermann, 1981), whereas fruiting bodies are formed only after about 6 months of incubation on a very complex medium containing saw dust, a variety of carbon and vitamin sources, and even milled bones (Holt et al., 1983).

The genetics of this organism has been scarcely explored. Analysis of the progeny of a fruiting body isolated from the forests revealed a bipolar mechanism of genetic control of fruiting-body formation as well as "Grenzhaut"-formation (Holt et al., 1983). The phenomenon of "Grenzhaut"-formation, which has not been analyzed with modern methodology, is a common one in wood destroying fungi (Jahn, 1979). It was first described for cultures of *Heterobasidion annosum* by Falck (1930).

Korhonen (1978), who analyzed 860 isolates of the fungus from different hosts and areas of the world, identified two different intersterility groups, P and S, with regard to fusion behavior of the mycelia on agar plates, which she could assign to the two main hosts of the fungus, P to Pine and S to Norway Spruce. There may, however, be more such intersterility groups of the fungus in the world (Korhonen, 1978). Studying the reaction of 31 strains (resp. isolates) of *H. annosum* Volger et al. (1977) demonstrated that these strains show conspicuous variations with regard to their growth on glucose and cellobiose at different pH values. Their results suggest a rather high genetic variability in this species. This was confirmed by more detailed physiological studies, including lignin-derivates as growth substances (Hüttermann et al., 1979). The same authors showed a high degree of variation in the isoenzyme pattern of two enzymes, lactate dehydrogenase and malate dehydrogenase, indicating a high degree of genetic variation in the loci coding for these enzymes. According to the classification of Johnson (1974), these two enzymes must be regarded, at least under the conditions inside the living tree, as nonregulatory enzymes, in that no key function in the metabolism can be ascribed to the reactions they catalyze. This situation is rather different for the enzyme polyphenole oxidase (laccase) examined in the same study. Here only one isoenzyme was found, exhibiting no variation in all 33 isolates studied (Hüttermann et al., 1979).

1.1. Infection Biology

Almost 150 species of trees, including some angiosperms, have been reported as being infected by *H. annosum*. The major hosts, however, are species of *Abies, Juniperus, Larix, Picea, Pseudotsuga,* and *Tsuga* (Sinclair, 1964). On most species of *Pinus* and *Juniperus*, the fungus causes only root rot and kills the tree within several months or few years after infection. More intriguing, especially for the regulated forests in central Europe and Scandinavia, is the course of the disease within the Norway Spruce (*Picea abies* Karst.). Here the fungus may remain inside the tree for 10 or more years without actually killing the host. This type of disease is discussed in this chapter.

Regardless of host and type of disease, there are two ways by which the spread of the fungus is accomplished—by spores and by contacts between healthy and infected roots.

1.2. Spores

As mentioned above, the fungus has two different types of spores for long-range dispersal, basiodiospores and conidiospores. Both types of sporophores have been found in nature and their spores are present at rather high densities in the air near forests in Central Europe (Dimitri, 1963).

Most workers have been unable to detect any considerable growth of mycelia of *H. annosum* in nonsterile soils (Bjorkman 1949; Molin, 1957). It therefore appears unlikely that the fungus is able to grow more than a few centimeters through nonsterile humus or litter. Thus the main pathway of infection is not via growth of mycelia through the soil, as is the case with most soilborne fungal diseases, instead of which the spore itself must reach a suitable target for infection. This can be either a wounded root of a healthy tree inside the soil or an infection of a stump after thinning or clearing out of the stand.

1.3. Direct Infection of Roots

Rishbeth (1951) demonstrated that spores are washed deeply into the soil; mineral soil taken from depths 90 cm below the forest floor still contained viable and infectious spores. It is now generally accepted that one way of infecting the spruce is the colonization of their roots by the fungus. If spores germinate in the vicinity of young or small-sized roots, the developing mycelium can in some cases penetrate the periderm (this was shown for roots with diameters up to 3 cm in the laboratory; Braun and Lulev, 1969, 1970; Peek et al., 1972). It is more likely, however, that the fungus will attack wounded roots. The wounds may be frost cracks or may have been caused by stones moved by the dynamics of the soil, owing to temperature differences during the four seasons. In addition to these "natural" causes, wounds may be applied to the roots by all kinds of modern forest practices like logging trees through the stand or simply driving heavy machinery over them.

1.4. Infection via Root Contacts and Through the Stump Surface

Hartig (1978) observed that because of the close contacts within the root system in our forests, the fungus is able to spread from an infected root to a healthy one. This pathway of infection is now considered the main and most important one, especially since Rishbeth (1951) showed that freshly cut stump surfaces are infected rather fast and easily by airborne spores. This observation, which has been confirmed by all subsequent studies, means that any open surfaces (stumps delivered by thinning of the stands or wounds in the trunk) will be infected rather fast by spores of *H. annosum*. The developing mycelium then spreads inside the trunk or roots rather efficiently with a speed of up to 2 m/yr (Rishbeth, 1957; Fowler, 1962). It eventually enters the trunk of the tree and colonizes the heartwood of the spruce up to a height of 10 m (Butin, 1983). This means that the most valuable part of the tree, the base of the trunk, is rendered useless.

The life of the fungus is not bound to living parts of the tree. It may survive for many years in dead trees or stumps. Low and Gladman (1960) and Rishbeth (1951) found the fungus in stumps up to 44 yr after felling.

To summarize the infection biology of the fungus, one has to state that this part of fungal biology apparently gives very little chance for any sort of control. The fungus does not grow in the soil. Its active life exclusively occurs inside the tree, either in the root system or in the trunk. Thus no biological control by antagonists (as in the systems described in this book) in the soil is possible; nor can systemic fungicidal agents be applied, since it is impossible to reach the complete root system with such agents. In addition, these agents need a long residence time to prevent the fungus from later out-growth. This fact alone would prevent any such means of control in our forests because of legal restriction. Therefore, one has to look for other means of controlling the fungus.

1.4.1. Attempts for Control of Fungus by Forestry Practices

In view of the economical importance of root and butt rot, especially in sustained-yield forests, strategies for avoiding infection by the fungus or at least reducing the amount of losses by appropriate silvicultural means or resistance breeding have been developed and tested in many countries during this century. The results of these attempts, however, have not been very promising.

In very careful investigations, Evers (1973) and Rehfuess (1973) studied the relationship between stand characteristics, especially soil chemistry, and the incidence of butt rot. These and other authors (Schlenker, 1976) came to the conclusion that an inverse relationship exists between the amount of calcium in the soil and the state of health of the trees. With increasing decalcification of the soils (acidification of topsoils), damage due to butt rot decreases significantly (Schlenker, 1976).

This phenomenon probably occurs because a higher infestation of antagonistic fungi is present in acid than in alkaline soils. This notion is supported by the fact that in new afforestations on acid soils, the incidence of butt rot is rather high, and on calcareous soils the percentage of diseased trees is highest on newly afforestated stands (Werner, 1973).

Haas (1971) reached the same conclusions by investigating the fungal sociology of these stands. He reported conclusive correlations between soil chemistry, composition of the macromycete flora, and incidence of butt rot. With increasing numbers of neutrophilic fungi, he found a higher incidence of butt rot, whereas with increasing frequence of acidophilic fungi, the portion of infected trees decreased (Haas, 1971).

Although these studies, which have been confirmed in other parts of the world, are very interesting with regard to the biology of the fungal–host interactions, they mean that virtually the best stands for growing spruce are exactly those with the highest risk of infection. Therefore the possibility of silvicultural means of control of this disease is almost nonexistent (Schmidt-Vogt and v. Schnurbein, 1976). Only under special conditions can a reduction of the disease be achieved; by reducing the area planted with spruce, an overall positive balance is expected (Moosmayer, 1971).

1.4.2. Resistance Breeding

The possibility of selecting and breeding resistance clones of spruce has been explored by several authors during the last decades. The most careful investigations were carried out by Dimitri in Hann.Münden, West Germany, and are summarized in his book (Dimitri, 1980b). Although these studies, which were carried out for more than one and a half decades, both in the field and in the laboratory, are an excellent piece of erudite scientific work, they yielded no practical result so far that would help foresters to prevent the disease. In addition, the criticism levelled against this concept has to be taken rather seriously. Unlike agricultural crops, which in general are genetically rather homogenous, if not even clones, forest stands are planted with seeds or trees that have a rather high variety in their genetical background. This concept has been very successful so far in the forestry practice when stands with long rotation times are to be established. The stands start with a population of some 5–20 if not 100,000 trees/ha, which number is reduced at a stand age of 120–200 yr to about 5–800 trees. During this time the stand has to cope with all kinds of soil conditions, climate, and air pollution. The stability of the stand as a whole, depends on the genetic variability of its individuals, which are selected and survive according to the stress conditions they meet.

Transferred to forestry practice this would mean that even if we could find one or several "superclones" that are completely resistant to both the attack and the spreading of the mycelia inside the tree, these clones would have rather limited practical value. No forester would establish and plant a new stand with an expected rotation time of about 100 yr on the population-genetics background of only a few clones. On the basis of 300 yr of experience in regulated sustained-yield forestry, this would be considered much too hazardous.

Research on both infection and spreading resistance is of high scientific value in understanding the host–parasite interaction and the genetics of both the host and the fungus (Ekman and v. Weissenberg, 1981), although the practical results for avoiding strategies against butt rot are missing so far.

Summarizing the results of research on control or reduction of root and butt rot, one unfortunately has to conclude that virtually all concepts presented thus far, have failed to yield any visible results.

1.5. Economical and Ecological Considerations for Prophylactic Treatment of Living Trees Against Fungi

From the biology and pathway of infection of the fungus as outlined above, it appears that the only strategy feasible for effective control of this fungus is the prophylactic treatment of healthy trees. This must be employed so as to block the access of the fungus to the trunk and to prevent further spreading, thus preserving the most valuable part of the tree from deterioration.

Although it might seem that such treatments would be rather expensive and thus uneconomical, this is not the case in the situation dealt with in this chapter. In a stand with a high risk of butt rot, which is almost always the case in the best and fastest growing spruce stands, protection for the last 10 or even 5 yr of the rotation time of the stand would result in a substantial gain for the forest service in charge. This can be demonstrated by the following example: in spruce stands of the highest yield class according to Kraft, the last 10 yr of rotation time will increase the value of an average trunk from about $25 to about $50, because during this time the stand will grow into the highest timber classifications, obviously yielding the highest prices per volume on the market. In central Europe the decision to let the stand grow and increase in value for another 10 yr may be a rather risky one. As outlined above, there is no means of detecting an infection of the fungus inside the root system. If, however, the stand is already infected within the root system, chances are very high that after another 10 yr most of the timber will be devalued and worthless because butt rot has deteriorated the lower 5–10 m of the trunk. In such a case, any prophylactic treatment at reasonable cost would result in a rather sound cost–benefit relationship.

1.6. First Attempts for Control of Fungi Inside the Tree

The idea of treating healthy trees with poisons to kill organisms connected with it, is surprisingly old. The first mention of this concept dates back to Leonardo da Vinci. He reports that, in the Renaissance, fruit trees were injected with arsenous compounds to poison the fruits of the trees, without leaving marks on the fruits themselves, which would have aroused suspicion. The overall "success" of this method, however, has not been reported, so that it was probably rather low (Leonardo da Vinci, quoted from Falck, 1935).

The first application of this concept in modern forestry was made by Richard Falck (1935), who was head of the Institute for Technical Mycology at the School of Forestry in Hann.Münden, Germany. His general approach was the same one that we use today. He injected the tree with solutions of antifungal substances, in the hope that this treatment would prevent the fungi from moving up into the trunk (Falck, 1935). The chemicals used by Falck were introduced by him for the protection of timber against fungal decay: arsenous compounds, respiratory poisons like dinitrophenole and dinitro-*o*-cresole, and chlorinated aromatic compounds such as dichlorobenzene, chlorotoluene, monochlornaphthalene, and chloroxylene.

In his description of the results of his studies, he relates that some of the agents, at least in the concentrations used by him, killed the trees almost immediately, especially when arsenous compounds were applied.

However, he also reports the cure of a very old linden tree (which was according to a legend planted by St. Bonifatius in the ninth century) by the application of a paste consisting of 50% monochloronaphthalene, 35% chloroxylene, and 3% dinitro-*o*-cresole in 12% flax varnish as a solvent. The treatment was applied in 1925 and the tree, apparently, was still flourishing in 1933, when Falck was forced to leave Germany. It was not possible to find out whether the tree is still alive; if this were so, it would be the first case of successful long-term cure using this method. Today, however, it would be impossible to use the chemicals employed by Falck, for obvious environmental reasons. Therefore, one has to start with an entirely different approach, and help the tree cope with the fungus by either interfering with its biochemistry in order to weaken the fungus or to enhance the defense mechanism the tree has established against the fungus.

2. BIOCHEMISTRY OF HETEROBASIDION ANNOSUM

In spite of the important role of *H. annosum* as a tree pathogen, the biochemical research on its mechanism of pathogenicity is still meagre. Only detailed knowledge of the metabolism of this wood-destroying fungus can lead to its control according to the two approaches mentioned above. To achieve this goal, it is necessary to exploit the biochemical differences between the host and the fungus as much as possible. Therefore, extensive studies of some of the main spheres of metabolism were initiated as a cooperative venture by the Institute of Forest Botany in Göttingen and the Department of Agriculture in Rehovot.

2.1. Metabolism of the Cell Wall

One very significant difference between the host and the pathogen is the cell wall. The cell wall of the tree contains cellulose and lignin as its main constituents, whereas fungal cell walls are built up from chitin and various glucans.

Cells walls of *H. annosum* contain basically the same structural components as other basidiomycetes, glucans and chitin (Chet and Hüttermann, 1980). Besides chitin, which contributes 16.4% of the wall, the main components are β-(1,3), β-(1,6)-glucans similar to those isolated from *Schizophyllum commune* by De Vries and Wessels (1975). The cell wall hydrolysate also contains mannose and xylose. A remarkable residue, however, remains insoluble after chemical and enzymatic hydrolysis, which is typical of walls of resting structures but not of other hyphal walls (Chet et al., 1967).

Another possible target for the biochemical control of the fungus is the enzyme systems involved in the cell-wall synthesis. The last and probably most important step of the chitin biosynthesis is catalyzed by chitin synthetase (E.C.2.4.1.16), which transfers *N*-acetylglucosamine units from uridine-diphospho (UDP) acetylglucosamine to acceptors and is markedly activated by *N*-acetylglucosamine.

The chitin synthetase from *H. annosum* has properties similar to the same enzyme from other fungal sources: it requires Mg^{2+}, proteolytic activation, and allosteric activation by *N*-acetylglucosamine. A total of 90% of the enzyme activity is located on membranes having high sedimentation coefficients (Flemming and Hüttermann, in press).

2.2. Metabolism During Differentiation

Differentiations of the pathogen that lead to dormant stages or to changes in the growth pattern of the hyphae may also be of interest for achieving biochemical control. Among the various chemicals that had morphogenetic effects in the basidiomycete *Sclerotium rolfsii* (Chet et al., 1966, Okon et al., 1972, 1973; Henis et al., 1973) L-threonine in a concentration of 10^{-2} *M* specifically induced the branching of hyphae and septum formation in the leading hyphae of the mycelium. Ethanol (2%) inhibited the formation of side hyphae and led to a very delayed appearance of hyphae. The formation of conidia could be stimulated by the combination of light and dark phases and starvation (Flemming and Hüttermann, 1981).

Growth and branching of hyphae as well as differentiation of the mycelium to conidiophores is intimately connected with protein turnover. Therefore, the intracellular digestion of protein by proteolytic enzymes such as aminopeptidases must play an important role in these processes. Extracts from both growing and threonine-induced branching mycelia exhibited high specific aminopeptidase activity, whereas a considerably lower activity was found in conidia. The changes in specificity were due to shifts in the isoenzyme spectra during differentiation. Density labeling experiments using deuterated amino acids revealed a *de novo* synthesis of leucine- and alanine aminopeptidase after 4 days during growth and threonine-induced branching (Haars et al., 1984).

2.3. Carbohydrate Metabolism

It is taken for granted that *H. annosum* has no competitive ability as a saprophyte in soils. Its aggressivity, as illustrated by its growth in wood, however, calls for complex adaptations in its biology. Some of these adaptations can be assumed to be correlated to glycosidic enzymes and their regulation according to the substrate. Hüttermann and Volger (1973b) found (using inhibitors of transcription and translation and the density labeling technique) that the increased activity of aryl-β-glucosidase, after the transfer of mycelia from glucose to cellobiose-containing media, is due to *de novo* synthesis of the enzyme. By isoelectric focusing it was shown that only one of four isoenzymes is inducible by cellobiose. The mycelial yields were 3 times higher on cellobiose than on glucose, which was attributed to the activity of the enzyme cellobiose phosphorylase (E.C.2.4.1.20), the first evidence for the presence of this enzyme in a fungus (Hüttermann and Volger, 1973a). Such growth stimulation on supplying di- or oligosaccharides instead of glucose is typical for bacteria that have enzymes that cleave the glycosidic bonds phosphorolytically rather than hydrolytically. The energy of the bond is conserved in glucose-1-phosphate and thus gives the system a much higher yield.

The white-rot fungus *H. annosum* is able to degrade cellulose and lignin under rather extreme conditions (low oxygen, high carbon dioxide). The first example of an enzyme that links the two areas of metabolism, phenols and cellulose, was described by Westermark and Eriksson (1974) in the fungus *Sporotrichum pulverulentum*, the cellobiose-quinone-oxidoreductase (or cellobiose dehydrogenase). It counteracts the action of phenoloxidase (E.C.1.14.18.1) by converting the quinones formed by phenoloxidase back to phenols, using the oxygen for the conversion of cellobiose to cellobiono-lactone, which is then converted to cellobionic acid by a lactonase (Westermark and Eriksson, 1974).

Hüttermann and Noelle (1982) found that *H. annosum* secretes high activities of the enzyme in the beginning of the growth phase. From the viewpoint of a participation of the enzyme in lignin metabolism this fact is rather difficult to interprete (see discussion below).

The physical properties of the cellobiose dehydrogenase of *H. annosum* are rather similar to the ones reported on enzymes of other organisms. An interesting observation was that its secretion is induced both by cellobiose *and* lignosulfonate, which had not been reported before.

The role of the enzyme in the degradation of wood still is not completely clear, although the presence of this enzyme linking cellulose and lignin metabolism has been discussed intensively (Crawford, 1981).

That lignosulfonate has an influence on carbohydrate catabolism was also shown by studies on the effect of L-threonine on oxalic acid synthesis by *H. annosum* (Hüttermann et al., 1980). This amino acid has not only a morphogenic effect on *H. annosum*, but it also causes an inhibition of the glyoxylate dehydrogenase activity and the synthesis of oxalic acid at a much lower concentration (1 m*M*), than is needed for the morphogenic effect. Malate synthetase was synthesized *de novo* in the presence of L-threonine. Lignosulfonate completely blocked the effect of the amino acid in both glucose or cellobiose-containing medium. The threonine effect was reported formerly for another pathogenic fungus, *S. rolfsii* (Kritzmann et al., 1976).

The secretion of oxalic acid by *H. annosum* when grown in the living tree was observed by von Aufsess (1972). The key biochemical reaction leading to the formation of oxalic acid is the oxidation of glyoxylic acid (derived from isocitrate) by the enzyme glyoxylate dehydrogenase (E.C.1.2.1.17). From this it is to be assumed that under the influence of L-threonine the glyoxylate shunt converts glyoxylate to malate. Apparently lignosulfonate blocks the activation of the glyoxylate cycle so that the fungus operates only on the tricarboxylic cycle, even in the presence of L-threonine. According to findings about the similarity between the physiology of *S. rolfsii* and *H. annosum* it is likely that oxalic acid is playing an important role in the pathogenicity of *Heterobasidion*. Inhibition of its production may help in the control of the pathogen.

2.4. Lignin Metabolism

Since the early work of Falck (1926), which has been confirmed repeatedly (Zycha, 1975), it is well known that lignin is preferentially degraded by *H. annosum* during its life inside the living spruce trees. Thus the content of lignin decreases continually after attack by *H. annosum*, until almost complete degradation, whereas cellulose degradation is considerably slower. It must be assumed that lignin degradation is of paramount importance during the infection of spruce by the fungus. The study of lignin metabolism, therefore, is useful because it yields information on the biochemical mechanisms of degradation of the second most important natural occurring substance. In addition, the studies on lignin biodegradation of *H. annosum* yield further insights about a very important part of the infection biology of the pathogen.

Our focus is on the following aspects:

1. Environmental parameters that influence and regulate lignin degradation.
2. Macromolecular mechanism of lignin degradation.
3. The mechanism of the ligninolytic system.
4. The role of phenol oxidase during lignin degradation.

1. As a typical white-rot fungus *H. annosum* is able to metabolize ^{14}C-DHP-lignin, labeled in all three main groups of the molecule: ring, side-chain, and methyl group (Trojanowski et al., 1984). It has been reported for other white-rot fungi that lignin degradation occurs in the stationary growth phase (Kirk et al., 1978). The metabolic regulation of lignin degradation by *H. annosum*, however, seems to be different from those of other white-rot fungi like *Phanaerochaete chrysosporium* (*Sporotrichum pulverulentum*). The main lignolytic activity was observed at the beginning of the growth phase. In many microorganisms secondary metabolism is initiated in response to several types of nutrient limitation. The ligninolytic system of *Phanaerochaete chrysosporium* is reported to be triggered by nitrogen depletion (Kirk et al., 1978). That the nitrogen source is important for the $^{14}CO_2$ release from labeled lignin was also evaluated by Reid (1983a,b). Recently, however, other authors found lignolytic enzyme production under nonlimiting nitrogen conditions (Buswell et al., 1984). Nitrogen limitation in cultures of *H. annosum* had no significant effect on lignin degradation (Hüttermann et al., 1983).

As was shown for other white-rot fungi (Eriksson et al., 1980) *H. annosum* needs a carbohydrate in addition to lignin as a carbon source, otherwise it is not able to degrade the lignin.

2. Whether lignin is degraded via low-molecular-weight aromatic products or by cleavage of the aromatic nuclei still bound to the polymer is not yet clear. The studies of Haars and Hüttermann (1980) indicated that *H. annosum* is unable to split the lignin molecule directly into intermediate-sized pieces (oligomers) and that lignin biodegradation *in vivo* follows the mechanism of exodegradation.

Using water-soluble fluorescence-labeled lignins (Mw > 1 milli-Dalton) as substrates for *H. annosum* it was found that the primary cleavage products accumulated in the culture are small molecules with a molecular weight below 1000 Dalton (Haars et al., 1982). Other studies support the idea of a cleavage of the aromatic ring still bound to the polymer (Ellwardt et al., 1981; Janshekar et al., 1982). It is, however, likely that both mechanisms are involved and that lignin degradation proceeds by a mixed function, that is, ring-fusion reactions in a largely intact polymer, accompanied by the release of some low-molecular-weight lignin fragments (Janshekar and Fiechter, 1983).

3. The ultimate goal of present-day lignin biochemistry is to find means of isolating "ligninase," the enzyme complex responsible for lignin degradation. Recent research using the white-rot fungus *Phanaerochaete chrysosporium* has made rapid progress toward elucidating the molecular mechanism of the so-called "ligninase" system which was isolated from culture fluid, purified, and characterized (Tien and Kirk, 1984). The hemoprotein catalyzes in the presence of H_2O_2 a variety of seemingly different oxidations in lignins. Molecular oxygen is incorporated into the products of C_α-C_β cleavage of the propyl side chain of lignin. Very recently the formation of cation radicals from methoxybenzenes was described (Kersten et al., 1985). Thus the mechanism in whole includes initial one-electron oxidation of a specific lignin subunit followed by subsequent breakdown reaction via radical cation intermediates (Schoemaker et al., 1985). Whether these mechanisms are also due to the lignin breakdown caused by *H. annosum* is currently under research in this laboratory (Trojanwoski et al., unpublished).

4. The opposite of "ligninase" phenoloxidases are copper-containing enzymes that oxidize phenols or lignins using O_2 as an electron acceptor. The question of whether these enzymes (E.C.1.14.18.1) are involved in lignin biodegradation has been the subject of numerous investigations. It is known for several white-rot fungi including *H. annosum* that the presence of lignin or phenols in the culture fluid induce extracellular phenoloxidase activity (Hüttermann et al., 1977; Haars et al., 1981). The extracellular increase of phenoloxidase activity upon addition of lignosulfonate in *H. annosum* cultures is due to *de novo* synthesis of the enzyme, as was shown by density labeling (Haars and Hütterman, 1983). The intracellular enzyme, on the contrary, remained unchanged.

The results obtained from genetic studies on *Phanaerochaete chrysosporium* by Ander and Eriksson (1976) and Gold et al. (1980, 1982) indicate that phenoloxidase plays an essential role in lignin biodegradation. Our results confirm that only when laccase activity was inhibited by the Cu-chelating substance thioglycolic acid could cleavage products of lignosulfonate be detected in the culture fluid. In the presence of phenoloxidase the detection of these low-molecular-weight intermediates was prevented by immediate repolymerization (Haars and Hüttermann, 1980). The results obtained by gel chromatography of water-insoluble lignins before and after attack of *H. annosum* (Hüttermann et al., 1980a,b) reveal that phenoloxidase is responsible for the polymerization of lignins. This simultaneous polymerization and repolymerization may be one reason why lignin degradation products are barely detectable even with aggressive white-rot fungi.

If this enzyme is not or only partially involved in lignin breakdown, what then is its main function? Some authors suggested a detoxifying function of phenoloxidases (Grabbe et al., 1968; Popoff et al., 1975), because a *p*-diphenoloxidase was capable of detoxifying some growth inhibiting lignins in the reaction zone. We however, doubt the detoxifying role of phenoloxidases; just the opposite was observed—catechol and guaiacol were converted to toxic substances in the presence of phenoloxidase (Haars and Hüttermann, 1980). It seems that not the phenol itself, but its oxidation product, the quinone, is toxic to the fungus. This observation is confirmed by Green (1977), who suggests that the enzyme glucose oxidase protects the fungal cell from toxic accumulations of quinoid intermediates. Johansson et al. (1976) reported that hydrolytic enzymes of *H. annosum* are more inhibited in the presence than in the absence of phenoloxidase. These facts seem to be interesting footnotes for the control of the pathogen.

2.5. Starting points for Combating the Pathogen

The results presented above lead to an improved understanding of the vulnerable sides of the pathogen. In other words, it becomes clear which parts of the metabolism can be damaged by a "biochemical weapon" without harming the host.

1. The differences in the cell-wall composition of the living tree and the pathogen offer two possibilities for a selective destruction of the fungal cell wall (1) by application of a cell-wall lytic enzyme complex and (2) by specific inhibition of the enzyme chitin synthetase with antibiotics (Polyoxin D).
2. Because many phenols that are naturally occurring substances in the tree (e.g., tannins) are toxic to the fungus, the possibility exists of increasing the resistance of the tree against *H. annosum* by applying these phenols as so-called "slow-release" formulations.

Theoretically, several other possibilities of interference with the fungal metabolism exist, for example, manipulation of the oxalic acid secretion, which enables the fungus to destroy the middle lamella of cells, or the inhibition of the ligninolytic system. However, these and other alternatives are either more difficult to installate in the tree or the biochemical background is not yet understood, as is the case with the ligninolytic mechanism. Therefore we focused our research on the two points mentioned above.

2.5.1. *Lysis of the Cell Wall by the Trichoderma harzianum Enzyme Complex*

For many years the fungus *Trichoderma harzianum* has been successfully used in the biological control of soilborne plant pathogens (Chet et al., 1979). The main mechanism involved in the antagonism of *T. harzianum* and the pathogenic fungus is the release of lytic enzymes, which are β-1,3-glucanase and chitinase (Elad et al., 1982, 1984). The enzymes are induced when the fungus is grown on living fungi or on chitin and glucosan-containing material. The cheapest source we found for the production of high-enzyme yields were cells of *Penicillium chrysogenium*, which we obtained as oven-dried, milled material from the Hoechst Company in Frankfurt, FRG. After scaling-up and optimization of the process the enzyme concentration was 25 g dry material/100 l. In a concentration of 1 mg/l the enzyme complex inhibited the growth of *H. annosum* in liquid culture (Cwielong, 1979; Hüttermann and Cwielong, 1982).

2.5.2. *Inhibition of the Biosynthesis of the Cell Wall by the Specific Antibiotic Polyoxin D*

As described in the previous section, the chitin–glucosan complex is the main structural backbone of the cell wall of *H. annosum*. During the growth of hyphal tips, two enzymes, the anabolic chitinsynthetase and the catabolic chitinase act together in a delicate equilibrium. Chitinsynthetase is localized on chitosomes having a higher sedimentation coefficient than the ones from other fungi (Flemming, 1979). Therefore, chitinsynthetase is a promising candidate for *in vitro* and *in vivo* inhibition with antibiotics such as Polyoxin D. This antibiotic (produced by the Kaken Company, Tokyo, Japan) is a structure analogue (competitive inhibitor) to the natural substrate of the enzyme, UDP-glucose-*N*-acetylglucosamine.

The affinity of chitinsynthetase toward the drug was 100 times higher than to the natural substrate. This means that the antibiotic can be applied in very low concentrations. A concentration of 100 μm/ml Polyoxin inhibited the chitinsynthetase by more than 80%, which led to a split of the cell wall and effusion of the cytoplasm. A combined treatment with threonine, which increases the branching of the hyphae (compare preceding chapter) and so the number of possible attack points for Polyoxin, may lead to an improved fungistatic effect. The addition of certain phenols, however, which also induced branching, compensated for the Polyoxin effect. The phenol probably competes with the drug for the binding site of the enzyme, or the change in the protein configuration, so that an enzyme-inhibitor (Polyoxin) complex cannot be formed.

2.5.2. *Increasing the Resistance of the Tree Against the Pathogen by Application of Phenolic Substances*

Phenolic and aromatic substances play a major role in the course of infection of a tree by fungi—as part of the defensive mechanism of the tree as well as products occurring during lignin degradation by the fungus. This was shown, for instance, by Shain and Hillis (1971) who found that the invasion of *H. annosum* to the sapwood of Norway spruce was inhibited by the accumulation of lignins. The so-called "reaction zone," a tissue produced as host response in advance of fungal invasion, contained considerably more hydroxymatairesinol than the heartwood.

Popoff et al. (1975) found that not hydroxymatairesinol but 15 different phenols—especially 4-methylcatechol—were the cause of the growth inhibition. They also reported (Johansson et al., 1976) on the effect of phenols on enzymes that are important for the metabolic activity of the fungus in wood. The enzymes cellulase, polygalacturonidase, and aryl-β-glucosidase were inhibited by partly different lignin fractions. The hydrolyzing enzymes were more inhibited by the lignins in the presence of fungal laccase than in the absence of laccase.

This effect was also found by Haars et al. (1981). The growth of *H. annosum* in liquid medium was inhibited by phenols and tannins that were oxidized by the fungus and/or induced high-laccase activities. If laccase was inhibited in the medium by the addition of thioglycolic acid, no toxic effect was exerted by the phenol. A correlation between the induction of laccase activity and toxicity was found in many cases, whereas the correlation of quinone formation and growth inhibition was not as strict.

Therefore, the structure–toxicity relationship of a variety of mainly phenolic substances, as well as the mechanism of toxicity and resistance of the fungus to certain phenols, were studied intensively (Haars et al., 1983), so that a choice of suitable substances for application in a "slow-release" formulation in the living tree is possible. The results of these studies are presented in the following sections.

2.6. Toxicity Mechanism

2.6.1. *Phenols*

The growth rate of the fungus exposed to the phenols as a function of increasing concentration followed the typical sigmoid curve, which can often be observed in nature and can be interpreted as follows:

Phase 1. In low concentrations only very sensitive cells are killed.

Phase 2. In higher concentrations, up to a specific level, a logarithmic inhibition of most cells takes place.

Phase 3. In concentrations above the specific level many resistant cells are present on which the substance has no effect.

Using the so-called "DR-curve" (Horsfall, 1956), which is a transformation of the sigmoid curves, enables a better comparison of substances with respect to their toxic mechanism. During the testing of 150 different aromates, the following correlations between toxicity and structure were observed:

1. The substances must be water-soluble, unpolar substances had no toxic effect.

2. Phenols with a C_{1-3} side chain were more toxic than phenols without a side chain.

3. The toxicity of phenols without a side chain can be increased by the introduction of methoxy groups and phenolic OH-groups. However, the toxic effect was compensated for when the methoxy groups were in the *o*-position in relation to the phenolic groups, because no *o*-cleavage of the ring was possible.

Similar results were obtained with conidia. Phenols that inhibited mycelial growth also had an inhibiting effect on conidia. The morphological appearance of the mycelia, especially the hyphal tips, was drastically changed in the presence of many phenols. An explanation of these morphological alterations is a disturbance of the cell wall synthesis by either an allosteric activation or an inhibited inactivation of chitinsynthetase, which leads to an unbalanced chitin:glucan relationship. Besides the effect on chitinsynthetase, changes in the permeability of the cell wall were also observed. In the presence of toxic phenol concentrations the uptake of radioactive amino acids was totally inhibited. The underlying mechanism is supposed to be an oxidation of lipids by the phenolic radicals leading to the formation of bonds between adjacent fatty acids.

The oxidation products of the phenols, the quinones, have stronger effects on the metabolism of the cell, which is responsible for the toxicity, than on the membranes and the cell wall; for example, all energy supplying processes, such as glycolysis, TCC, and PCC, are nearly completely inhibited. The effect on the oxidoreductases isocitratdehydrogenase and glucose-6-phosphate dehydrogenase is especially strong. Other enzymes (e.g., acid phosphatase and aminopeptidase) are also affected. Laccase activity is not depressed, but rather is increased, as was described above. The mechanism is supposed to be an indirect decoupling of the oxidative phosphorylation (Grabbe et al., 1968; Johansson and Hägerby, 1974).

Other fungicidal substances include chlorinated aromates (e.g., pentachlorophenol, phenylmercuriacetate, or the phytohormone indolylacetic acid), which were strong inhibitors of mycelial growth. The last substance especially will, of course, affect the physiology of the tree, so that it cannot be applied in practice. Another suitable inhibitor is Sparassol from *Sparassis crispa* (Falck, 1923).

2.7. Resistance Mechanisms of *H. annosum* Against Fungicidal Substances

The prior condition for long-term application of fungicidal substances in the tree is proof that the fungus does not develop resistance against the compound. The problem of resistance always appears when an organism is exposed for a long time to sublethal doses of the toxic substance. Such conditions are expected in a fungicide depot inside the living tree because a concentration gradient is formed by the diffusion of the phenols.

Such a situation was simulated with different phenols in liquid culture for 1 yr. An adaptation of the fungus was not observed in any case.

Besides adaptation, another form of resistance known as spontaneous mutation exists. Ultraviolet irradiation did not induce the resistance of *H. annosum* to the four phenols, whereas exposure to the chemical mutagen *N*-methyl-*N*-nitro-*N*-nitrosoguanidine (NMG) induced low resistance against some phenols which, however, was lost after transfer from solid to liquid media. The causes for these different effects in solid and liquid media are not yet understood. Probably the higher diffusion velocity in liquid medium, the different composition of the media, or different osmotic pressure was responsible for this effect.

Since a detoxifying effect is attributed to laccase by some authors (see discussion above), an attempt was made to isolate a laccase-deficient mutant. This would probably elucidate questions about whether this enzyme is really involved in the toxicity mechanism of the phenols and whether the phenols act in reduced or oxidized form. Using two types of mutagenesis (UV and NMG), researchers succeeded in isolating several auxotrophic and thermosensitive mutants, but not a laccase-less clone. An explanation for this phenomenon can be the findings of Hüttermann et al. (1979) and the possibility that two genes code the synthesis of the laccase protein.

It can be stated that *H. annosum* has no "repair" mechanism to detoxify the substances or achieve resistance against them during long-term exposure. This seems to be a hopeful aspect for applications in "depot-formulation," which is described in the next section.

2.8. Control of *H. annosum* Inside the Tree with "Slow-Release" Formulations of Fungicidal Substances

The classical methods of applying fungicides (e.g., in the soil or at critical locations of infection) are, of course, not possible in the case of *H. annosum*. Since direct control of the fungus in the forest so far seems to be impossible, two other methods of coping with the problem have been tried:

1. Selection and breeding of resistant clones of spruce (Dimitri, 1980a,b).
2. Control of *H. annosum* inside the tree with "slow-release" formulations (Dimitri et al., 1983).

On the basis of the data obtained in liquid culture, the best phenols to use for treatment of trees against *H. annosum* are phenols with side chains (see Section 2.6.1). Following this, we tested some phenols and tannins on spruce stem slices, a more natural substrate. Some phenols that were toxic in liquid medium lost their toxicity on wood, thus indicating that there are some interactions between phenol and wood (Haars et al., 1981). The observed differences in the sensitivity of *H. annosum* in different environments indicate the necessity of screening such agents for potential fungistatic action under conditions as similar as possible to the situation of the living tree. The substances that proved to be toxic on spruce stem slices were transformed to slow-release formulations by the Hoechst Company (Frankfurt, FRG). These so-called "dispensers" should release their active ingredients over a reasonable period of time in order to establish a protective screen within the tree. They were delivered by Hoechst in the form of cylindrical rods of 12-mm diameter or flexible bars of 6-mm diameter. The dispensers were tested on agar as 4-mm slices and in field experiments. Some 60- and 84-yr-old spruce stands in different forest districts were infected (Dimitri 1980a,b). About 10 cm above the inoculum hole the dispenser was introduced. The application of the dispenser was performed in two series of experiments—one 4 wk before and one 6 wk after infection by the fungus. The trees were harvested after 12 months and cut in disks that were stored at room temperature for 10–12 days. After that time the mycelium and conidiospores of *H. annosum* were easily visible with the dissecting microscope.

The horizontal and vertical diffusion of the phenols along the stem was monitored by spraying with appropriate chemicals, which gave a color reaction with the phenol. It is obvious that diffusion towards the horizontal direction can only be low. Transport into the vertical direction was much better, but has to be improved by multiple treatments or a better dispersion method. This was probably the reason why only one of the three tested dispenser types had a significant effect on the infection rate; compared to the control (81% infected stems), where only 63% infections were counted in treated stems. The application of the dispenser 6 wk after infection resulted in 25% better production (Dimitri et al., 1983).

3. CONCLUSIONS

In summary, it can be stated that there are several hopeful aspects in the fight against *H. annosum* using fungistatic agents. However, the data available so far, indicate that there are still some problems to be solved:

1. A better horizontal diffusion of the chemical inside the tree has to be achieved.
2. The uptake of the chemicals by the xylem water transport system must be prevented.
3. The stability of the chemicals in the "slow-release" formulation has to be tested over a longer time period.

The improvement of this "slow-release" formulation will not easily be achieved, because our results show that both the formulation and the fungicide must have certain properties which cannot be tested on agar or spruce stem slices. Indeed, some of the expectations were fulfilled after transformation of the substance from the liquid media to agar and then to spruce stem slices and the living tree. In most cases, however, it was not possible to predict the behavior of the fungicidal agent because of the change of environmental conditions and the complex reactions in the living tree.

Therefore, a screening program for effective chemicals in a long-term formulation will only be successful if performed either in the living tree itself or under very similar conditions.

If the problems mentioned above can be solved, this method may help to save old valuable stands toward the end of the rotation period.

REFERENCES

Ander, P., and K. E. Eriksson (1976). The importance of phenoloxidase activity in lignin degradation by the white-rot fungus *Sporotrichum pulverulentum*. *Arch. Microbiol.*, **109:**1–8.

Aufsess, H. v. (1972). Einige licht-und rasterelektronenmikroskopische untersuchungen an rotfaulem fichtenholz. *Forstwiss. Cbl.*, **91:**98–105.

Bjorkman, E. (1949). Soil antibiotics acting against the root fungus (*Polyporus annosus* Fr.). *Physiol. Plant.*, **2:**1–10.

Braun, H. J. and J. Lulev (1969). Infektion unverletzter, fingerstarker fichtenwurzeln durch den wurzelschwamm *Fomes annosus* (Fr.) Cooke. I. Bau des oberflächen-periderms und erste infektionsbefunde. *Forstwiss. Cbl.*, **88:**327–338.

Braun, H. J. and J. Lulev (1970). Infektion unverletzter, fingerstarker fichtenwurzeln durch den wurzelschwamm *Fomes annosus* (Fr.) Cooke. II. Infektionswege und ausbreitung des pilzes kurz nach der infektion. *Forstwiss. Cbl.*, **89:**269–275.

Buswell, J. A., B. Mollet, and E. Odier (1984). Ligninolytic enzyme production by *Phanerochaete chrysoporium* under conditions of nitrogen sufficiency. *FEMS Microbiol. Lett.*, **25:**295.

Butin, H. (1983). *Krankheiten der Wald- und Parkbäume*. Thieme, Stuttgart, pp. 115–117.

Chet, I., and A. Hüttermann (1980). Chemical composition of hyphal walls of *Fomes annosus*. *Eur. J. For. Pathol.*, **10:**65–70.

Chet, I., Y. Henis, and R. Mitchell (1966). The morphogenetic effect of sulphur-containing aminoacids, glutathione and iodoacetic acid on *Sclerotium rolfsii* Sacc. *J. Gen. Microbiol.* **45:**541–546.

Chet, J., Y. Henis, and R. Mitchell (1967). Chemical composition of hyphal and sclerotial walls of *Sclerotium rolfsii* Sacc. *Can. J. Microbiol.*, **13:**137–141.

Chet, I., Y. Hadar, Y. Elad, J. Katan, and Y. Henis (1979). Biological control of soil-borne plant pathogens by *Trichoderma harzianum*. In *Soil-Borne Plant Pathogens* (B. Schippers and W. Gams eds.). Academic, New York.

Crawford, R. L. (1981). Lignin biodegradation and transformation. Wiley, New York.

Cwielong, P. (1979). Optimierung und produkton lytischer enzyme von *Trichoderma harzianum* zum direkten abbau von *Fomes annosus*. Dipl. Arbeit, Forstl. Fakultät Göttingen.

Dimitri, L. (1963). Untersuchungen über die ausbreitung von *Fomes annosus* (Fr.) (Cooke). *Phytopath. Z.*, **48:**349–369.

Dimitri, L. (1980a). *Proceedings of the Vth International Conference on Problems of Root and Butt Rot in Conifers*. Kassel.

Dimitri, L. (1980b). *Untersuchung über die Widerstandsfähigkeit der Fichte* (*Picea abies* Karst) *gegenüber dem Wurzelschwamm Fomes annosus (Fr.) Cooke*. Sauerländers Verlag, Frankfurt, 126 pp.

Dimitri, L., G. Schumann, and A. Hüttermann (1983). Control of *Heterobasidion annosum* (Fr.) Bref. (*Fomes annosus* Cooke) inside the tree with slow release formulations of fungicidal substances. *Proceedings of the IUFRO Conference*. Australia, 1983.

Ekman, R., and K. v.Weissenberg (1981). Association between some silvicultural, physical, and chemical properties of *Picea abies* and spread of *Fomes annosus*. *Acta Acad. Aboensis Ser. B.*, **41:**1-22.

Elad, Y., I. Chet, and Y. Henis (1982). Degradation of plant pathogenic fungi by *Trichoderma harzianum*. *Can. J. Microbiol.*, **28:**719-725.

Elad, Y., R. Barak, and I. Chet (1984). Parasitism of sclerotia of *Sclerotium rolfsii* by *Trichoderma harzianum*. *Soil. Biol. Biochem.*, **16:**381-386.

Ellwardt, P. Ch., K. Haider, and L. Ernst (1981). Untersuchungen des mikrobiellen ligninabbaus durch ^{13}C-NHR-spektroskopie an spezifisch ^{13}C-angereichertem DHP-lignin aus coniferylalkohol. *Holzforschung*, **35:**103-109.

Eriksson, K. -E., A. Grünewald, and L. Vallander (1980). Studies of growth conditions in wood for three white-rot fungi and their cellulase-less mutants. *Biotechnol. Bioeng.*, **22:**363-376.

Evers, F. H. (1973). Zusammenhang zwischen chemischen bodeneigenschaften und kernfäulebefall in fichtenbeständen. *Mitt. Ver. Forst. Standortskde. Forstpflanzenzücht*. **22:**65-71.

Falck, R. (1923). Über ein krystallisiertes stoffwechselprodukt von *Sparassis ramosa* schäff. *Chem. Ber.*, **56:**2555.

Falck, R. (1926). Über korrosive und destruktive holzzersetzung und ihre biologische bedeutung. *Ber. Dt. Bot. Ges.*, **44:**652-664.

Falck, R. (1930). Neue mitteilungen über die rotfäule. *Mitt. Forstwirtsch. Forstwiss.*, **1:**525-570.

Falck, R. (1935). The application of respiration poisons in the cure and prophylactic treatment of the living tree. *Pal. J. Bot. Horticult. Sci.* **1:**70-92.

Flemming, S. (1979). Entwicklungsgenetische Untersuchungen an *Fomes annosus*. Untersuchungen über Enzymaktivitäten des Chitinstoffwechsels während der Morphogenese. Dissertation, University of Göttingen.

Flemming, S. and A. Hüttermann (1981). Induction of branching and conidiogenesis in *Fomes annosus*. *Eur. J. For. Pathol.*, **11:**170-178.

Flemming, S. and A. Hüttermann (1986). Chitinsynthetase from *Fomes annosus*. *Arch. Microbiol.* (in press).

Fowler, M. E. (1962). *Fomes annosus* in northeastern United States. In *Conference and Study Tour on Fomes annosus. Scotland, 1960, 20-22*. IUFRO, Firenze.

Gold, M., T. Cheng, K. Krishnanghura, M. Mayfield, and L. Smith (1980). Genetics and biochemcial studies on *P. chrysosporium* and their relation to lignin degradation. In *Lignin Biodegradation*, Vol. II (T. K. Kirk, T. Higuchi, and H. M. Chang, eds.). CRC, Boca Raton, Fl, p. 65.

Gold, M., T. Cheng, and M. Mayfield (1982). Isolation and complementation studies of auxotrophic mutants of the lignin-degrading basidiomycete *Phanerochaete chrysosporium*. *Appl. Environm. Microbiol.* **44:**996-1000.

Grabbe, K., R. Koenig, and K. Haider (1968). Die bildung der phenoloxydase und die stoffwechselbeeinflussung durch phenole bei *Polystictus versicolor*. *Arch. Mikrobiol.*, 63:133-153.

Green, T. R. (1977). Significance of glucose oxidase in lignin degradation. *Nature*, **268:**78-80.

Haars, A. and A. Hüttermann, (1980a). Macromolecular mechanism of lignin degradation by *Fomes annosus*. *Naturwissenschaften*, **67:**39.

Haars, A. and A. Hüttermann (1980b). Function of laccase in the white-rot fungus *Fomes annosus. Arch. Microbiol.*, **125:**233–237.

Haars, A. and A. Hüttermann (1983). Laccase induction in the white-rot fungus *Heterobasidion annosum* (Fr.) Bref (*Fomes annosus* Fr. Cooke). *Arch. Microbiol.*, **134:**309–313.

Haars, A., I. Chet, and A. Hüttermann (1981). Effect of phenolic compounds and tannin on growth and laccase activity of *Fomes annosus. Eur. J. For. Pathol.*, **11:**67–76.

Haars, A., O. Milstein, S. Lohner, and A. Hüttermann (1982). Rapid, sensitive, and inexpensive assay for the determination of lignin degradation using different fluorescence labelled lignin derivatives. *Holzforschung*, **36:**85–91.

Haars, A., P. Cwielong and A. Hüttermann, (1983). Effect of phenolic and other aromatic substances on growth morphology and certain enzymatic activities of *Fomes annosus* in: G. Kile (ed.): Proc. Vth Int. Conf. *Root- and Butt-Rot of Forest Trees*, pp. 207–220, Melbourne: CSIRO.

Haars, A., D. Lange, W. Jaszcynski, and A. Hüttermann (1984). Regulation of aminopeptidases in the white-rot fungus *Heterobasidion annosum. Microbiol. Lett.*, **27:**103–112.

Haas, H. (1971). Makromycetenflora und kernfäulebefall älterer fichtenbestände auf der schwäbischen alb. *Mitt. Ver. Forstl. Standortskde. Forstpflanzenzücht.*, **20:**50–59.

Hartig, R. (1878). *Die Zersetzungserscheinungen des Holzes der Nadelholzbäume und der Eiche in Forstlicher, Botanischer und Chemischer Richtung*. Springer, Berlin, 151 pp.

Holt, C. E., H. Gockel, and A. Hüttermann (1983). The mating system of *Fomes annosus. Eur. J. For. Pathol.*, **13:**174–181.

Horsfall, J. G. (1956). *Principles of Fungicidal Action*. Massachusetts Chronica Botanica Company, Waltham, MA.

Hüttermann, A. and P. Cwielong (1982). The cell wall of *Fomes annosus* (Heterobasidion annosum) as target for biochemical control. I. Direct lysis of the cell wall by lytic enzymes from *Trichoderma harzianum. Euro. J. For. Pathol.*, **12:**238–245.

Hüttermann, A. and A. Noelle (1982). Characterization and regulation of cellobiose dehydrogenase in *Fomes annosus. Holzforschung*, **36:**283–286.

Hüttermann, A. and C. Volger (1973a). Cellobiose phosphorylase in *Fomes annosus. Nat. New Biol.*, **245**:64.

Hüttermann, A. and C. Volger (1973b). Induction of aryl-β-glucosidase in *Fomes annosus* by cellobiose. *Arch. Mikrobiol.*, **93:**195–204.

Hüttermann, A., M. Gebauer, Ch. Volger, and Ch. Rösger (1977). Polymerisation und abbau von natrium-ligninsulfonat. *Holzforschung*, **31:**83–89.

Hüttermann, A., C. Volger, R. Schorn, G. Ahnert, and H.-K. Ganser (1979). Studies on isoenzyme polymorphisms in *Fomes annosus. Eur. J. For. Pathol.*, **9:**265–274.

Hüttermann, A., Ch. Herche, and A. Haars (1980a). Polymerization of water-insoluble lignins by *Fomes annosus. Holzforschung*, **34:**64–66.

Hüttermann, A., W. E. Kuhl, and I. Chet (1980b). The effect of L-threonine on oxalic acid synthesis by *Fomes annosus. Eur. J. For. Pathol.*, **10:**339–344.

Hüttermann, A., J. Trojanowski, A. Haars, and U. Böttcher (1983). Physiology and mechanisms of lignin degradation by *Heterobasidion annosum* in: G. Kile (ed.): Proc. Vth. Int. Conf. *Root- and Butt-Rot of Forest Trees*, pp. 257–268, Melbourne: CSIRO.

Jahn, H. (1979). *Pilze, Die an Holz Wachsen*. Busse, Herford, 268 pp.

Janshekar, H. and A. Fiechter (1983). Lignin: Biosynthesis, application, and biodegradation. Adv. Biochem. Eng. Biotechnol., **27:**119–178.

Janshekar, H., C. Brown, Th. Haltmeier, M. Leisola, and A. Fiechter (1982). Bioalteration of pine kraft lignin by *Phanaerochaete chrysosporium. Arch. Microbiol.*, **132:**14–21.

Johansson, M. and E. Hägerby (1974). Influence of growth conditions, metabolic inhibitors, and phenolic compounds on the ATP pool in *Fomes annosus. Physiol. Plant.*, **32**:23–32.

Johansson, M., T. Popoff, and O. Theander (1976). Effect of spruce root constituents on extracellular enzymes of *Fomes annosus. Physiol. Plant.*, **37**:275–282.

Johnson, G. B. (1974). Enzyme polymorphism and metabolism. *Science*, **184:**28–37.

Kersten, P. J., M. Tien, B. Kalyanaraman, and T. K. Kirk (1985). The ligninase of *Phanerochaete chrysosporium* generates cation radicals from methoxybenzenes. *J. Biol. Chem.*, **260:**2609–2612.

Kirk, T. K., E. Schulte, W. J. Connors, L. F. Lorenz, and J. G. Zeikus (1978). Influence of culture parameters on lignin metabolism by *Phanaerochaete chrysosporium. Arch. Microbiol.*, **117:**277–285.

Korhonen, K. (1978). Intersterility groups of *Heterobasidion annosum. Comm. Inst. For Fenn.*, **94:**1–25.

Kritzman, G., I. Chet, and Y. Henis (1976). Metabolism of L-threonine and its relationship to Sclerotium formation in *Sclerotium rolfsii. J. Gen. Microbiol.*, **95:**78–86.

Low, D. J. and R. J. Gladman (1960). *Fomes annosus* in Great Britain. An assessment of the situation in 1959. *For. Rec.*, **41**, 22 pp.

Molin, N. (1957). The infection biology of *Fomes annosus. Statens Skogforskningsinst.Medd.*, **47**, 36 pp.

Moosmayer, H. U. (1971). Zur waldbaulichen planung im nordostteil der schwäbischen alb. *Allg. Forst. J. Ztg.*, **4:**108–113.

Okon, Y., I. Chet, and Y. Henis (1972). Lactose-induced synchronous sclerotium formation in *Sclerotium rolfsii* and its inhibition by ethanol. *J. Gen. Microbiol.*, **71:**465–470.

Okon, Y., I. Chet, and Y. Henis (1973). Effects of lactose, ethanol and cycloheximide on the translocation pattern of radioactive compounds and on Sclerotium formation of *Sclerotium rolfsii. J. Gen. Microbiol.*, **74**:251–258.

Peek, R.-D., W. Liese, and N. Parameswaran, (1972). Infektion und Abbau der Wurzelrinde durch *Fomes annosus. Eur. J. Forest Pathol. 2:*104–115.

Popoff, T., O. Theander, and M. Johansson (1975). Changes in sapwood of roots of Norway spruce attacked by *Fomes annosus*. Part II: Organic chemical constituents and their biological effects. *Physiol. Plant.*, **34:**347–356.

Rehfuess, K. E. (1973). Kernfäulebestand und ernährungszustand älterer fichtenbestände (Picea abies Karst.) im wuchsgebiet baar-wutach. *Mitt. Ver. Forstl.Standortskde. Forstpflanzenzücht.*, **22:**9–26.

Reid, I. D. (1983a). Effects of nitrogen supplements on degradation of Aspen wood lignin and carbohydrate components by *Phanaerochaete chrysosporium. Appl. Environ. Microbiol.*, **45:**830–837.

Reid, I. D. (1983b). Effects of nitrogen sources on cellulose and synthetic lignin degradation by *Phanaerochaete chrysosporium. Appl. Environ. Microbiol.*, **45**:838–842.

Rishbeth, J. (1951). Observations on the biology of *Fomes annosus*, with particular reference on East Anglian pine plantations. II. Spore production, stump infection and saprophytic activity in stumps. *Ann. Bot. (Lond.) NS*, **15:**1–22.

Rishbeth, J. (1957). Some further observations on *Fomes annosus* Fr. *Forestry*, **30:**69–89.

Schlenker, G. (1976). Einflüsse des standorts und der bestandesverhältnisse auf die rotfäule der fichte. In *Der Wurzelschwamm (Fomes annosus) und die Rotfäule der Fichte* (Picea abies) (H. Zycha, ed.). Parey, Hamburg, 83 pp.

Schmidt-Vogt, H. and U. v. Schnurbein (1976). Maβnahmen zur bekämpfung der rotfäule. In *Der Wurzelschwamm* (Fomes annosus) *und die Rotfäule der Fichte* (Picea abies) (H. Zycha, ed.). Parey, Hamburg, 83 pp.

Schoemaker, H. E., P. J. Harvey, R. M. Bowen, and J. M. Palmer (1985). On the mechanism of enzymatic lignin breakdown. *FEBS Lett.* **183:**7–12.

Shain, L. and W. E. Hillis (1971). Phenolic extractives in Norway Spruce and their effects on *Fomes annosus*. *Phytopathology*, **61**:841–845.

Sinclair, W. A. (1964). Root- and butt rot of conifers caused by *Fomes annosus*, with special reference to inoculum dispersal and control of the disease in New York. *Mem. Cornell Univ. Agr. Exp. Sta.*, No. 391, 54 pp.

Stalpers, J. A. (1974). Revision of the genus *Oedocephalum* (Fungi imperfecti). *Proc. Kon. Ned. Akad. Wet., Ser. C*, **77**:383–401.

Tien, M. and T. K. Kirk (1984). Lignin degrading enzyme from *Phanaerochaete chrysosporium*: purification, characterization and catalytic properties of a unique H_2O_2-requiring oxygenase. *Proc. Natl. Acad. Sci. USA*, **81**:2280.

Trojanowski, J., K. Haider, and A. Hüttermann (1984). Decomposition of ^{14}C-labelled lignin holocellulose and lignocellulose by mycorrhizal fungi. *Arch. Microbiol.*, **139**:202.

Volger, C., C. Rösger, and A. Hüttermann (1977). Untersuchungen zur physiologischen variabilität von *Fomes annosus*-isolaten. *Eur. J. For. Pathol.*, **7**:262–282.

Vries, O. M. H. de and J. G. A. Wessels (1975). Chemical analysis of cell wall regeneration and reversion of protoplasts from *Schizophyllum commune*. *Arch. Microbiol.*, **102**:209–218.

Werner, H. (1973). Untersuchungen über die Einflüsse des Standortes und der Bestandesverhältnisse auf die Rotfäule (Kernfäule) in Fichtenbeständen der Ostalb. *Mitt. Ver. Forstl. Standortskde Forstpflanzenzüecht.*, **22**:27–64.

Westermark, U. and K.-E. Eriksson (1974). Carbohydrate-dependent enzymic quinone reduction during lignin degradation. *Acta Chem. Scand.*, **B28**:204–208.

Zycha, H. (Herausg.) (1976). Die Rotfäule der Fichte. Forstwiss. Forschungen, Heft 36.

14

PERSPECTIVES IN THE CONTROL OF PLANT VIRUSES

J. DUNEZ
Station de Pathologie Végétale, Institut National de la Recherche Agronomique, Centre de Bordeaux, France

The origin and history of viruses is still a matter of speculation. Since viruses are obligate pathogens, it is difficult to see how they could have developed before microorganisms or other more highly developed forms of life. Some of the earliest publications dealing with virus diseases refer to a human disease which probably was smallpox. The earliest illustrations of virus diseases in plants are paintings from the 17th century in which several Dutch masters beautifully illustrated breaking in the flowers of tulips and carnations. Preliminary results demonstrating the transmissibility of tobacco mosaic and the ultrafiltrability of the responsible agent were reported at the end of the 19th century. But it is only in the 20th century, especially the beginning of the second half, that the science of virology actually started. Since that time, the biochemical and physical characteristics of many viruses have been precisely defined. In general, a virus particle consists of a nucleic acid (ribonucleic acid, RNA, or deoxyribonucleic acid, DNA) and a coat made up of several protein subunits. Certain virus particles are surrounded by a membrane that is partially derived from the infected cell. The viral genome is made up of 1 molecule of nucleic acid (monopartite genome) or several molecules (multipartite genome) with molecular weights in the range of 10^6–10^8 daltons; plant virus RNAs fall within the range of 10^6–10^7 daltons. In contrast with higher organisms and other plant pathogens, viruses contain only one type of nucleic acid (RNA or DNA). Because of the small size of their genome, which carries only a few genetic codes [for coat protein(s), replicase, translocation factor, etc.], viruses are obligate pathogens entirely dependent on cell constituents and molecules for their own synthesis. In 1971, Diener described a new type of pathogen, a viroid which, up to the present time, has been isolated only from plants. A viroid consists of a small infectious RNA molecule of 0.12×10^6 daltons, which is devoid of protein and does not seem to have any encoding capacity.

The impact of viruses in agriculture is well known. To point out a few examples, we could mention over 30 million trees killed in Brazil alone by citrus tristeza virus and losses of 70–100% of the yield in some cultivars of potato, beet, and fruits. Sometimes the growth is not significantly affected or the yield reduced, but the quality is severely altered; for example, plum pox virus infects millions of *Prunus* trees in Europe and causes the fruit of some cultivars to be totally unmarketable, even for canning or distillation. Most of the severe virus diseases spread naturally and often have a wide host range. Various modes of spread are known: besides the vegetative propagation of infected plant material, some viruses are seed- or pollenborne; many are transmitted by animals, especially anthropods and nematodes; some are soilborne by soil fungi; and a few are transmitted mechanically by contact or via infected tools. These modes of spread permit the survival of viruses and short- or middle-distance spread. Long-distance spread occurs from the worldwide movement of plant material. With the development of international exchanges of plant materials, control is becoming more and more difficult and, even when quarantine stations exist, there are many possibilities to circumvent the quarantine regulations. Most of the common viruses are distributed worldwide and occur wherever susceptible plants are grown. Only a few of the most devastating diseases have been successfully restricted to a given area (e.g., plum pox in Europe).

Efforts to control virus diseases started over 50 yr ago. In contrast to other plant pathogens, chemical control of plant viruses has been almost totally ineffective, perhaps because virus replication is so dependent upon the normal metabolism of the host cells. Plant virus control has been and still is based primarily on prevention and sanitation practices, that is, producing and propagating virus-free plants and protecting these plants against reinfection. We discuss briefly this very conventional technique and focus especially on the diagnostic aspect which is a critical factor in any sanitation scheme or control program. We attempt to point out the extent to which new procedures have revolutionized the diagnosis of plant virus diseases. The availability of new technology in molecular biology and genetic engineering now makes it possible to consider new approaches to plant virus disease control. The future, in terms of plant virus control, very likely will involve genetic engineering techniques to introduce into plants specific genes or genetic codes not only from plants, as is done now in various breeding programs for resistance, but also from the viral genome itself.

1. PRODUCTION OF VIRUS-FREE PLANTS—A CONVENTIONAL TECHNIQUE FOR THE CONTROL OF PLANT VIRUSES—PAST, PRESENT, AND FUTURE

Sanitation schemes for the production of virus-free plants have been practised for more than 30 yr; they have proved to be an efficient means of control and will, undoubtedly, remain in use for many years. Healthy stock can be obtained by visual selection of virus-free plants from a plant population or by the use of appropriate detection procedures in the case of symptomless infections. It frequently occurs that some cultivars, especially those belonging to species that are vegetatively propagated, are uniformly infected. In such cases, two techniques can be used to obtain virus-free material from an infected plant—heat treatment and meristem-tip culture. Very few major modifications have been made to these techniques during the past 30 yr.

The effect of heat treatment was first described in 1936 by Kunkel. Since then many different cultivars belonging to a diverse group of genera and species have been successfully heat treated. Initially, the technique consisted of treating plant material in hot water (50–55°C) for a few minutes. Problems arose because such temperatures may be lethal to the host tissue and survival of the treated plant material was often very poor. Virologists found that another technique based on heat treatment in warm air (37–38°C) for several weeks resulted in better host survival and various modifications of this method are still in use. With a few exceptions (strawberry plants), the virus is never totally eliminated from the treated plant. Following heat treatment, virus is not present in certain parts of the plants, especially the distal parts of the shoots, which developed during heat treatment. Although the mechanism of thermotherapy is not clearly understood, heat treatment seems to adversely affect both virus replication and translocation. The distal parts or individual buds are removed from the treated plant and propagated. Virus-free clones of a wide variety of important cultivated species and genera have been obtained by this technique.

In 1952 Morel and Martin reported the possibility of obtaining virus-free dahlias by growing meristem tips of dahlia mosaic virus-infected plants on a synthetic nutrient medium. The technique was then successfully applied to potatoes and later to various other species and genera and has permitted the production of large numbers of virus-free plants. There are two sorts of difficulties associated however with both of these techniques. First, *in vitro* meristem-tip culture of certain plants, especially woody plants, is difficult. Improvements have been introduced recently which may make it possible to use the technique with a wider range of species. Instead of being placed on the nutrient medium, the excised meristem tip is micro-grafted onto a young virus-free rootstock grown *in vitro*. This shoot-tip grafting recently allowed the production of virus-free citrus and peach plants (Navarro et al., 1975, 1982). Another serious problem involves plants infected with viroids. Viroids seem to be more invasive than viruses, making it more difficult to eliminate them from the meristematic tissue. Since viroids multiply preferentially at higher temperatures (30–35°C) than most plant viruses, thermotherapy is not effective and production of viroid-free plant viruses requires meristem culture. For species that are infected with both viruses and viroids, viroid-free plants are much more difficult to obtain than virus-free plants. The percentage of viroid-free plants usually is low, but the efficiency can be increased by using the meristem-grafting technique, which permits the use of a smaller piece of the meristem tip.

The proportion of plants that are actually freed of viruses and viroids after heat treatment or meristem culture varies from 5 to 85. Plants that are still infected generally contain a low level of virus, which generally is undetectable if the plants are assayed soon after propagation. Reliable testing of plants from heat treatment or meristem-tip culture requires several successive assays based on sensitive detection methods. New *in vitro* propagation methods result in the production of very large numbers of plants. With some species, millions of plants can be produced during a 12-month period from a single mother plant. It is imperative, therefore, that virus-infected plants be identified and not used as a propagation source. It is obvious that accurate diagnosis is a key factor in plant virus control, including testing the virus-free stock, detecting and eliminating reinfections, surveying nurseries, and enforcing various plant introduction and quarantine regulations. For many years, plant virus diagnosis was based on biological indexing, but in the past 10 yr, especially the last 2–5 yr, a tremendous change has taken place in plant virus disease diagnosis.

1.1. The Diagnosis of Plant Virus Infections—From Biological Indexing to the Use of Biotechnology

1.1.1 Biological and Serological Indexing

Of course, traditional biological indexing using mechanical or graft transmission still remains in use. The technique is highly sensitive and has the advantage of detecting transmissible diseases, the causal agents of which are still unknown. In addition, new techniques based on the immunological properties of the virus or on the characteristics of its nucleic acid have been developed or are in progress.

Viruses are immunogenic and with only a few exceptions, the antigenically active part of the virus particle is the protein. Upon injection into an animal, virus particles induce synthesis of antibodies, which can be collected in the serum fraction of the blood and will specifically recognize the corresponding virus. Serological indexing has been used for more than 25 yr. In 1959 Cornuet reported that the technique was already being used for the detection of some potato viruses. Until about 1976, serological indexing remained of limited use, because of a lack of sensitivity compared to bioassay. Clark and Adams (1977) demonstrated that immunoenzymatic techniques, already applied to human and animal diseases, were readily applicable to plant viruses, and there has since been a tremendous development of these techniques. The enzyme linked immunosorbent assay (ELISA) is now applied to the detection of most known viruses in leaves, stems, roots, and seeds (Clark and Bar Joseph, 1984). The ELISA test is 1000× more sensitive than the precipitin type serological techniques used previously. In addition, it is very rapid, simple, and well adapted to routine testing and mass indexing.

Various modifications and improvements have been introduced that give the test a higher degree of sensitivity and better versatility. The HADAS (Heterologous Antibody Double Antibody Sandwich) technique utilizes specific antibodies produced in two different animal species (Bar Joseph and Malkinson, 1980). The advantage of this modification results from the amplification and use of a nonspecific conjugate instead of the specific conjugate used in the standard sandwich ELISA test. Another possibility is to use Fab or $F(ab')_2$ fragments of antibodies (devoid of the Fc fragment). The virus is trapped between these fragments coated on the plate and the whole antibodies. A nonspecific anti-Fc conjugate is used and, once again, amplification results in increased sensitivity. The list of possible modifications is endless. We can assume that in the coming years new developments in the immunoenzymatic diagnosis of plant viruses will involve the avidine–biotin technique and the use of fluorescent substrates instead of colored substrates. An advantage of avidine–biotin is the extreme avidity of these two molecules. In the modified ELISA test the virus is trapped between the coating antibodies and biotinylated antibodies; avidine molecules are then trapped by the biotin molecules, and finally these avidine molecules are used to trap a nonspecific biotin–enzyme conjugate. Even though this procedure involves additional steps, the sensitivity is significantly increased (Kendall et al., 1983). The use of fluorescent substrates does not modify or complicate the normal ELISA test. Only the usual substrates are modified. Antibodies are conjugated with alkaline phosphatase or galactosidase and a fluoroegenic substrate is used. The technique developed in animal and human immunochemistry already has been suggested for plant viruses (Torrance and Jones, 1982).

Various immunoelectron microscopy techniques (IEM) have also been developed (Milne and Lesemann, 1984). In general the techniques involve the observation of virus particles trapped by antibodies adsorbed on the grids and surrounded by a halo of antibodies fixed on these particles. In contrast to the ELISA technique, IEM does not require purified antibodies; crude diluted antiserum is suitable. In locations where an electron microscope is available, the IEM procedure can provide very rapid results with a sensitivity of the same order as ELISA.

When applied to purified virus suspensions, the precipitin techniques detect about 1 μg/ml and ELISA and the immunoelectron microscopy technique about 1–10 ng/ml. With the improvements described above one can expect serological detections as low as 0.05–0.1 ng of virus.

1.1.2. Biotechnology and Virus Diagnosis

Monoclonal antibodies. Until recently serological diagnosis was carried out with crude antisera or antibodies purified from the crude serum. These antisera contained a mixture of different specific antibodies, depending upon the number of antigenic determinants on the injected antigen. These polyclonal antibodies were satisfactory for most routine diagnostic procedures. Sometimes it is of interest, however, to be able to detect a specific serotype. Specific antisera for detecting a single serotype can be obtained by cross-absorption of polyclonal antiserum with the purified virus of other serotypes, but it is possible now to produce monoclonal antibodies (Prabhakar et al., 1984). When dealing with viruses that are difficult to obtain in a higher purified state, the antisera produced generally contain a mixture of antibodies against the virus and normal antigenic cell components. The contaminating antibodies can be eliminated by cross-absorption with preparations from healthy plants, but it is better to use the monoclonal antibody technique. There may be certain disadvantages to the monoclonal antibodies: because of the high specificity of each monoclonal antibody it is necessary at times, in routine diagnostic tests, to use a mixture of different monoclonal antibodies. Production of monoclonal antibodies is a complicated and time-consuming procedure. Generally speaking, it is preferable to use a polyclonal antiserum for routine diagnostic procedures, but when the virus is difficult to purify or when precise serotyping is required for virus classification, it is better to turn to the monoclonal technique. Monoclonal antibodies will become more widely available in the future and will permit a higher degree of sensitivity in serological assays. Monoclonal antibodies will play an even more significant role in basic virus research as we seek to gain a better knowledge of the biochemical and serological properties of plant viruses.

Molecular hybridization. With the discovery (Diener, 1971) of the unique characteristics of viroids (RNA devoid of a protein coat) it became apparent that normal serological techniques could not be used to detect viroids in infected plants. Viroids had been detected, as had most viruses, by transmission into susceptible host plants. A rapid electrophoretic technique consisting of a short, simplified purification procedure was proposed for potato spindle tuber viroid (PSTV) (Schumann et al., 1978) and chrysanthemum stunt viroid (CSV) (Macquaire et al., 1981). The technique was used for a couple of years, but has largely been replaced by other techniques, although the electrophoretic procedure is still in use in the search for dsRNA, which is discussed in the next section. In the meantime, the possibility of reverse transcription of viroids was demonstrated and this allowed the application of molecular hybridization for viroid and virus detection. By using reverse transcriptase it is possible to transcribe the viroid RNA into a complementary DNA (cDNA) which is an excellent probe for the detection of viroid RNA (Owens and Diener, 1981). This probe can be labeled upon transcription from the viroid RNA template, but obtaining reasonable amounts of labeled probe would require repeated viroid purifications and transcriptions. It is better therefore, to use the transcribed single-stranded cDNA to synthesize the complementary strand, then introduce these double-stranded molecules into a plasmid. The recombinant plasmid is then used to transform bacteria (Owens and Cress, 1980; Van Wezenbeek et al., 1982). The cloned recombinant plasmids containing the viroid insert are produced in high amounts by selected bacteria from which they can be easily purified. Plasmids are then labeled by nick translation and used for hybridization. The dot-blot hybridization technique on nitrocellulose or a gene screen is especially suitable for routine detection (Macquaire et al., 1984). There are two advantages to using a cloned cDNA: (1) unlimited amounts can be obtained from bacteria, and (2) the whole plasmid, including the PSTV insert, is labeled and therefore the sensitivity of the test is amplified. So far the plasmids were labeled by radioactive markers, especially ^{32}P; nick translation using the $\alpha^{32}P$ dCTP as the marker provides a high specific radioactivity and a high sensitivity to detection by autoradiography. Encouraging results have been obtained with biotinylated probes that might be developed in the future. The comparable sensitivity of the two techniques, in terms of purified viroid detected, are 10 ng by electrophoresis (similar to bioassay) and 5–50 pg by molecular hybridization with ^{32}P-labeled probes. Of course, molecular hybridization is also applicable to virus detection and several examples have been proposed. Usually it is not necessary to use very long transcripts to produce a probe providing a high sensitivity (Rosner et al., 1983). For those viruses or viroids that have been totally or partly sequenced and that could be difficult to transcribe, it is now possible to use complementary synthetic oligonucleotides as probes; 60–100 nucleotides are enough to provide a suitable probe (Bar Joseph et al., 1985). Another special interest in such probes is that their preparation does not require previous purification of the pathogen. This is very useful when detection techniques are needed for a pathogen that is not present in a particular area and should not be introduced into the area even for scientific purposes.

Presence of double-stranded RNAs, a characteristic of virus infection. It has been claimed by several authors (Morris and Dodds, 1979; Morris et al., 1983) and is generally accepted now that dsRNAs, especially molecules with high molecular weight, do not normally exist in detectable amounts in most healthy plants. If this is, in fact, correct, the presence of dsRNA molecules in plant extracts could indicate a virus infection. According to Dodds et al. (1984), the dsRNA detected could be the genome of a dsRNA virus or the replication form of a single-stranded RNA virus or some plasmid-like molecule of a class of pathogen not yet identified. Several rapid methods have been described for extracting and identifying dsRNA in plant extracts and numerous applications have been reported (Bar Joseph et al., 1983). The advantage of this technique is that it does not require expensive equipment or knowledge of the identity of the suspect virus. Identification of dsRNA in extracts from plants infected by an unidentified pathogen is a good indication of virus infection and permits subsequent orientation of the research to identify the causal agent of the disease.

Virologists today have a wide range of diagnostic techniques, in addition to the traditional bioassay which has served as the standard procedure in the past. Few virologists would have imagined the tremendous development that has taken place in the past few years in the area of immunology and serology with the introduction of new immunoenzymatic procedures. Who would have predicted 10 yr ago that molecular hybridization would be developed and would have an impact on disease diagnosis?

The advantages of having different detection methods are obvious; they result not only in increased sensitivity but also expand the range of applications. Each detection procedure is directed toward a special characteristic of the virus—transmissibility, serological properties, nucleic acid characteristics, and sequences. These detection techniques will increase the reliability of virus-free production and will be valuable tools for basic research on viral replication and pathogenesis.

1.2. Maintainance of the Virus-Free Stocks—Control of Reinfections

The ultimate success of any virus-free production program is the potential for preventing natural reinfection, that is, the presence and efficiency of natural vectors of the virus. Of the 27 taxonomic groups of plant viruses, 20 include viruses that are arthropod or nematode-transmitted. They include the great majority of viruses causing the most severe virus diseases (barley yellow dwarf, citrus tristeza, grapevine fan leaf, potato virus Y, plum pox, etc). Control of virus diseases involves production of virus-free plants, eradication of all virus sources (wild and crop species), and control of vectors. Efficiency of this last control method depends on the manner in which the virus is borne in the vector. The spread of viruses that need periods of hours or days to be acquired and inoculated (persistent) is easier to prevent than the spread of stylet-borne viruses, which are acquired and inoculated within a few minutes (nonpersistent). This topic has been reviewed recently by Harrison (1983).

Nematicides, applied as soil fumigants before planting, are generally quite effective for nematode control; more than 99% of *Xiphinema* and *Trichodorus* are killed to a depth of 70 cm (Cooper and Thomas, 1971), but reinfestation from the deep soil levels may occur within 5 yr. Systemic nematicides also can be applied as granules just before planting. They have low phytotoxicity, but have only a temporary nematicidal effect.

The range of effective insecticides has increased in the past 10 yr. Pyrethroids act rapidly and provide very good vector control of several virus diseases, including semipersistent and nonpersistent viruses (potato leaf roll, beet yellows, and potato virus Y). In most cases, however, insecticides can only be expected to slow down virus spread. The present trend is toward a more discriminated use of pesticides, in combination with other alternative techniques intended to prevent insects from reaching susceptible plants at the time they are the most defenseless. Light-reflecting material may be used to prevent aphids from landing on plants. Mulches of grey plastic sheeting decreased the spread of potato virus Y and cucumber mosaic virus in peppers by more than 90%. Mulching of tomato crops with polyethylene sheets also gives a good control of the white flies vectors of tomato leaf curl virus (Cohen and Malamed Madjar, 1978). Of course, such methods are efficient only as long as the plastic is not covered with plant foliage. The use of screens in combination with insecticides proved to be very efficient, but such methods are very expensive and can only be of use for high-value crops. Another control technique is to prevent aphid vectors from feeding on plants by the use of repellents. Alarm pheromones reduce settling and reproducing of *Myzus persicae* on treated leaves and inhibit for several days the acquisition of beet yellows and potato virus Y. It was demonstrated in 1962 by Bradley et al. that oils sprayed on leaves interfere with the acquisition and/or inoculation of potato virus Y. The technique has been shown to be effective for several other viruses.

The mode of action of oil inhibition is still unclear. The treatment may simply delay infection to such a degree that yield is greatly increased. Similar results of delaying infection and spread can be obtained by changing cultural practices. Gibbs and Harrison (1976) showed that advancing the sowing date of peas enables the plants to reach a more mature and less susceptible stage of growth by the time vectors are active. In contrast, delaying the planting date of maize until the temperature reaches 24°C decreases the incidence of maize rough dwarf virus (Harpaz, 1982), because the virus is not able to complete its cycle in the planthopper vector at high temperature.

None of these procedures gives a total control of virus spread; the most effective way to reduce the level of virus spread is to combine different techniques into an integrated control. In addition to the control techniques discussed above, one should point out the advantages of incorporating resistance to virus transmission into plant breeding programs. Some cultivars have been described which are readily infected by grafting or mechanical inoculation but cannot be infected by aphids. For example, there exists, in muskmelon, resistance governed by a single dominant gene which confers resistance to infestation by *Aphis gossypii* and prevents this aphid species from colonizing plants (Pitrat and Lecoq, 1980). This resistance is aphid species specific. Similar results have been reported in peach cultivars resistant to plum pox transmission by *Myzus persicae* (Massonié and Maison, 1982).

One should not overlook the importance of seed transmission as a means by which viruses survive and are disseminated. In 1981, Mandahar listed 119 viruses transmitted to progeny seedlings, most of which also are pollenborne. The level of seed infection varies with the virus and the host plant; it may sometimes be as low as 0.01–0.1%. Under these conditions virus detection is difficult and requires techniques that are highly sensitive and well adapted to mass indexing. Considerable improvement in seed testing will result from the application of new technologies (ELISA and molecular hybridization). Examples of genetic resistance to virus transmission through the seeds and plants carrying this resistance are now available, for example, barley and lettuce that are resistant to seed transmission of barley stripe mosaic and lettuce mosaic viruses, respectively.

We have only dealt with the major means of plant virus spread. Most of the control procedures we have discussed are more or less conventional and limited in their effectiveness. Very often they are expensive and some insecticides can induce the appearance of vector strains that are resistant to the insecticides. Integrated control programs that include chemical controls for forecasting virus epidemics, adaptation of different cultural practices, and breeding for resistance will contribute to slowing down the spread of plant viruses. It should also be possible to devise better ways of producing virus-free plant material with the use of the rapid new *in vitro* propagation techniques. Nevertheless, it is likely that production of virus-free plants and control measures associated with the production and protection of these virus-free plants will remain one of the standard means of plant virus control. New developments in virus control involve the use of chemicals which, so far, have produced rather disappointing results, and breeding for virus resistance by using new techniques in genetic engineering.

2. CHEMICAL DRUGS FOR THE CONTROL OF PLANT VIRUSES

For many decades chemotherapy has been considered a method with great potential but, so far, results have been rather disappointing. Several chemicals have been tested. Some substances are reported to inhibit infection, but none are able to eradicate the virus from a systemically infected plant. Multiplication of the virus seems to be decreased when the infected plant is treated with chemical analogues of purine or pyrimidine bases of nucleic acid such as 2-thiouracil or 8-azaguanine. More recent attempts were concentrated on the use of virazole (= Ribavirin = 1-β-D-ribofuranosyl-1,2,4-triazole-carboxamide), an analogue of guanine. In 1977, Shepard showed that this substance did not eradicate potato virus X (PVX) when administrated to infected mesophyll protoplasts of tobacco or to proliferating calli raised from these protoplasts. However, treating the same calli with virazole on a medium promoting onset of shoot morphogenesis resulted in the absence of PVX in most regenerated plants. It seems, therefore, that virazole has a definite therapeutic potential for certain *in vitro* propagation systems. Nevertheless, in the present case these results should be interpreted carefully because virus-free plants are often obtained naturally from infected calli without any chemical treatment. Renaudin et al. (1982) demonstrated that virazole is able to inhibit turnip mosaic virus replication in infected cabbage protoplasts without affecting synthesis of the cellular ribosomal or transfer RNAs. Hansen (1979) reported that this drug was able to inhibit apple chlorotic leaf spot virus in *Chenopodium quinoa* plants. Mancino and Agrios (1984), working with cowpea chlorotic mottle-infected cowpea plants, also observed a suppression of symptoms after virazole treatment, even though viral antigen concentration in the treated plants did not appear to be altered. Virazole also has been applied to woody perennial plants (Hansen, 1984); the drug was sprayed on leaves of 2-yr old *Prunus serrulata* infected with a virus-like agent causing the green ring mottle disease. Nine weekly applications prevented symptom expression on the newly developed foliage and seemed to eliminate the infectious agent from the previously infected wood. Back-indexing confirmed the elimination of the pathogen; one year after virazole treatments were stopped, no symptom could be detected. In parallel experiments carried out with peach trees infected with Prunus necrotic ring spot virus, infected trees continued to display symptoms even after more than twenty virazole applications.

On the basis of these reports, virazole seems to have some effect on certain viruses, but results are unconsistent. In fact, this drug is known to have a broad spectrum of activity to both RNA and DNA viruses (Sidwell et al., 1972), but its mode of action is still under study. Some authors have suggested an inhibition of the pathway for GTP synthesis or inhibition of RNA polymerase as with influenza virus. More recently, however, it was proposed that the virazole 5′ phosphate is a competitor of the *in vitro* capping guanylation of the vaccinia virus mRNAs. This could account for the broad spectrum of activity against RNA and DNA viruses where capped RNA are involved. In the absence of capped RNA, antiviral activity would not occur (e.g., uncapped poliovirus RNA). This hypothesis does not explain some of the results that have been reported, in particular the lack of effect on cowpea chlorotic mottle virus which has a capped RNA, and Prunus necrotic ring spot which very likely also has a capped RNA.

Much more information is needed on the mechanisms of replication and expression of viruses before some effective chemotherapeutants can be developed. The discovery and use of drugs having a broad spectrum of activity is unlikely, since the mechanism of virus replication is so integrally associated with general cell metabolism that secondary cytotoxic effects would result. The desired objective would be to find a chemical compound that has very specific activity against the target virus, but little or no effect on normal cell metabolism. A better knowledge of the virus genome will facilitate the development of efficient chemotherapeutants. The more complex the genome is, the better will be the chance to discover a special function sensitive to a given molecule and therefore to produce efficient drugs. Despite previous disappointing results in the field of the chemical control of plant viruses, drugs should not be overlooked in regard to future possibilities of virus control and in their present utilization in eliminating some viruses, for example, in meristem tip or *in vitro* culture.

3. BREEDING FOR RESISTANCE: FROM THE USE OF SPECIFIC RESISTANCE GENES TO THE DISCOVERY OF ANTIVIRAL ACTIVITIES IN PLANTS

In the future, control of virus diseases in plants very likely will rely on developments taking place in the area of genetic engineering. Up to the present time, breeding for resistance to plant pathogens has been based on identification of certain traits of resistance in a given cultivar or species and their subsequent introduction by genetic crossing into a crop species. This technique has been successful, as evidenced by the virtual disappearance of viruses such as bean common mosaic virus. With certain crops it is quite difficult, however, to combine virus resistance with desirable horticultural characteristics, this being especially true with woody plant species. To illustrate the difficulties of breeding for resistance we can briefly mention two examples. The case of tobacco mosaic virus (TMV) in tomatoes is an interesting situation. In tomato, there are two dominant alleles, Tm_2 and Tm_2^2, both from *Lycopersicum peruvianum* (Laterrot, 1973, 1977) that govern the hypersensitivity type of resistance to TMV. The system has two disadvantages: no strain is adapted to Tm_2 but one TMV strain is adapted to Tm_2^2. Moreover, in the case of homozygosis, plants exhibit low fertility and yields. On the contrary, heterozygosis can result in severe necrosis on leaves and fruits under conditions of high temperature and light or high inoculum pressure. This clearly illustrates the difficulties usually encountered in breeding programs. A dominant gene Tm_1 isolated from *Lycopersicum hirsutum* confers resistance to viral multiplication, but, once again, homozygosis is associated with low fertility and yields, indicating the existence of a genetic linkage between virus resistance and poor agronomic production. Presently, breeding tomato for resistance leads to the selection of plants containing, in an heterozygous condition, both Tm_1 and Tm_2, which confer a total protection against the most frequent TMV strains. Of interest is also the resistance of the Charentais type muskmelon to cucumber mosaic virus. Two types of resistance have been introduced. The first type is the resistance to infection by the common strains of CMV, which is oligogenic, recessive (Risser et al., 1977), and ineffective against CMV strains of the "Song" pathotype (Leroux et al., 1979). The second type, already mentioned in a previous section, prevents transmission of both common and Song CMV strains by *Aphis gossypii*, but is ineffective against other species of aphid vectors.

Many characteristics of resistance have been transferred between genetically compatible species and we hope to control, by this conventional breeding method, some of the very severe virus diseases. For example, two apricot CVS have been discovered that apparently are resistant to plum pox infection; both belong to the *siberica* type of *Prunus armeniacae*, which may be used in a future breeding program for plum pox resistance.

Two research groups investigating the hypersensitivity reaction of tobacco plants carrying the N resistance gene to TMV have identified an antiviral factor AVF (Sela, 1981) or an inhibitor of virus replication (IVR) (Loebenstein and Gera, 1981; Gera et al., 1983). These substances have been purified and the IVR was characterized as proteinaceous in nature and was demonstrated to inhibit replication of TMV in protoplasts and tomato leaf disks, and CMV and potato virus X replication in various host tissues.

The AVF also is a type of protein. It is a glycoprotein that was reported by Sela (1981) to function as an interferon-like molecule. The same research group also reported that animal interferon is able to inhibit plant virus replication (Orchanski et al., 1982; Reichman et al., 1983). The existence and respective role of these substances is discussed in another chapter of this book. It is interesting to find that a nonspecific control mechanism of virus replication may exist in plants. This opens new perspectives for a control of plant viruses by substances associated with active resistance phenomena. Identification at the genome level of the responsible gene(s) could provide new possibilities of breeding for nonspecific virus resistance by genetic engineering comprising cloning of this(these) gene(s) and transfer to plants devoid of such resistance gene(s).

4. USE OF VIRAL GENES OR VIRAL INFORMATIONS IN BREEDING FOR RESISTANCE TO PLANT VIRUSES

4.1. The First Steps: Cross-Protection

In the previous section we discussed briefly the use of plant genes in breeding for virus resistance. It was demonstrated more than 50 yr ago that viruses themselves could be used in plant virus control.

In 1929 MacKinney demonstrated that tobacco plants infected with a mild strain of tobacco mosaic virus (TMV) that produces a light green mosaic, showed no additional symptoms following inoculation with a variant that caused yellow mosaic when inoculated to healthy plants. Since this first report of cross-protection between virus strains, many other examples have been reported with various crops and viruses. There have been several reviews written on virus interactions (Bennett, 1953; Kassanis, 1963) and possible applications of the cross-protection phenomenon are still frequently discussed (Sequeira, 1984).

Cross-protection between virus strains has at least two important applications. First, it is a valuable tool for virus classification. Except for a few questionable reports of protection between different viruses, cross-protection is restricted to strains of a given virus and always indicates a certain degree of relationship. It should be mentioned, however, that cross-protection may not occur with every strain of a given virus.

The second application of cross-protection involves field control of virus diseases. The best examples are the protection of tomato and citrus plants against tobacco mosaic virus and tristeza virus, respectively. Before 1971, 50–60% losses in tomato due to TMV infection were not uncommon in Europe. The virus is spread mechanically by contact and its spread is difficult to control. It has been common knowledge since the early reports of MacKinney (1929) that mild TMV strains could be used to protect against severe strains of the virus. Several natural mild variants were isolated and some were artificially induced by mutagenesis. Some were selected and propagated for their ability to confer cross-protection when challenged with severe strains (Marrou and Migliori, 1974). One of these strains obtained in the Netherlands by mutagenesis is especially well known and has been used commercially in several European countries during 1975–1980 (Fletcher and Rowe, 1975); in France alone more than 10 million tomato plants were protected. After 1980, the number of protected plants decreased because of the availability of resistant cultivars and some unexpected breakdowns of protection.

The second successful application of cross-protection concerns citrus tristeza virus (CTV). The virus was introduced in Brazil in the 1920's. At this time practically all orange groves consisted of the Pera cultivar grafted onto sour orange. High populations of the aphid *Toxoptera citricidus*, later demonstrated to be an efficient vector of the tristeza virus, were also present in the groves. The CTV induces incompatibility between the sweet orange scion grafted onto sour orange rootstock and, within a few years, more than 30 million orange trees were killed by the virus. In highly infected areas, a few trees did not decline, even though the results of indexing tests showed that they were CTV infected. The CTV isolate present in these nondeclining trees was then challenged with severe strains of the virus and appeared to confer a high degree of protection (Muller and Costa, 1977). This mild isolate was propagated and more than 30 million orange trees are now protected (Costa and Muller, 1980) from the debilitating effects associated with the severe strain of CTV.

Many other applications of cross-protection have been reported, but appear to be of limited use because of low stability of the mild strains, difficulties encountered in propagating these strains, or failure to get uniform distribution of the protectant strain in the inoculated plant, because of no or low vertical transmission of the mild strain to progeny. The mechanisms of cross-protection are still unclear and a better understanding of the process is needed before envisaging new developments and new applications. Several hypotheses have been proposed, but none is totally acceptable. Some researchers explain interference by postulating a limited supply of essential cell constituents for virus increase. Others suggest that the virus replicates at certain specific sites in the host cell; if there are a limited number of these replication sites and if the mild strain has a special affinity for these sites, all might be occupied by the mild strain, thus preventing the severe one from becoming established. Others propose that, when the virulent strain enters the cell, the virus genome is uncoated, replicated, and encapsulated in the mild strain coat protein, and further uncoating will not occur. Hence the virulent genome trapped in the mild strain coat protein will be progressively eliminated or will remain at a very low level. Most of these hypotheses attempt to explain the absence of symptoms in protected plants by the absence of the severe strain, but it is now known that protection can exist in the presence of the severe strain. In 1977, Cassels and Herrick investigated cross-protection in TMV-infected tomato plants and they observed that both the mild and the severe strains were present in the symptomless doubly-infected plants. The concentration of the mild strain in these plants was identical to that observed in plants inoculated only with the mild strain; the concentration of the severe strain was less. On some apical leaves of the "protected" plants symptoms occasionally would appear, indicating a breakdown in cross protection. Analysis of the virus strain content of these apical leaves showed an increase in the titer of the severe strain. It would appear, therefore, that symptom development was correlated with an increase of the titer of the severe strain. In this particular situation, protection clearly depends on the severe: mild strain ratio. Doz et al. (1982, 1983) investigated cross-protection with another plant virus combination. They reported that *Chenopodium quinoa* plants inoculated with a mild tomato black ring virus (TBRV) strain are protected against a severe TBRV strain. Quantitation of the strain content by the immunoenzymatic ELISA test applied to crude extracts as well as virus suspensions purified from doubly-infected plants, showed that both strains were present. Their respective titers were simi lar, if not identical, to those observed in plants singly inoculated with each strain. The authors concluded that replication of the severe strain was not altered and that the mild strain simply controlled symptom expression. These results, which seem to contradict theories claiming that the absence of symptoms correlates with the absence of the severe strain, probably illustrate the existence of multiple mechanisms that are still to be clarified, as proposed by Zaitlin in 1976. Cross-protection has been reported primarily with viruses. In 1978 Niblett et al. reported, however, that mild strains of potato spindle tuber viroid protected against severe strains of the viroid; this report seems to refute the involvement of the coat protein, as suggested by Zoeten and Fulton (1975). More recently, it has been suggested (Zaitlin, personal communication) that regulation could come from the formation of a hybrid between the mild and the severe genome. Such a hypothesis fits well with recent results we discuss in the following section involving the formation of RNA–RNA hybrids in cell regulation phenomena. Even though many questions regarding cross-protection remain unanswered, it is clear that the phenomenon exists and that a plant can be efficiently protected by preinoculation with a mild virus strain. As we mentioned before, besides various difficulties, including the selection of a stable mild protective strain, the two major limitations are (1) uniformity of virus distribution in the plant (it is generally accepted that protection needs a uniform distribution of the mild strain), and (2) vertical transmission to progeny. These are major difficulties because the virus often is unevenly distributed, especially in perennial plants, and, even when seed transmission exists, it occurs at a low and inconsistent level.

4.2. Possibilities of Propagating Copies of Protective Viral Genome in the Plant Genome

Recent progress in the fields of molecular biology and genetic engineering shed new light on this matter and offers hope that protective elements from the viral genome might be introduced into the cell nuclear genome. The three problems that must be dealt with are (1) the nature of the viral genome, (2) the vehicle, and (3) the expression of the viral genome introduced in the cell genome.

4.2.1. Nature of the Viral Genome

Of course the problem only exists for RNA viruses and viroids. For single-stranded (ss) DNA viruses, the DNA molecule can be easily duplicated into a double-stranded (ds) DNA molecule which may, in theory, be transferred to the cell genome using an appropriate vector. This can be done directly with the dsDNA viruses. For the RNA viruses the problem is to obtain a DNA copy comprising all the information of the viral RNA, or at least, all the information needed for establishing and propagating a protecting state. We have previously mentioned, in the section dealing with diagnosis and molecular hybridization, that it is now possible to transcribe an RNA molecule into a complementary cDNA. This single-stranded cDNA molecule is then duplicated into a double-stranded molecule that may be introduced into a plasmid. Finally, it is possible to obtain a recombinant plasmid containing a partial or full-length copy of the viral RNA genome. In 1983 Cress et al., working with potato spindle tuber viroid (PSTV), constructed a plasmid containing a full-length PSTV insert with two contiguous fragments from two cloned partial-length PSTV cDNA. The full-length fragment was then subcloned immediately downstream a promoter sequence. Plasmids containing monomeric and dimeric PSTV cDNA and RNA were isolated from *E. coli* cells and tested for infectivity. Plasmids containing dimers and RNA transcripts containing the PSTV sequence were infectious. Similarly, Ohno et al. (1983) constructed tandemly repeated clones of hop stunt viroid; transcripts directed by a promoter were demonstrated to be infectious. Recently Tabler and Sanger (1984) constructed a set of monomeric and oligomeric PSTV-specific cDNA forms representing complete DNA copies of the circular PSTV RNA genome in the pBR 322 plasmid and in bacteriophage M13. Both double-stranded and single-stranded PSTV cDNA were infectious upon inoculation to tomato plants. All dimeric and multimeric forms were infectious, irrespective of their polarity (ssDNA) or regardless of their orientation in the vector DNA (dsDNA). In addition, certain cloned monomeric cDNA also are infectious as are subgenomic DNA fragments, representing together the complete molecule, which themselves are noninfectious, when inoculated as a mixture.

These results obtained with plant viroids confirm results previously reported with QB bacteriophage and poliovirus. The DNA copies of a viral RNA genome are infectious and can be transcribed into infectious RNA. The problem is to know if full copies are needed or if partial ones are enough. In terms of cross-protection it is still impossible to answer this question because the molecular basis of cross-protection is not fully understood, nor is the molecular basis of symptom expression and pathogenicity. Full-length copies of plant virus RNAs have now been obtained. A full-length cDNA of tobacco mosaic virus genomic RNA was assembled from partial clones (Dawson et al., 1985). Ahlquist et al. (1985) constructed cDNA clones of the three genomic RNAs of brome mosaic virus and the capped *in vitro* transcripts of these cDNAs could be used to efficiently infect whole plants and protoplasts. Another significant contribution has been the construction of a plasmid containing a full-length copy of the M.RNA of cowpea mosaic virus which was transcribed *in vitro* into complete plus-strands of M.RNA. Applying site-directed mutagenesis to this full-size DNA copy contributed to a further understanding of the genetic organization of the M.RNA (Vos et al., 1984).

4.2.2. Transfer of Copies of the Viral Genome into the Plant Genome—Plant Vehicles

The previous discussion dealt only with the possibility of constructing viral cDNA copies that are infectious upon inoculation to plants and are transcribed into infectious RNA. The second question is whether it is feasible to transfer the viral genetic information into the cell DNA. The point is to have the appropriate molecule or vehicle that could be used as a vector for introducing foreign genetic information into plants. This has been recently reviewed by Howell (1982). Several potential vehicles exist; they may be derived from plant genome components or from naturally occurring plant vectors and pathogens. The aim is to transform the recipient host plant so that it can express the newly acquired genes with a high degree of stability. Some years ago, cauliflower mosaic virus (CaMV) DNA appeared as a potential attractive vector. The advantage of this viral dsDNA is that it could be manipulated *in vitro* as a bacterial plasmid and then reintroduced to plants. Several problems exist, the first of which is the restricted host range of the virus, even considering all the caulimoviruses. In addition, there is little room in the CaMV DNA to insert foreign DNA; most modifications of the genome destroy infectivity. Another major disadvantage is the lack of integration and the lack of vertical transmission. Nevertheless, CaMV DNA has been used successfully to introduce foreign genes into plants.

The most attractive vectors have been discovered in soil bacteria, *Agrobacterium tumefasciens* and *A. rhizogenes.* These bacteria possess plasmids that have a morphogenic influence on the infected plant. These plasmids have been identified first in *A. tumefasciens*, where they have been called Ti plasmids. They have been characterized by the major type of amino acid derivative, or opine, produced in the induced tumors. These Ti plasmids are able to direct the synthesis of specific opines in the tumors because a piece of the Ti plasmid called T-DNA is transferred from bacteria to the plant and integrated into the plant nuclear genome (Chilton et al., 1977; White et al., 1982, 1983). Similar infection of plants by *A. rhizogenes* results in the development of adventitious roots directed by the Ri plasmid. Transformed tissues, tumors, or adventitious roots can be induced to regenerate into plants that still possess the T-DNA. Significant progress is being made in the regeneration of the transformed plants. In the case of plants transformed with the Ti plasmid, shoots that could regenerate from certain tumors are probably not transformed (Van Slogteren et al., 1983). A Ti plasmid mutant was constructed by Zambryski et al. (1983) in which all the oncogenic functions of the T-DNA have been deleted and replaced by pBR 322. This Ti plasmid still mediates efficient transfer and stabilization of its truncated T-DNA into plant cells; integration and expression of this minimal T-DNA does not interfere with normal plant-cell differentiation. Moreover, a DNA fragment, e.g. a viral DNA fragment, cloned in a pBR vector can be inserted in this Ti plasmid upon a single recombination.

In the case of *Agrobacterium rhizogenes* the original Ri T-DNA induces little or no differentiated cell proliferation (Moore et al., 1979). Adventitious roots regenerate readily into plants that transmit the Ri T-DNA to their progeny (Tepfer, 1984; David et al., 1984). The resulting regenerated plants show some abnormal phenotypic characteristics which very likely will be removed in the near future by the elimination of the responsible fragment of the T-DNA.

4.2.3. Expression of Foreign Genes in Plants

In summary, the *Agrobacterium* plasmids undoubtedly will be used in the coming years for introducing various foreign genes, including virus genomes, into the nuclear plant cell; but another question is how much these foreign genes can be expressed. Clear demonstration that a viral gene can be transcribed in plants has been given by Schewmaker et al. (1985). Using the T-DNA of *Agrobacterium tumefasciens* they introduced a full-length copy of CaMV into various plant genomes. Analysis of turnip galls revealed two transcripts derived from the integrated genome; the 1.9 kb transcript derived from the region VI of the CaMV genome is found in CaMV-infected leaves. A large 7.5-kb transcript initiates at, or near, the other known promoter which, in infected cells, directs the synthesis of the full-length transcript. That viral promoters do function when integrated in the plant genome was confirmed by demonstrating that both the full-length and the region VI promoters can direct expression of a bacterial kanamycin-resistant gene. Several other reports demonstrate that foreign genes introduced into plant genomes can, indeed, be expressed. One of them involves the introduction, without integration, of a foreign gene with the CaMV DNA as vector. The CaMV genome possesses seven open reading frames (ORF); two of which are not essential for infection and can be detected or expanded by a small insertion of foreign DNA. Brisson et al. (1984) reported the replacement of CaMV ORF II by the R67 plasmid-encoded dehydrofolate reductase (dhfr); this gene confers resistance to methothroxate in *E. coli.* The chimeric gene was propagated in turnip plants and the dhfr gene was expressed, producing a functional enzyme. Four different coding sequences inserted behind the nopaline synthase gene promotor have been properly expressed following the infection of plant cells with Ti plasmid carrying these genes; one was octopine synthase and the three others were for bacterial antibiotic resistance (Herrera-Estrella et al., 1983a,b).

It is obvious that the tremendous developments taking place in genetic engineering, especially the results obtained with the *Agrobacterium* plasmids, provide plant virologists with new tools and will allow new approaches in the control of plant virus diseases. One is based on the use of "protective viral genes" as a continuation of cross-protection and the second involves the possible use of viral antisense RNA.

4.3. Use of Viral Genes for the Control of Plant Diseases

When discussing the problem of cross-protection, we mentioned that several difficulties that are encountered (e.g., uneven distribution of the protective strain and lack of vertical transmission) could be overcome if the protective genome could be integrated into the cell nuclear genome. The possibility of manipulating all virus genomes, including viral RNA, as DNA molecules, and the existence of suitable vectors, open up new areas for cross-protection. The remaining question is to know what part of the genome or of its copy should be used and introduced. Several virus genomes have been thoroughly investigated, especially from the point of view of their replication and expression of certain genes but, so far, very little is known concerning the origin and molecular basis of their pathogenicity. Dickson (1981) suggested that viroids can compete with the small nuclear RNA in the processing of the hnRNA and interfere with splicing. Such regulation is based on the possibility of RNA–RNA hybrid formation and correlates with recent theories explaining regulation by such a process. If this theory is accurate, it would appear that a small part of the molecule is involved in pathogenicity and, perhaps, also in cross-protection. Being able to obtain full-length cDNA copies and site-specific mutagenesis will certainly be a major factor in understanding the origins of virus pathogenicity. Returning to the question of cross-protection with viruses or viroids, since it is not possible to identify the part of the genome responsible for the phenomenon, the first step could be to introduce the complete copy of the mild strain genome into the plant nuclear genome via an *Agrobacterium* plasmid, then regenerate the plant and see what happens. The main questions to be answered are:

1. Will the genome be transcribed into viral RNA transcripts?
2. Will these transcripts show the 3′ and 5′ ends of the viral RNA (when dealing with RNA viruses)?
3. Will they be translated into proteins, especially into the coat protein, which could be used as a marker because of its easy identification?
4. Will the transformed plants be protected against the severe virus strain?

Depending on the size and structure of the viral genome, a full-length copy may be difficult to produce. When dealing with a multipartite genome, virus engineering a part of the genome can be enough. In their work on tomato black ring virus, a bipartite genome virus, Doz et al. (1982) showed that the pathogenicity of the severe strain (expressed as top necrosis on *Chenopodium quinoa*) was associated with the RNA_2, the smaller viral RNA (1.5×10^6 daltons mol. wt.), which seems also to be involved in cross-protection. In such a case, manipulation to introduce a viral information into the cell genome could be restricted to this RNA_2.

Very recently, Powell Abel et al. (1986) showed that expression of the tobacco mosaic virus (TMV) coat protein gene integrated in the nuclear genome of tobacco plants interfered with symptom development when those plants were inoculated with TMV.

Viroids or some of the small satellite RNAs might be excellent experimental models to study the control of virus diseases with copies of viral sequences. Several satellites have been identified in plant viruses and some are associated with special symptom expression. This is the case of cucumber mosaic virus strains, particles of which contain various satellite RNAs of 0.1×10^6 daltons. On the basis of symptom expression (i.e., their capacity to induce lethal necrosis of tomatoes), Jacquemond and Lot (1981) distinguished the necrotic and nonnecrotic types of satellite RNA. In addition, the nonnecrotic type protects the tomato plants against the necrosis normally associated with the necrotic type (Jacquemond, 1982). The 300 nucleotide molecule of the nonnecrotic satellite could be easily cloned and introduced into a tomato plant as a protective gene to determine whether it would confer protection from the necrosis associated with the necrotic satellite.

Full-length copies of different satellite RNAs of CMV have been obtained in several laboratories. In particular, obtention of a cloned copy of the nonnecrotic CMV strain will soon enable it to be transferred into tomato plants as a protective gene to determine whether it can confer protection from the necrosis associated with the necrotic satellite (Baulcombe et al. 1986).

4.4. The Antisense RNAs: A Potential Mode for the Control of Plant Viruses

It is well known that the expression of certain genes is regulated at the level of transcription either negatively (repressors) or positively (activators) by protein factors. Protein factors also regulate the translation of specific mRNAs. In addition, it was recently demonstrated that RNA molecules are also involved in regulating the expression of certain genes. A classic example was produced by Lacatena and Cesarini (1981) with the regulation of the plasmid Col E1. A short antisense transcript is believed to form a RNA–RNA duplex with the 5′ end of the replication primer and, consequently, to inhibit replication. Another example of regulation of gene expression is provided by the membrane proteins of *E. coli.* There are two major outer membrane proteins called OmpF and OmpC which are essential for diffusion of small hydrophilic molecules. They are encoded by two genes ompF and ompC, the regulation of which is controlled by the osmolarity of the culture medium. Mizuno et al. (1984) reported evidence for a unique regulatory mechanism of gene expression mediated by a small RNA they called "mRNA interfering complementary RNA" (mic RNA). This mic RNA is produced from an independent transcriptional unit (the mic F gene) located immediately upstream the ompC gene, but transcribed in the opposite direction.

This small 174-base mic RNA appears to block the translation of the OmpF mRNA by hybridizing to it in its 5′ end. This natural regulation (i.e., the ability of antisense RNA to inhibit gene expression) has been investigated and confirmed by genetic engineering. Coleman et al. (1984) constructed plasmids that produce a complementary RNA to the *E. coli* lipoprotein (lpp) mRNA. Construction of this gene necessitated isolation of the lpp gene region comprising the Shine Delgarno sequence for ribosome binding and the coding region for the first 20 residues of the lipoprotein. This fragment was inserted into a plasmid in the opposite direction from the normal lpp gene; the resulting plasmid is able to produce a mic-lpp-mRNA upon induction. Production of lpp in cells harboring this plasmid is reduced 16-fold. In addition, plasmids were constructed with 2 mic-lpp-genes; cells harboring these plasmids showed a production of lpp reduced 31-fold, demonstrating that production of mic RNA inhibits lipoprotein synthesis. Inhibition of gene expression by mic RNA also occurs in eucaryotes. Izant and Weintraub (1984) used the Herpes Simplex virus thymidine kinase (TK) gene, which is efficiently expressed when introduced into mouse cell by direct needle injection. A fragment containing the TK structural gene was excised from a parental plasmid and reinserted into a promotor-containing vector in the opposite orientation. The DNA containing such flipped gene construction was coinjected with the cloned wild-type gene into mouse lacking the TK gene. The appearance of TK gene activity was significantly reduced. The Herpes Simplex virus-TK flipped gene constructions do not inhibit the activity of chicken-TK expression, suggesting that the antimessage production can provide a specific molecular immunity to the expression of subsequently transcribed genes and indicating that the phenomenon is sequence specific.

Of course, as suggested by Travers (1984), the extent to which this novel form of regulation of gene expression is utilized naturally in prokaryotes and eukaryotes remains to be established. Nevertheless, it provides a very attractive potential mode of regulation, especially in the field of virus control.

In contrast with the proposals given in the section above (introduction of viral genes), the point is not to clone and to introduce in the plant genome and in the normal orientation a copy of the mild-strain genome, but to start with a virulent pathogenic strain and to clone and transfer in the opposite orientation a part of the viral genome copy into the nuclear cell genome, using the same vectors. Regulation can involve replication of the viral genome or more likely its expression. Indeed, it is most likely that the mic RNA acts primarily by binding to the mRNA and probably prevents interaction with ribosomes. In such a case it should be of interest to clone a short sequence complementary to the 5′ end of the viral genome. The expression of some virus genomes has been elucidated. It appears that the genome or the different pieces of the multipartite genome are translated into a polyprotein that is further processed into several proteins, replicase, and coat protein. Inhibition of translation will inhibit synthesis of this polyprotein and, therefore, block the virus synthesis. When the viral genome is made up of two pieces, as observed for example with cowpea mosaic virus, expression of the two virus RNA species is closely integrated—the processing of the polypeptide encoded by one RNA species depending on a protein synthesized from the second RNA species (Franssen et al., 1984). In this case, using an antisense message from the adequate genome species could also block completely the virus synthesis.

In the case of the regulation of the *E. coli* membrane lipoprotein (Coleman et al., 1984), the mic RNA induced a reduction in the amount of lipoprotein mRNA and the authors proposed several models for explaining such a regulation: (1) reduced stability of the mRNA when not entirely traversed by ribosomes; (2) hybrid formation between the mic and the mRNAs causing premature termination of the transcription; and (3) destabilization of the mRNA. Alternatively, the mic RNA may directly inhibit initiation of transcription or cause pausing of mRNA elongation. In fact, several types of mechanisms may occur. In attempts to use antisense RNA for virus control it will be interesting to use copies of different small parts of the viral genome. This may also help to elucidate some steps which are still unclear, concerning the replication and expression of the viral genomes, one of the possibilities being that the mic RNA interfere with the part of the genome directly involved in symptom expression. If the mic RNA acts only by binding to a mRNA, thus preventing initiation of protein synthesis, the use of such a regulation system very likely will be ineffective with viroids, the genome of which does not appear to encode for any protein. Once again, use of mic antisense RNA will provide basic information on the replication, expression, and molecular basis of the pathogenicity of viruses and viroids. There is now some evidence that processing of the hnRNA through splicing depends on snRNA, which have 10–15 nucleotides near their 5′ end complementary to the intervening sequences. The existence and role of mic RNA provide additional evidence for the role of hybrid RNA-RNA in cell regulation. If pathogenicity of viroids and viruses appears (Dickson, 1981) as competition with the snRNA with which they share some sequence homology, there will be another approach, using antisense RNA complementary with the interacting regions of the virus or viroid genome, to investigate origins of their pathogenicity.

As early as 1978, Zamecnik and Stephenson reported that synthetic oligodeoxynucleotide complementary to Rous Sarcoma virus terminal repeats decreased virus infection and could inhibit viral translation *in vitro*. Indeed there seems to be some possibility of testing the effect of the mic RNA on the translation of the viral RNA *in vitro* before trying to transform the plant by *Agrobacterium* harboring a Ti or Ri plasmid containing an antisense sequence for the viral RNA.

5. CONCLUDING REMARKS

Attempts to control plant viruses began as soon as researchers recognized the role of these pathogens in plant disease, certainly long before their physical chemical characteristics were described. Control has been primarily based on sanitation, that is, prevention of disease by the production and subsequent protection of virus-free plants. After more than 30 yr, this method remains the key to plant virus control. The efficiency and reliability of certification schemes are increasing as a result of significant improvements in multiplication methods (*in vitro* propagation of plant material) and disease diagnosis (immunoenzymatic techniques using polyclonal or monoclonal antibodies, molecular hybridization, etc.). Very recent results, within the past 5 yr and in some cases within the past few months, open new areas and new perspectives. Owing to the discovery of appropriate plant vehicles (e.g., *Agrobacterium* plasmids), rapid introduction of new traits into the plant nuclear genome can be now envisaged. The possibility exists for the introduction not only of plant resistance genes but also of information about viral origin through copies of mild virus strains or antisense viral sequences for the production of mic RNAs.

Nonspecific control of plant viruses could result from the use of genes governing the synthesis of antiviral, interferon-like molecules or inhibitors of virus replication. Such genes, which seem to exist only in certain plants, could be identified, isolated, cloned, and introduced into important crop plants that do not naturally contain such genes.

Exciting possibilities based on chemical or genetic control also exist in the area of specific virus control. As we mentioned previously, if antiviral drugs are still to be discovered, they most likely will be active against specific viruses or specific viral functions. For instance, an antiviral factor efficient against Herpes virus was recently identified. It acts specifically on a virus encoded kinase different from the normal cell kinases; most viruses do not have specific kinase and therefore are not inhibited by the drug. Viruses of the Potyvirus group contain several genes; drugs may be found that interfere with inclusion proteins (some of them may be involved in virus particles translocation) or the small protein located at the 5′ end of the viral RNA.

Recent discoveries of plant vehicles (*Agrobacterium* plasmids) that make possible the integration of foreign information into the cell nuclear genome allow two types of manipulation:

1. Introduction of a protective genome or cloned copies of mild virus strains.
2. Introduction in the opposite orientation of small viral sequences that could be transcribed into antisense mRNA, thereby providing regulatory properties against the infecting virus.

We need much more information on the origin and expression of pathogenicity, but it is likely that these techniques will provide specific control of given viruses. Dickson (1981) proposed that the expression of pathogenicity in viroids may result from 12–15 nucleotide sequence homology between viroid RNA and the small nuclear RNAs (snRNAs) which could be involved in the splicing of hnRNAs. Pathogenicity of viroids could therefore result in a competition between the snRNAs and the viroid RNAs at the splicing level. Changes in 1–2 of these 12–15 nucleotides could greatly influence the competitiveness of viroids and, therefore their pathogenicity. Similarly, regulation by mic RNAs involves short sequences and changes in a few nucleotides should result in a decreased interference, that is to say, that newly appeared strains differing by a few nucleotides could partly or totally escape regulation.

In the future, genetic engineering of the plant genome will provide exciting new possibilities for controlling severe plant diseases. Some limitations come from the characteristics of the Ti and Ri plasmids themselves, from the host range of *Agrobacterium spp.*, and from the difficulty of transforming and regenerating some types of plants. The use of plant transposable elements for cloning (O'Reilly et al., 1985) and transfer, will be a way to overcome these difficulties and broaden the field of application of these techniques of genetic engineering to introduce foreign information into the plant genome. It is quite obvious that these, apparently sophisticated, methods will reduce the adverse effects of virus infection and also reduce the general level of virus infection in commercial crops. As far as we know at the present time, they will not provide complete resistance and, therefore, will never completely replace the standard techniques of classical plant breeding for resistance and the conventional sanitation programs. As a result of the introduction of new technology in plant multiplication and disease diagnosis, certification schemes will provide increasingly higher-quality plant material and will remain the foundation of plant virus disease control. For this reason some emphasis has been placed on these conventional, but highly effective control methods, while at the same time recognizing the exciting possibilities that may come about as a result of the rapid progress being made in the areas of genetic engineering and biotechnology.

REFERENCES

Ahlquist P., R. French, M. Janda, and Y. Antignus (1985). Replication and expression of homologous and foreign genes in engineered RNA viruses. 1st International Congress of Plant Molecular Biology. Abstracts, p. 149.

Bar Joseph, M. and M. Malkinson (1980). Hen egg yolk as a source of antiviral antibodies in the enzyme linked immunosorbent assay (ELISA): A comparison of two plant viruses. *J. Virol. Methods*, **1:**179–183.

Bar Joseph, M., A. Rosner, M. Moscovitz, and R. Hull (1983). A simple procedure for the extraction of double stranded RNA from virus infected plants. *J. Virol. Methods*, **6:**1–8.

Bar Joseph, M., D. Segev, S. Twizer, and A. Rosner (1985). Detection of avocado sunblotch viroïd by hybridization with synthetic oligonucleotide probes. *J. Virol. Methods*, **10:**69–73.

Baulcombe, D. C., Saunders, G. R., Bevan, M. W., Mayo, M. A., and Harrison, B. D. (1986). Expressions of biologically active viral satellite RNA from the nuclear genome of transformed plants. *Nature*, **324:**446–449.

Bennett, C. W. (1953). Interaction between viruses and virus strains. *Adv. Virus Res.*, **1:**39–67.

Bradley, R. H. E., C. W. Wade, and F. A. Wood (1962). Aphid transmission of potato virus Y inhibited by oils. *Virology*, **18:**327–329.

Brisson, N., J. Paszkowski, J. R. Penswick, B. Gronenborn, I. Potrykus, and T. Hohn (1984). Expression of a bacterial gene in plants using a viral vector. *Nature*, **310:**511–514.

Cassels, A. C. and C. C. Herrick (1977). Cross protection between mild and severe strains of tobacco mosaic virus in doubly inoculated tomato plants. *Virology*, **78:**253–260.

Chilton, M. D., M. H. Drummond, D. J. Merlo, D. Sciaky, A. L. Montaya, M. P. Gordon, and E. W. Nester (1977). Stable incorporation of plasmid DNA into higher plant cells: The molecular basis of crown gall tumorigenesis. *Cell*, **11:**263–271.

Clark, M. F. and A. N. Adams (1977). Characteristics of the microplate method of enzyme-linked immunosorbent assay for the detection of plant viruses. *J. Gen. Virol.*, **34:**475–483.

Cohen, I. and V. Malamed Madjar (1978). Prevention by soil mulching of the spread of tomato yellow leaf curl virus transmitted by *Bemisia tabaci* in Israel. *Bull. Entomol. Res.*, **68:**465–470.

Coleman, J., P. J. Green, and M. Inouye (1984). The use of RNAs complementary to specific mRNAs to regulate the expression of individual bacterial genes. *Cell*, **37:**429–436.

Clark, M. F. and M. Bar Joseph (1984). Enzyme immunosorbant assays in plant virology. In "Methods in Virology." (K. Maramorosh and H. Koproski, eds.) *Academic Press*, **VII:**51–87.

Cooper, J. I. and P. R. Thomas (1971). Chemical treatment of soil to prevent transmission of tobacco rattle virus to potatoes by *Trichodorus spp. Ann. Appl. Biol.*, **69:**23–24.

Cornuet, P. (1959). Maladies de la pomme de terre. In *Maladies à Virus des Plantes Cultivées et Méthodes de Lutte.* INRA, Paris, pp. 131–132.

Costa, A. S. and G. W. Muller (1980). Tristeza control by cross protection: A U.S.-Brazil cooperative success. *Plant Dis.*, **64:**538–541.

Cress, D. E., M. C. Kiefer, and R. A. Owens (1983). Construction of infectious potato spindle tuber viroid cDNA clones. *Nucleic Acid Res.*, **11:**6821–6835.

David, C., M. D. Chilton, and J. Tempe (1984). Conservation of the T-DNA in plants regenerated from hairy root cultures. *Biotechnology*, **2:**73–76.

Dawson, W. O., G. L. Garantham, D. A. Knorr, and D. L. Beck (1985). Synthesis of full length cDNA clones of tobacco mosaic virus. 1st International Congress of Plant Molecular Biology. Abstracts, p. 13.

Dickson, E. (1981). A model for the involvement of viroids in RNA splicing. *Virology*, **115:**216–221.

Diener, T. O. (1971). Potato spindle tuber viroid: a replicating low molecular weight RNA. *Virology*, **45:**411–428.

Dodds, J. A., T. J. Morris, and R. L. Jordan (1984). Plant viral double stranded RNA. *Ann. Rev. Phytopathol.*, **22:**151–168.

Doz, B., R. Delbos, and J. Dunez (1982). Prémunition: une compétition entre souches faibles et souches fortes pour une voie commune d'expression des symptômes. In *La Sélection des Plantes pour la Résistance aux Maladies. Colloq. INRA*, **11:**29–44.

Doz, B., M. Jacquemond, G. Marchoux, and J. Dunez (1983). Interaction entre souches de virus; application à la protection des plantes. In *Les Antagonismes Microbiens. Colloq. INRA*, **18:**58–70.

Fletcher, J. T. and J. M. Rowe (1975). Strain changes in populations of tobacco mosaic virus from tomato plants. *Ann. Appl. Biol.*, **81:**409–412.

Franssen, H., N. Leunisse, R. Golbach, G. Lomonossof, and D. Zimmern (1984). Evidence that the 32,000 dalton protein encoded by bottom component of cowpea mosaic is a proteolytic processing enzyme. *J. Virol.*, **50:**183–190.

Gera, A., G. Loebenstein, and S. Shabtai (1983). Enhanced tobacco mosaic virus production and suppressed synthesis of a virus inhibitor in protoplasts exposed to antibiotics. *Virology*, **127:**475–478.

Gibbs, A. and B. D. Harrison (1976). *Plant Virology, the Principles.* E. Arnold, London.

Hansen, A. J. (1979). Inhibition of apple chlorotic leaf spot virus in *Chenopodium quinoa* by ribavirin. *Plant Dis. Rep.*, **63:**17–20.

Hansen, H. J. (1984). Effect of ribavirin on green ring mottle causal agent and necrotic ring spot virus in Prunus species. *Plant Dis.*, **68:**216–218.

Harpaz, I. (1982). In *Pathogens, Vectors and Plant Diseases: Approaches to Control.* (K. M. Harris and K. Maramorosh, eds.). Academic, New York, pp. 1–21.

Harrison, B. D. (1983). Progress and problems in the control of arthropod, nematode and seedborne plant viruses. In *Control of Virus Diseases.* (E. Kurstak and K. G. Marusyk, eds.). Dekker, New York, pp. 265–299.

Harrison, B. D. and P. D. Winslow (1963). Laboratory and field studies on the relation of Arabis mosaic virus to its nematode vector *Xiphimena diversicaudatum. Ann. Appl. Biol.*, **49:**621–633.

Herrera-Estrella, L., M. De Block, E. Messens, J. P. Hernalsteens, M. Van Montagu, and J. Schell (1983a). Chimeric genes as dominant selectable markers in plant cells. *EMBO J.*, **2:**987–995.

Herrera-Estrella, L., A. Depicker, M. Van Montagu, and J. Schell (1983b). Expression of chimeric genes transferred into plant cells using a Ti plasmid-derived vector. *Nature*, **303:**209–213.

Howell, S. H. (1982). Plant molecular vehicles: Potential vectors for introducing foreign DNA into plants. *Ann. Rev. Plant Physiol.*, **33:**609–650.

Izant, J. G. and H. Wentraub (1984). Inhibition of thymidinekinase gene expression by antisense RNA: a moleculàr approach to genetic analysis. *Cell*, **36:**1007–1015.

Jacquemond, M. (1982). Phénomènes d'interférence entre les deux types d'ARN satellites du virus de la Mosaique du Concombre. Protection des Tomatoes vis-à-vis de la nécrose léthale. *CR Acad. Sci., Paris*, **294:**991–994.

Jacquemond, M. and H. Lot (1981). L'ARN satellite du virus de la Mosaique du Concombre. I. Comparaison de l'aptitude à induire la nécrose de la Tomate d'ARN satellites isolés de plusieurs souches de virus. *Agronomie*, **1:**927–932.

Kassanis, B. (1963). Interactions of viruses in plants. In *Advances in Virus Research*, Vol. 10 (Smith and Lauffer, eds.). Academic, New York, pp. 219–255.

Kendall, C., I. Ionescu-Mathiu, and G. R. Dreesman (1983). Utilization of the biotin-avidin system to amplify the sensitivity of the enzyme linked immunosorbent assay (ELISA). *J. Immunol. Methods*, **56:**329–339.

Kunkel, L. O. (1936). Heat treatment for the cure of yellows and other virus diseases of peach. *Phytopathology*, **26:**809.

Lacatena, R. M. and G. Cesareni (1981). Base pairing of RNA I with its complementary sequence in the primer precursor inhibits Col E1 replication. *Nature*, **294:**623-626.

Laterrot, H. (1973). Résistance de la tomate au virus de la mosaique du tabac. Difficultés rencontrées pour la sélection de variétés résistantes. *Ann. Amélior. Plantes*, **23:**287-313.

Laterrot, H. (1977). Résistance de la tomate au virus de la mosaique du tabac. Etat actuel de la sélection. *PHM Revue Hortic.*, **175:**3-7.

Leroux, J. P., J. B. Quiot, H. Lecoq, and M. Pitrat (1979). Mise en évidence et répartition dans le Sud-Est de la France d'un pathotype particulier du virus de la mosaique du Concombre. *Ann. Phytopathol.*, **11:**431-438.

Loebenstein, G. and A. Gera (1981). Inhibitor of virus replication released from tobacco mosaic virus infected protoplasts of a local lesion responding tobacco cultivar. *Virology*, **114:**132-139.

MacKinney, H. H. (1929). Mosaic diseases in the Canary Islands. *J. Agric. Res.*, **39:**557-558.

Macquaire, G., M. Monsion, J. C. Bachelier, C. Faydi, and J. Dunez (1981). Detection of Chrysanthemum stunt and chlorotic mottle viroids by slab gel electrophoresis. *Acta Hortic.*, **110:**321-328.

Macquaire, G., M. Monsion, C. Mouches, T. Candresse, and J. Dunez (1984). Spot hybridization: Application to viroid identification. *Ann. Virol. (Inst. Pasteur)*, **135E:**219-230.

Mancino, L. J. C. and C. N. Agrios (1984). Effects of antiviral compounds on symptoms and infectivity of Cowpea chlorotic mottle virus. *Plant Dis.*, **68:**219-222.

Mandahar, C. L. (1981). In *Plant Diseases and Vectors: Ecology and Epidemiology* (K. Maramorosh and K. F. Harris, eds.). Academic, New York, pp. 241-292.

Marrou, J. and A. Migliori (1974). La prémunition entre souches de Phytovirus. *Ann. Phytopathol.*, **6:**486-491.

Massonie, G. and P. Maison (1982). Résistance du pêcher *Prunus persica* à la transmission du virus de la sharka par *Myzus persicae. Colloq. INRA*, **11:**65-74.

Milne, Q. G. and D. E. Lesemann (1984). Immunosorbent electron microscopy in plant virus studies. In *Methods in Virology* (K. Maramorosh and M. Koproski, eds.). Academic, New York, pp. 85-101.

Mizuno, T., M. Y. Chou, and M. Inouye (1984). A unique mechanism regulating gene expression: Translational inhibition by a complementary RNA transcript (mic RNA). *Proc. Natl. Acad. Sci., USA*, **81:**1966-1970.

Moore, L., G. Warren, and G. Strobel (1979). Involvement of a plasmid in the hairyroot disease of plants caused by *Agrobacterium rhizogenes. Plasmid*, **2:**617-626.

Morel, G. and C. Martin. (1952). Guérison de dahlias atteints d'une maladie à virus. *CR Acad. Sci., Paris*, **235:**1324-1325.

Morris, T. J. and J. A. Dodds (1979). Isolation and analysis of double stranded RNA from virus infected plant and fungal tissue. *Phytopathology*, **69:**854-858.

Morris, T. J., J. A. Dodds, B. Hillman, R. L. Jordan, S. A. Lommel, and S. J. Tamaki (1983). Viral specific dsRNA. Diagnostic value for plant virus diseases identification. *Plant Mol. Biol. Rep.*, **1:**27-30.

Muller, G. W. and A. S. Costa (1977). Tristeza control in Brazil by preimmunization with mild strains. *Proc. Inst. Soc. Citric.*, **3:**315-316.

Navarro, L., C. N. Roistacher, and T. Murashige (1975). Improvement of shoot tip grafting *in vitro* for virus-free citrus. *J. Am. Soc. Hortic. Sci.*, **100:**471-479.

Navarro, L., G. Llacer, M. Cambra, J. M. Arregui, and Z. Juare (1982). Shoot tip grafting *in vitro* for elimination of viruses in peach plants (*Prunus persica*). *Acta Hortic.*, **130:**185-192.

Niblett, C. L., E. Dickson, K. F. Fernow, R. K. Horst, and M. Zaitlin (1978). Cross protection among four viroids. *Virology*, **91:**198-203.

Ohno, T., M. Ishikawa, N. Takamatsu, T. Meshi, Y. Okada, T. Sano, and E. Shikata (1983). *In vitro* synthesis of infectious RNA molecules from cloned Hop stunt viroid complementary DNA. *Proc. Jpn. Acad., Ser. B*, **8:**251–254.

Orchanski, P., M. Rubinstein, and I. Sela (1982). Human interferons protect plants from virus infection. *Proc. Natl. Acad. Sci., USA*, **79:**2278–2280.

O'Reilly, C., N. S. Shepherd, A. Pereira, Z. Schwarz-Sommer, I. Bertram, D. S. Robertson, P. A. Peterson, and H. Saedler (1985). Molecular cloning of the al locus of *Zea mays* using the transposable elements En and Mu1. *EMBO J.*, **4:**877–882.

Owens, R. A. and D. E. Cress (1980). Molecules cloning and characterization of PSTV cDNA sequence. *Proc. Natl. Acad. Sci., USA*, **77:**5302–5306.

Owens, R. A. and T. O. Diener (1981). Sensitive and rapid diagnosis of potato spindle tuber viroid disease by nucleic acid hybridization. *Science*, **213:**670–671.

Pitrat, M. and H. Lecoq (1980). Inheritance of resistance to cucumber mosaic virus transmission by *Aphis gossypii* in Cucumis melo. *Phytopathology*, **79:**958–961.

Prabhakar, B. S., M. V. Haspel, and A. L. Nokkins (1984). Monoclonal antibody techniques applied to viruses. In *Methods in Virology* Vol. VII (K. Maramorosh and H. Koprowski, eds.). Academic, New York, pp. 1–18.

Powell, Abel P., Nelson, R. S., De, B., Hoffman, N., Rogers, S. G., Fraley, R. T., and Beachy, R. N. (1986). Delay of disease development in transgenic plants that express the tobacco mosaic virus coat protein gene. *Science*, **232:**738–743.

Reichman, M., Y. Devash, R. J. Suha Dolnik, and I. Sela (1983). Human leucocyte interferon and the antiviral factor (AVF) from virus infected plants stimulate plant tissues to produce nucleotides with antiviral activity. *Virology*, **128:**240–244.

Renaudin, S., C. Mouches, and J. M. Bove (1982). Inhibitory effects of virazole and Actinomycin D on turnip yellow mosaic virus replication in chinese cabbage protoplasts. *Ann. Virol. (Inst. Pasteur)*, **133E:**295–314.

Risser, G., M. Pitrat, and J. C. Rode (1977). Etude de la résistance du melon au virus de la mosaique du concombre. *Ann. Amélior. Plantes*, **27:**509–522.

Rosner, A., I. Ginzburg, and M. Bar Joseph (1983). Molecular cloning of complementary cDNA sequences of citrus tristeza virus RNA. *J. Gen. Virol.*, **64:**1757–1763.

Schewmaker, C. K., J. R. Caton, C. M. Houk, and K. O. Gardner (1985). Transcription of cauliflower mosaic virus integrated into plant genomes. *Virology*, **140:**281–288.

Schumann, G. L., H. D. Thurston, R. K. Horst, O. Kawamotos, and G. I. Nemoto (1978). Comparison of tomato bioassay and slab gel electrophoresis for detection of potato spindle tuber viroid in potato. *Phytopathology*, **68:**1256–1259.

Sela, I. (1981). Interferon like factor from virus infected plants. *Perspect. Virol.*, **11:**129–139.

Sequeira, L. (1984). Cross protection and induced resistance: Their potential for plant disease control. *Trends Biotechnol.*, **2:**25–29.

Shepard, J. F. (1977). Regeneration of plants from protoplasts of potato X infected tobacco leaves. II. Influence of virazole on the frequency of infection. *Virology*, **78:**261–266.

Sidwell, R. W., J. H. Huffman, G. G. Khare, L. B. Allen, J. T. Witkowski, and R. R. Robins (1972). Broad spectrum antiviral activity of virazole: 1-β-ribofuranosyl 1,2,4-triazole-3-carboxamide. *Science*, **177:**705–706.

Tabler, M. and H. L. Sanger (1984). Cloned single and double stranded copies of potato spindle tuber viroid (PSTV) RNA and coinoculated subgenomic DNA fragments are infectious. *EMBO J.*, **3:**3055–3062.

Tepfer, D. (1984). Transformation of several species of higher plants by *Agrobacterium rhizogenes:* Sexual transmission of the transformed genotype and phenotype. *Cell*, **37:**959–967.

Torrance, L. and R. A. C. Jones (1982). Increased sensitivity of detection of plant viruses obtained by using a fluorogenic substrate in ELISA. *Ann. Appl. Biol.*, **101**:501-509.

Travers, A. (1984). Regulation by antisense RNA. *Nature*, **311**:470.

Van Slogteren, G., J. Hoge, P. Hoykaas, and R. Schilperoort (1983). Clonal analysis of heterogeneous crown gall tumor tissues induced by wild-type and shooter mutant strains of *Agrobacterium tumefasciens.* Expression of T-DNA genes. *Plant Mol. Biol.*, **2**:321-323.

Van Wezenbeek, P., P. Vos, J. Van Boom, and A. Van Kammen (1982). Molecular cloning and characterization of a complete DNA copy of potato spindle tuber viroid RNA. *Nucl. Acid Res.*, **10**:7947-7957.

Vos, P., J. Verver, P. Van Wezenbeek, A. Van Kammen, and R. Goldbach (1984). Study of the genetic organization of a plant viral RNA genome by *in vitro* expression of a full length DNA copy. *EMBO J.*, **3**:3049-3053.

White, F., G. Ghidossi, M. Gordon, and E. Nester (1982). Tumor induction by *Agrobacterium rhizogenes* involves the transfer of plasmid DNA to the plant genome. *Proc. Natl. Acad. Sci., USA*, **79**:3193-3197.

White, F., D. Garfinkel, E. Huffman, M. Gordon, and E. Nester (1983). Sequences homologous to *Agrobacterium tumefasciens* in the genomes of uninfected plants. *Nature*, **301**:348-350.

Zaitlin, M. (1976). Viral cross protection: More understanding is needed. *Phytopathology*, **66**:582-583.

Zambryski, P., H. Joos, C. Genetello, J. Leemans, M. Van Montagu, and J. Schell (1983). Ti plasmid vector for the introduction of DNA into plant cells without alteration of their normal regeneration capacity. *EMBO J.*, **2**:2143-2150.

Zamecnik, P. C. and M. H. Stephenson (1978). Inhibition of rous sarcoma virus replication and cell transformation by a specific oligodeoxynucleotide. *Proc. Natl. Acad. Sci., USA*, **75**:280-284.

Zoeten de, G. A. and R. W. Fulton (1975). Understanding generates possibilities. *Phytopathology*, **65**:221-222.

15

GENETIC TRAITS TO BE MANIPULATED IN PLANTS FOR VIRUS DISEASE RESISTANCE

ILAN SELA
Virus Laboratory, The Hebrew University of Jerusalem, Faculty of Agriculture, Rehovot, Israel

1. INTRODUCTION

The Ti plasmid of *Agrobacterium tumefaciens* has been developed as a vehicle that enabled the introduction of foreign DNA fragments into plant cells (Schell et al., 1985). Furthermore, in a limited number of cases, foreign, nonplant genes, when brought under control of the nopaline synthase promoter in T-DNA-transformed cells, became expressible (Herrara-Estrella et al., 1983a). The introduction of the bacterial gene aminoglycoside phosphotransferase into a T-derived plasmid, and its successful expression, turned the transformed cells resistant to kanamycin and brought into the system a selectable marker (Herrara-Estrella et al., 1983b). In some cases it was possible to regenerate whole plants with foreign DNA sequences (DeBlock et al., 1984; Horsch et al., 1984). At the present stage, it seems that the technical aspects of plant cell transformation are sufficiently advanced, but more study is needed regarding the various factors of expression regulation. However, it is apparent that soon the transformation technology of plant cells will become a common tool in plant genetic engineering.

With this expectation, one wonders what genetic traits one would clone into a plant when this technique becomes routinely possible? The cloning of many desirable traits, such as crop yield, fruit shape, and shock resistance, does not seem presently feasible, since many genes are involved in determining these characteristics. At the present initial stage traits controlled by a single, preferably dominant gene, should be sought. For practical uses, gene activity may be amplified by DNA recombination technique, genes many be transferred from one plant to another (circumventing the *genus* barrier for classica sexual fertilization), or one may envisage the successful introduction of a nonplant gene as a trait in a certain crop. In many cases manipulations in the regulatory elements may suffice, and thus gene transfer may be avoided.

Our laboratory has specialized in the protection mechanisms of plants toward viral infection. The title of this chapter implies a rather theoretical approach, and hitherto not a single resistance gene has been engineered into a plant, except by sexual crossing. This chapter presents a concise, updated description of the relevant topics in the field of resistance, with the minimum references necessary for understanding the speculative aspects that follow. For broader review articles, extension to other related aspects, and back references the reader is referred to Sela (1981a, 1984) and Lobenstein et al. (1984).

2. VIRUS LOCALIZATION

In many cases virus infection is confined to relatively small leaf areas and the virus cannot spread in the tissue. In the case of tobacco mosaic virus (TMV) infection in plants of the genus *Nicotiana*, localization is determined by a single dominant plant gene (the N-gene). In this particular virus–host interaction, as well as in many (but not all) other local infections, necrotic local lesions develop: the rest of the plant is virus-free and is completely or partially protected from subsequent viral infections.

Initial TMV infection in N-gene-carrying *Nicotiana* spp. proceeds at a rate comparable to that of TMV infection in a systemic host. However, about 30–48 h after infection, an abrupt cessation of virus replication occurs and the virus remains confined to those sections of the tissue to which it has spread until that point. Hence the initiation of localization is a relatively late event in a viral infection, and, by inference, some infection products should accumulate sufficiently to trigger localization.

2.1. The N-Gene

The N-gene has been transferred by conventional breeding techniques from *N. glutinosa* to a number of tobacco varieties, creating tobacco cultivars differing from the type variety by (presumably) only this genetic element. The mode of the N-gene-related activities and their connection to TMV localization are not yet known. However, some evidence connecting the N-gene with active antiviral substances has been presented (Antignus et al., 1975, 1977; Sela et al., 1978;

Loebenstein and Gera, 1981; Gera and Loebenstein, 1983). Two sets of phenomena related to the N-gene are of great interest. First, in all natural TMV infections necrotization takes place besides virus localization. In fact the letter "N" stands for "Necrosis." The development of necrotic local lesions is the visual phenotypic expression of TMV localization. Necrotization and virus localization do not fit into a cause and consequence situation (Sela, 1981; Loebenstein et al., 1984), mainly because the virus lesion expands beyond the physical boundary of the necrotic lesion, and also because, even in the case of TMV and the N-gene-carrying host, it is possible to prevent the process of necrotization while virus localization remains unaffected. Necrotization and localization are, therefore, uncoupled phenomena that are linked in their appearance.

The second point of interest is the outstanding stability of control conferred on viral infection by the N-gene. Usually, 2–3 independent resistance genes are bred into a plant for the production of a commercial, virus-resistant cultivar. Even so, viral mutagenesis eventually evolves a virus strain that circumvents these protective mechanisms. In contrast, the single dominant N-gene has remained unaffected in this respect since the evolution of the ancestor of *Nicotiana* and none of the countless TMV strains and artificial mutants has been able to overcome its localizing effect. It seems that the factor in TMV infection that triggers the N-gene is so essential to viral infection that any considerable change leads to a lethal mutation.

How does the N-gene exert its activity? It is apparent that the N-gene controls more than one metabolic pathway. The simplest interpretation would be that the unsegregative genetic entity defined as the N-gene is, in fact, a locus of closely linked genes. Thus a number of unrelated phenomena are expressed concurrently. Another explanation may be that the situation is a classic one gene–one protein case, but the "N-protein" represents a general type of activity (protein kinase, protein glycosylase, protease, or any other modifying activity). In such a case the enzyme directed by a single gene may have multiple effects. Still another possibility is that the N-gene is a genetic regulatory element affecting the expression of a number of unrelated genes.

The situation relevant to the topic of this chapter is even more complicated, since (as described in Section 3.2) plants that carry the recessive allele "n" also express resistance to viral infection in the form of "green islands," and the difference between the effect of the two alleles appears to be merely quantitative.

2.2. Antiviral Substances Associated with the N-Gene

Plant-sap fractions from virus-infected plants exhibit antiviral activity in excess of that of corresponding control saps, as first demonstrated by Sela and Applebaum in 1962 (Sela and Applebaum, 1962). It was assumed that infected plants develop an antiviral factor (AVF), which was found in a number of plant-virus combinations (for more elaborate background information the reader is referred to recent review articles; Sela, 1981, 1984).

As it turned out, the system of choice for pursuing AVF study was localized

TMV infection in *N. glutinosa*. At a certain stage of this study, an AVF fraction could be demonstrated upon polyacrylamide gel electrophoresis. Only a small proportion of the AVF activity (then considered negligible) was detected when the same experiment was repeated in the TMV–tobacco system, which allows systemic infection. However, when the tobacco plants were replaced by an N-gene-carrying cultivar of the same variety, the *N. glutinosa* situation was observed (Antignus et al., 1975, 1977). The appearance of AVF was, therefore, closely associated with the presence of the N-gene in the plant. Later, some evidence pointed to the possibility of the N-gene being (directly or indirectly) involved in the processing of a precursor of AVF to its active form and the involvement of phosphorylation in that process (Sela et al., 1978).

Only minute amounts of AVF are produced in plants and, when purified, they are still active at the picogram level. In fact, it was estimated that AVF exerts its antiviral activity at a level of a few molecules per cell. Even with relatively crude preparations of AVF, the end point of its activity lies at a few nanograms of protein per milliliter. Furthermore, recent experiments in my laboratory have confirmed previous observations that AVF is heterogeneous and that the low amount of active material is divided among a number of subspecies of AVF. Evidence such as loss of activity by degrading enzymes or binding to concanavalin A led to the postulation that AVF is a phosphorylated glycoprotein (Mozes et al., 1978). However, since the evidence was indirect and could result from a residual activity of some of this subspecies, it would be best, at present, to assume that some of the AVF species are phosphorylated and/or glycosylated.

A similar situation was found in tobacco protoplasts. Loebenstein and coworkers (Loebenstein and Gera, 1981, 1981–1982, 1982; Gera and Loebenstein, 1983) reported that N-gene-carrying protoplasts infected with TMV secreted a substance(s) with antiviral activity to the medium. They termed the responsible antiviral agent "Inhibitor of Virus Replication" (IVR), whose activity was reduced if protoplasts were pretreated with actinomycin D or chloramphenicol (Gera et al., 1983). Again, the association of an antivirally active substance with the N-gene was demonstrated. Antisera raised against semipurified IVR cross-reacted weakly with AVF (Loebenstein, personal communication), but the relatedness of the two substances should yet be confirmed.

Other recent papers have indicated the involvement of active substances in N-gene-related resistance. Wieringa-Brants (1983) demonstrated that intercellular fluids from TMV-infected N-gene-carrying tobacco were antivirally active when introduced to another tobacco. Roberts (1983) reported that progeny seedlings of N-gene-carrying tobacco, obtained by selfing individual plants that had been repeatedly inoculated with TMV, are somewhat protected from TMV-infection. This indicated that the activation of the resistance mechanism is seed-transmissible. Since these were selfed seedlings and the antiviral effect did not pass to the second generation, the involvement of a dilutable active substance was implied. Antiviral factors in other localized infections were also reported by Faccioli and co-workers (Faccioli and Capponi, 1983; Capponi et al., 1984).

2.2.1. *AVF-Induced Activities in Plants*

The extremely low levels of AVF, though still sufficient for antiviral activity, indicate that after being initiated, the antiviral activity is amplified. It is probable that AVF turns on a number of metabolic pathways, which lead to the actual antiviral phenomenon.

Two sets of activities were induced by AVF in plants. Both activities were found in AVF-treated plant tissues, regardless of whether they carried the N-gene or virus infection, and they seem to be related.

We know that TMV-RNA can be acylated with histidine at its 3′ terminus. A certain factor (DF), capable of removing the histidine from the charged TMV-RNA, was found to be induced in plant tissues by AVF (Sela et al., 1976; Devash et al, 1981). The DF activity was stimulated by a low-molecular-weight fraction from virus-infected plants. This led to the discovery of the second phenomenon—the stimulation, in AVF-treated plant tissues, of an ATP-polymerizing enzyme (Devash et al., 1981, 1985) whose activity was dependent on the presence of double-stranded RNA (dsRNA). The oligoadenylates produced in this system were antivirally active (Reichman et al., 1983). Though the resemblance to the 2′, 5′-oligoadenylate system of interferon (Williams and Kerr, 1980, and Section 4.2) is apparent, plant oligonucleotides are not identical to animal ones (Devash et al., 1985). The AVF alone was capable of priming plant tissue by stimulating the production or activity of the ATP-polymerizing enzyme—which, however, would not become operative before virus infection took place. Presumably, the viral replicative form, dsRNA, is required for this activation (Reichman et al., 1983).

3. THE SYSTEMIC INFECTION

Tobacco mosaic virus infection in *Nicotiana spp.* that carry the recessive "n" allele is systemic, that is, the virus spreads all over the plant. However, death of tissue as a result of viral infection is a rare event. Normally, some kind of delicate balance is maintained, restricting the virus from multiplying beyond a certain level in many systemic infections. This balance takes the visual form of a green-yellow mosaic pattern, where the virus is confined to the yellow areas and the so-called "green islands" remain virus-free (Atkinson and Matthews, 1970), are resistant to subsequent viral infection (Loebenstein et al., 1977), and contain antivirally active substances (Kluge, 1976) resembling AVF (Ratner, 1981) and IVR (Loebenstein, personal communication). According to Ratner (1981), AVF was also found in the yellow areas, but a quantitative comparison between the yellow and green areas was not possible. The AVF appeared late and at relatively low levels in systemic infections (Antignus et al., 1977) nevertheless, the n-gene-carrying plant reacted to infection by producing AVF, hence the effect of the N(n)-allele seems to be merely quantitative.

4. ANTIVIRAL FACTORS OF NONPLANT ORIGIN AFFECTING VIRAL INFECTION IN PLANTS

4.1. Interferons

Various human interferons were applied to TMV-infected tobacco-leaf disks and the virus titers were monitored by ELISA. Surprisingly, almost complete protection from TMV was observed (Orchanski et al., 1982). Furthermore, human leukocyte interferon was about 100–1000 times more effective in plants than in its animal reference cells.

Later, TMV infection was followed up in tobacco protoplasts (Rosenberg et al., 1985). This is a more precise system where infection is synchronized, the exact number of cells is determined, the timing of interferon application is easily manipulated, and cells can be readily lysed for virus determination. Tobacco mosaic virus replication was followed up, not only serologically, but also by dot-blot hybridization with a probe specific to TMV-RNA. A few interesting observations were made from this study.

1. A number of human interferons, including some recombinant species (isolated from transformed bacteria) were active in reducing TMV infection.
2. Interferon could be applied between 24 hr prior to 1 hr after TMV inoculation, earlier or later applications minimized its effect. This indicated that the cells were primed for protection by the interferon treatment, namely, interferon conferred on the cells an "antiviral state."
3. Interferon affected the level of TMV-RNA in infected cells.
4. The antiviral effect of interferon is transient, fading away within 48–96 hr after application.

Interferon-treated plant tissues reacted similarly to AVF-treated tissues by stimulating ATP-polymerization to oligonucleotide with antiviral activity (Devash et al., 1985; Riechman et al., 1983).

4.2. Oligoadenylates

In interferon-treated animal cells, ATP is polymerized to a series of oligoadenylates in which the AMP monomers are linked by 2′,5′-phosphodiester bonds rather than by regular 3′,5′-bonds. These oligonucleotides are cumulatively referred to as 2′,5′-A (Williams and Kerr, 1980). The 5′-phosphorylated trimers and the longer 2′ -5′A's activate a constitutive nuclease, which degrades polyribosomes and inhibits protein synthesis *in vitro* (Revel, 1984). However, when applied to cells the 5′-dephosphorylated 2′,5′-A "cores" are also active in inhibiting protein synthesis and conferring antiviral activity (Baglioni et al., 1981).

Although the plant oligonucleotides mentioned in Section 2.2.1 are not identical to 2′,5′-A and they do not activate any endonuclease (Devash et al., 1985), the 5′-dephosphorylated 2′,5′-A trimer inhibited TMV replication in tobacco-leaf disks (Devash et al., 1982). Furthermore, a series of synthetic analogues of 2′,5′-oligoadenylates including 2′,5′-ionosine, 2′,5′-A-A-araA, 2′,5′-A-A-tubercidine, and 2′,5′-cordycepin inhibited TMV multiplication in tobacco-leaf disks, protoplasts, and intact plants (Devash et al., 1984). While all these analogues were antivirally active in plants, their corresponding 3′,5′-counterparts, or possible degradation products, were not. The antiviral effect of 2′,5′-A was transient, and by 60 hr after application TMV resumed a normal (but lagged) rate of multiplication. The synthetic analogue 2′,5′-cordycepin, however, was much more stable, conferring its antiviral activity for the duration of the experiment (96 hr), with no indication of any lift.

5. GENETIC TRAITS TO BE CONSIDERED FOR CLONING FOR VIRUS RESISTANCE

5.1. General Considerations

Hitherto, phaseolin and Zein are the only plant genes and some bacterial antibiotic resistant—the only foreign ones successfully cloned and expressed in plants (Schell et al., 1985). The technique of introducing foreign genes into plants is still in its infancy and species, or even tissue-specific factors, are involved. Since the system is very little understood, many of the considerations and suggestions discussed below are speculative.

Based on the background information given in the previous sections, a few genetic targets emerge as candidates for cloning or other genetic manipulations. Some are native genes that should be transferred, amplified, or regulated in plants, and others may be foreign DNA segments that may also lead to more resistant plants.

In my opinion, the value of merely introducing into a plant a gene that is already carried by that plant, is questionable. Although one may argue for an effect obtained by gene amplification, it is almost certain that the relevent gene is under expressional control and random cloning would not help. The limiting factors in all of the following suggestions is the manipulation of the controlling elements, to enable us to turn them on and off at will. The reader is requested to bear this in mind when considering these suggestions.

5.2. Genetic Manipulations with the N-Gene

The N-gene is the most intriguing of the genetic entities discussed in this chapter. This single gene is central, since it sets in motion a number of activities related to resistance and controls the resistance phenomenon at large. However, no protein can, as yet, be identified as the N-gene product.

Transferring the N-gene among species of the genus *Nicotiana* by DNA recombination is of no advantage, because this can be achieved by conventional breeding. However, this gene should be located, studied, and cloned. It may be manipulated to respond to viruses other than TMV, the recessive allele may be regulated to react more strongly, or it may even be introduced to plant genera besides *Nicotiana*.

The mystery in this situation is the basic central question— "What is the N-gene?" The N(n)-locus has been identified, but are there any structural differences between the "N" and "n" alleles? As already mentioned (Section 3) the difference between their effects is probably of a quantitative nature and may well be due to regulation of transcription. If so, the search is for a regulatory element (promotor, enhancer, repressor, etc.) whose transfer to a non-*Nicotiana* plant may be worthless.

One cannot propose any possible genetic manipulations with the N-gene before its activity is understood. On the other hand, without DNA manipulations we will not understand the system and this, in turn, requires the location and isolation of the DNA segment termed "N" (or "n").

Since no N-protein is known, it is impossible to proceed from protein to mRNA to DNA. What then can be done to break the circle? Central to any attempt to approach the N-gene problem is the preparation of genomic and cDNA libraries from dual tobacco cultivars differing "only" in their carrying the "N" versus "n" alleles. The cDNA libraries should also be constructed of TMV-infected and noninfected N-gene-carrying plants, giving rise to libraries where this gene is turned off or on.

The variation among the host DNA-related mRNA's in TMV-infected n- or N-carrying tobacco cultivars should be minimal and result from the difference between these two alleles. Bearing in mind the possible complexity mentioned in Section 2.1, an approach similar to the isolation of clones specific to T-cell membrane proteins could still be attempted (Hedrick et al., 1984). Single stranded cDNA's from TMV-infected (turned on) tissue would be hybridized with mRNA's from a noninfected (turned off) N-gene carrying cultivar. Only a few unmatched cDNA species should occur, representing the N-related sequences. These can now serve as probes for selecting N-gene related sequences from cDNA libraries as well as from genomic libraries. The few selected sequences can then be compared with n-libraries to study the relatedness between the two alleles.

Another approach for locating and isolating the N-gene is by insertion–mutation (Zoller and Smith, 1982). The employment of a random agent of recombination, such as a T-DNA derived vector with a selectable marker, may result in some successfully transformed tobacco cells, which lost their N-gene effect as a result of this insertion–mutation in the "N" locus. Another laborious selection is then required to determine the N-mutated individuals and the pertinent sequences can be isolated by probing with T-DNA.

To minimize the background "noise" in the procedures suggested above, *in-*

situ hybridization is recommended, selecting only those libraries which hybridize to the chromosome known to carry the N-gene.

5.2. Genes for AVF

As mentioned before, AVF(s) appears in plants only at extremely low levels. Recently, we were able to purify a subspecies of AVF to a degree of a single band upon polyacrylamide gel electrophoresis. The protocol now requires the accumulation of sufficient amounts of this particular protein for some sequence analysis and for the raising of specific antibodies. The isolation of the respective mRNA then follows and is (or its cDNA) used as a probe for screening libraries for the respective gene. All this seems rather technical, but a possible basic problem should be taken into consideration. If indeed, as we have assumed (Sela et al., 1978), the induction of AVF activity may be a result of some post-translational processing, its mRNA would not be stimulated and its isolation would become much more difficult.

The AVF bears some similarities to interferon in structure and activity (Sela, 1981b). There is a slim chance that some sequence homology occurs between their genes and that cloned interferon sequences may serve as probes for tobacco AVF genes.

5.3. The ATP Polymerizing Enzyme

The approach discussed above for AVF seems to be also appropriate in this case. However, *a priori*, the isolation of this enzyme seems an easier task than that of AVF, because of the availability of a convenient assay monitoring the progress of purification.

5.4. Cloning of Genes for Interferon and 2′, 5′-Oligoadenylates Synthetase in Plants

Interferons and 2′,5′-oligoadenylates suppress TMV infection in tobacco, even though they are foreign to the plant system. Both respective genes from human cells have been cloned in bacteria. Would the introduction of these genes to tobacco plants render them resistant? At least in one case a human interferon gene has been cloned in tobacco—but so far has not been expressed (J. Schell, personal communication). This exemplifies the type of expected problems dealt with in the introduction to this section.

However, interferons and 2′,5′-oligoadenylates both exert harsh conditions on the affected cells (for instance inhibition of protein synthesis) and their release in their native systems is under control, particularly following viral infection. If their expression in plants is achieved one wonders whether the plant can survive when either of the products of these genes is constantly released.

ACKNOWLEDGMENT

The contributions of the author's laboratory were made possible by grants from the Richard Allan Shankman Memorial Fund, Inc. and from the Israel–Germany Biotechnology Fund.

REFERENCES

Antignus, Y., I. Sela, and A. Hauschner (1975). A phosphorous containing fraction associated with antiviral activity of *Nicotiana* spp. carrying the gene for localization of TMV infection. *Physiol. Plant Pathol.*, **6:**159–168.

Antignus, Y., I. Sela, and I. Harpaz (1977). Further study on the biology of an antiviral factor (AVF) from virus-infected plants and its association with the N-gene of *Nicotiana* species. *J. Gen. Virol.*, **35:**107–116.

Atkinson, P. H. and R. E. F. Matthews (1970). On the origin of dark green tissue in tobacco leaves infected with tobacco mosaic virus. *Virology*, **40:**344–356.

Baglioni, C., S. B. D'Allesandro, T. W. Nilsen, J. A. J. Jen Hartog, R. Crea, and J. H. Van Boin (1981). Analogs of (2′-5′) oligo (A). Endonuclease activation and inhibition of protein synthesis in intact cells. *J. Biol. Chem.*, **256:**3253–3257.

Capponi, R., G. Faccioli, and S. Montanari (1984). Produzione di Fattori antivirali (AVFs) in piante virus infette. *Atti Giorante Fitopatol.*, **3:**361–372.

Schell, J., L. Herrera-Estrella, P. Zambryski, M. De Block, H. Joos, L. Eillmizer, P. Eckes, S. Trosahl, and M. Van Montagu (1985). Genetic engineering of plants. In *The Impact of Gene Transfer Techniques in Eukaryotic Cell Biology*. Springer-Verlag, Berlin, pp. 73–90.

De Block, M., L. Herrera-Estrella, M. Van Montagu, J. Schell, and P. Zambryski (1984). Expression of foreign genes in regenerated plants. *EMBO J.*, **3:**1681–1684.

Devash, Y., A. Hauschner, I. Sela, and K. Chakraburtty (1981). The antiviral factor (AVF) from virus-infected plants induces discharge of histidinyl-TMV-RNA. *Virology*, **111:**103–112.

Devash, Y., S. Biggs, and I. Sela (1982). Multiplication of tobacco mosaic virus in tobacco leaf-disks is inhibited by (2′-5′) oligoadenylate. *Science*, **216:**1415–1416.

Devash, Y., A. Gera, M. Reichman, W. Pfleiderer, R. Charubala, I. Sela, and R. J. Suhadolnik (1984). 5′-Dephosphorylated 2′-5′-adenylate and its analogs: inhibition of tobacco mosaic virus (TMV) replication in TMV-infected leaf discs, protoplasts and intact tobacco plants. *J. Biol. Chem.*, **259:**3482–3486.

Devash, Y., M. Reichman, I. Sela, and R. J. Suhadolnik (1985). Plant oligoadenylates: Enzymatic synthesis, isolation and biological activity. *Biochemistry*, **24:**593–599.

Faccioli, G. and R. Capponi (1983). An antiviral factor present in plants of *Chenopodium amaranticolor* locally infected by tobacco necrosis virus: Extraction, partial purification, biological and chemical properties. *Phytopathol. Z.*, **106:**289–301.

Gera, A. and G. Loebenstein (1983). Further studies of an inhibitor of virus replication from tobacco mosaic virus-infected protoplasts of a local lesion-responding tobacco cultivar. *Phytopathology*, **73:**111–115.

Gera, A., G. Loebenstein, and S. Shabtai (1983). Enhanced tobacco mosaic virus production and suppressed synthesis of a virus inhibitor in protoplasts exposed to antibiotics. *Virology*, **127:**475–478.

Hedrick, S. M., D. I. Cohen, E. A. Nielsen, and M. M. Davis (1984). Isolation of cDNA clones encoding T-cell-specific membrane-associated proteins. *Nature*, **308:**149–151.

Herrera-Estrella, L., A. Depicker, M. Van Montagu, and J. Schell (1983a). Expression of chimeric genes transferred into plant cells using a Ti-plasmid-derived vector. *Nature*, **303:**209–213.

Herrera-Estrella, L., M. De Block, E. Messens, J. P. Hernalsteens, M. Van Montagu, and J. Schell (1983b). Chimeric genes as dominant selectable markers in plants cells. *EMBO J.* **2:**987–995.

Horsch, R., R. Fraley, S. Rogers, P. Sanders, A. Lloyd, and N. Hoffmann (1984). Inheritance of functional foreign genes in plants. *Science*, **223:**496–498.

Kluge, S. (1976). Protein content in yellow green and dark green tissue of mosaic-diseased tobacco plants. *Biochem. Physiol. Pflanzen*, **170:**91–95.

Loebenstein, G. and A. Gera (1981). Inhibitor of virus replication released from tobacco mosaic virus-infected protoplasts of a local lesion-responding tobacco cultivar. *Virology*, **114:**132–139.

Loebenstein, G. and A. Gera (1981–1982). Inhibitor(s) of virus replication (IVR) and localization of viruses in plants. *Texas Rep. Biol. Med.*, **41:**213–218.

Loebenstein, G. and A. Gera (1982). Resistance mechanisms to viruses in plants associated with inhibitor(s) of virus replication (1982). *Colloq. l'INRA*, **11:**19–28.

Loebenstein, G., J. Cohen, S. Shabtai, R. H. A. Coutts, and K. R. Wood (1977). Distribution of cucumber mosaic virus in systemically infected tobacco leaves. *Virology*, **81:**117–125.

Loebenstein, G., A. Gera, and A. Stein (1984). Plant defense mechanisms against viral infections. In *Control of Virus Disease*. (E. Kurstak, ed.) Marcel Dokker, New York, pp. 375–391.

Mozes, R., Y. Antignus, I. Sela, and I. Harpaz (1978). The chemical nature of an antiviral factor (AVF) from virus-infected plants. *J. Gen. Virol.*, **38:**241–249.

Orchanski, P., M. Rubinstein, and I. Sela (1982). Human interferons protect plants from virus infection. *Proc. Natl. Acad. Sci. USA*, **79:**2279–2280.

Ratner, I. (1981). A study on the induction of antiviral activities specific to virus-infected plants. M.Sc. Thesis, The Hebrew University of Jerusalem.

Revel, M. (1984). The interferon system in man: nature of the interferon molecules and mode of action. In *Antiviral Drugs and Interferon, The Molecular Basis of Their Activity*. (Y. Becker, ed.). Martinus Nijhoff, Boston, pp. 357–4333.

Reichman, M., Y. Devash, R. J. Suhadolnik, and I. Sela (1983). Human leukocyte interferon and the antiviral factor (AVF) from virus-infected plants stimulate plant tissues to produce nucleotides with antiviral activity. *Virology*, **128:**240–244.

Roberts, D. A. (1983). Acquired resistance to tobacco mosaic virus transmitted to the progeny of hypersensitive tobacco. *Virology*, **124:**161–163.

Rosenberg, N., M. Reichman, A. Gera, and I. Sela (1985). Antiviral activity of natural and recombinant human leukocyte interferons in tobacco protoplasts. *Virology*, **140:**173–178.

Sela, I. (1981a). Plant virus interactions related to resistance and localization of viral infections. *Adv. Virus Res.*, **26:**201–237.

Sela, I. (1981b). Interferon-like factor from virus-infected plants. *Perspect. Virol.* **11:**129–139.

Sela, I. (1984). Interferon-like peptides in plants. In *Antiviral Drugs and Interferon: The Molecular Basis of Their Activity* (Y. Bocker, ed.). Martinus Nijhoff, Boston, pp. 335–356.

Sela, I. and S. W. Applebaum (1962). Occurrence of antiviral factor in virus-infected plants. *Virology*, **17:**543–548.

Sela, I., S. E. Grossberg, A. H. Mehler, and J. J. Sedmak (1976). Discharge of aminoacyl viral RNA by a factor from interferon-treated cells. *Science*, 194:527–529.

Sela, I., A. Hauschner, and R. Mozes (1978). The mechanism of stimulation of the antiviral factor (AVF) in *Nicotiana* leaves. *Virology*, **89:**1–6.

Wieringa-Brants, D. H. (1983). A model to stimulate acquired resistance induced by localized virus infections in hypersensitive tobacco. *Phytopath. Z.*, **106:**369–372.

Williams, R. B. G. and I. M. Kerr (1980). The 2-5A (pppA2′p5′A2′p5′A) system in interferon-treated and control cells. *TIBS*, **4:**138–140.

Zoller, M. J. and M. Smith (1982). Oligonucleotide-directed mutagenesis using M-13-derived vector: an efficient and general procedure for the production of point mutants in any fragment of DNA. *Nuc. Acid Res.*, **10:**6475–6484.

16

AN IMPROVED SYSTEM FOR BIOLOGICAL CONTROL OF DAMPING-OFF BY USING PLASMIDS IN FUNGI

TERUYOSHI HASHIBA
Faculty of Agriculture, Tohoku University, Sendai, Japan

One exciting application of the genetic engineering technique is biological control of plant diseases. Although plasmids are easily found in prokaryotes, only a few example of plasmids, other than viruses and organelle DNAs, are known in eukaryotes. There is even less experimental data for the presence of plasmids in phytopathogenic fungi. Nevertheless, genetic and molecular evidence accumulated to date encourage further study for plasmids in phytopathogenic fungi.

We searched for plasmids among three weakly pathogenic isolates of *Rhizoctonia solani* and found a linear DNA plasmid, pRS64 (molecular weight $1.68 \pm 0.06 \times 10^6$) (Hashiba et al., 1983, 1984; Hashiba and Yamada, 1984). Weakly pathogenic isolates containing plasmids showed abnormally slow growth and low virulence to Japanese radish seedlings, whereas pathogenic isolates that lacked the pRS64 plasmid showed normal growth and high virulence to radish seedlings. Thus it seems highly probable that the weakly pathogenic character of *R. solani* isolate may be under the control of the pRS64 plasmids, and this system seems to offer an example for biological control of damping-off to Japanese radish seedlings. In this chapter I describe these results and outline some of the difficulties that must be surmounted before biological control of plant diseases is feasible.

1. DNA PLASMIDS IN PHYTOPATHOGENIC FUNGI

1.1. DNA Plasmids in the Fungus *Rhizoctonia solani*

1.1.1. Factors in the loss of Pathogenicity in Rhizoctonia solani

Several reports have been published on degenerative diseases of *R. solani*. Castanho et al. (1978) reported that a degenerative decline occurred in a severely diseased isolate of *R. solani* that had been associated with an assortment of double-stranded RNA (dsRNA) segments of three different sizes (2.2, 1.5, and 1.1 megadalton), but so far no virus particles have been observed. The decline was transmissible from one strain to another strain.

The relationship between the presence of dsRNA in *R. solani* and virulence was examined (Finkler et al. 1983). There seems to be no direct correlation between hypovirulence and the presence of dsRNA in *R. solani*, as implicit from the report by Castanho and Butler (1978).

More recently, Zanzinger et al. (1984) investigated the role of dsRNA on virulence of *R. solani*. Sixty-three isolates containing one to several segments of dsRNA exhibited a wide range of virulence from hypovirulence to high virulence. They suggested that the dsRNA is commonly found in a given population of field isolates of *R. solani*, and that the presence of dsRNA is not associated with hypovirulence.

1.1.2. Decline of Damping-Off of Japanese Radish Seedling

Suppressiveness to damping-off of Japanese radish was induced by the addition of living mycelium of *R. solani* anastomosis group 4 every 2 wk into soil (Homma et al., 1981, 1986). Disease suppressiveness could be induced by adding living mycelia to soil, but not by dead mycelia. When the incidence of damping-off was estimated in the suppressive soil pasteurized at 40–100°C by aerated steam, the suppressiveness was slightly less following treatments of the soil at 40 and 50°C, and completely absent after soil treatments at 60°C or above.

The weakly pathogenic isolates 1668 RI-1, 1271 RI-64, and 1271 RI-1 were isolated from soil initially infested with the strongly pathogenic isolates 1668, 1271, and 1272 of *R. solani*, and were planted weekly to successive crops of radishes (Homma et al., 1981) (Fig. 1).

1.1.3. Properties of the Weakly Pathogenic Isolates of R. solani

Properties of weakly pathogenic isolates differ from the strongly pathogenic isolates in five aspects (Table 1).

1. Weakly pathogenic isolates of *R. solani* showed abnormally slow growth, hyphal distortion, and lysis of hyphal cells.

2. The weakly pathogenic isolates produced a large amount of oxalic acid in culture medium containing SO_4^{2-}, but the strongly pathogenic isolates produced no oxalic acid in culture medium (Hyakumachi et al., 1984).

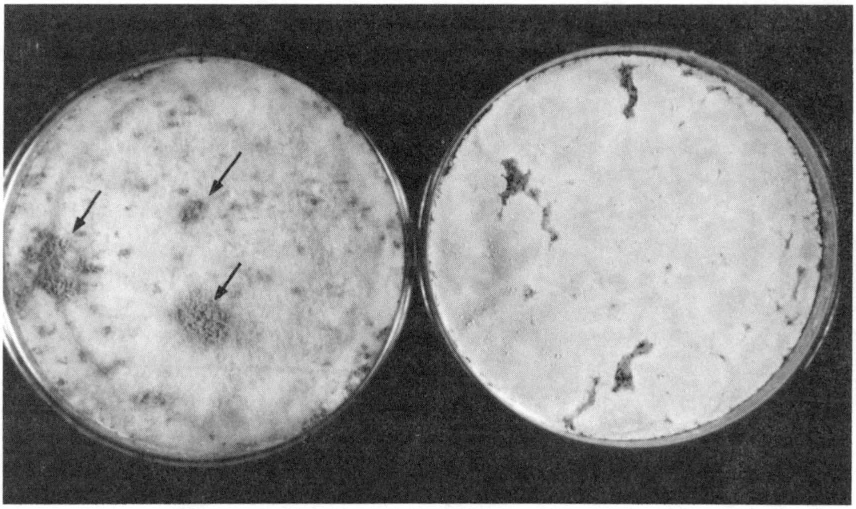

Fig. 1. Plate cultures of a strongly pathogenic isolate (left) and weakly pathogenic isolate 1271 RI-64 (right) of *Rhizoctonia solani* anastomosis group 4 (AG-4). Arrows indicate sclerotia.

3. The strongly pathogenic isolates showed high virulence to radish seedlings, whereas the weakly pathogenic isolates showed low virulence to radish seedlings, though this varies from high to low (Fig. 2).

4. The weakly pathogenic isolates formed no sclerotia, but the strongly pathogenic isolates were able to form sclerotia (Fig. 1). Our previous study indicated that the addition of cyclic AMP in the culture medium induced sclerotial formation in the isolates of *R. solani* anastomosis group 1 (AG-1), which are normally unable to form sclerotia (Hashiba and Ishikawa 1978; Hashiba, 1982). The levels of cyclic AMP in the strongly pathogenic isolate 1271 were considerably higher in comparison with the weakly pathogenic isolates that are unable to form sclerotia. The weakly pathogenic isolates that contained lesser amounts of cyclic AMP had either low adenylate cyclase or high phosphodiesterase activity and failed to form sclerotia, but had approximately similar levels of cyclic AMP-binding protein and cyclic AMP-dependent protein kinase activity (Hashiba et al., 1984; Uno et al., 1985a,b) (Table 1). Thus it is concluded that cyclic AMP may play an important role in the induction of sclerotia.

5. The weakly pathogenic isolates contained the plasmids, but the strongly pathogenic isolates contained no detectable plasmids.

1.1.4. Protoplast Formation

Protoplasts were prepared according to the method of Hashiba (Hashiba, 1982a,b; Hashiba and Yamada; 1982) (Fig. 3). Mycelium (1 g fresh weight) was suspended in 10 ml of the enzyme mixture (cellulase "Onozuka" RS, 20 mg/ml; macerozyme R-10, 5 mg/ml; and β-glucuronidase, 0.06 ml/ml) in a 100-ml

Fig. 2. Disease incidence of Japanese radish seedlings by the strongly pathogenic (left) and weakly pathogenic isolate (right).

TABLE 1. Properties of the Weakly Pathogenic and Strongly Pathogenic Isolate of Rhizoctonia solani AG-4

Isolate	*Sclerotium Formation*	*Plasmid*	*Oxalic Acid Production*	*Adenylate Cyclase (units/mg protein)*	*Phosphodiesterase (units/mg protein)*	*cAMP Content (pmol/mg protein)*	*8-N_3-cAMP Bound (pmol/mg protein)*
1271	+	−	−	15.9	33.0	17.0	0.80
RI-64	±	+	+	0.6	3.4	1.2	0.32

Erlenmeyer flask. The flask was placed on a reciprocal shaker at 75 strokes/min at 34°C for 3 hr. The culture was filtered through a 150-μm steel sieve to remove mycelial fragments, and the filtrate was centrifuged at least twice at 700xg for 5 min to remove the enzymes. Intact protoplasts were further separated from mycelial fragments and cell debris by an aqueous two-phase system which consisted of 0.6 *M* sucrose and 0.6 *M* mannitol. Two milliliters of the protoplast suspension were layered onto 4 ml of 0.6 *M* sucrose, 1 m*M* $CaCl_2$, and 5 m*M* HEPES-KOH buffer (pH 7.0) and centrifuged at 200xg for 5 min. The intact protoplasts were located at the interphase of the two-phase system (Fig. 4).

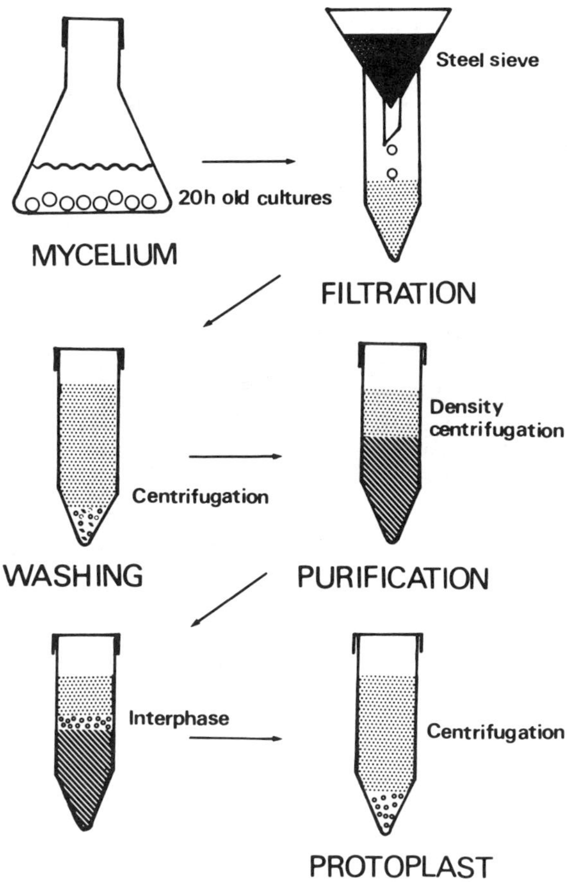

Fig. 3. Procedure for isolation of protoplasts from *R. solani* mycelium.

1.1.5. Preparation of DNA from Protoplasts

The protoplasts were washed and suspended in 4 ml of 0.1 *M* citrate-phosphate buffer (pH 6.0) containing 0.6 *M* mannitol and 0.01 *M* EDTA and then lysed by adding 0.5 ml of 10% sodium dodecyl sulfate solution. After the addition of 5 *M* NaCl at a final concentration of 1 *M*, the mixture was kept on ice overnight and subjected to centrifugation at 25,000xg for 30 min. To the supernatant were added 2 volumes of cold (−20°C) ethanol and 0.1 volume of 3 *M* sodium acetate. The mixture was incubated at −20°C for 2 hr. The precipitate was collected by centrifugation at 25,000xg for 20 min and dissolved in 100 μl of TES buffer (30 m*M* Tris-HCl, pH 8.0; 50 m*M* NaCl; 5 m*M* EDTA). After the addition of RNase A at a final concentration of 50 μl/ml, the DNA samples were subjected to agarose gel electrophoresis. Plasmid DNA bands were visualized

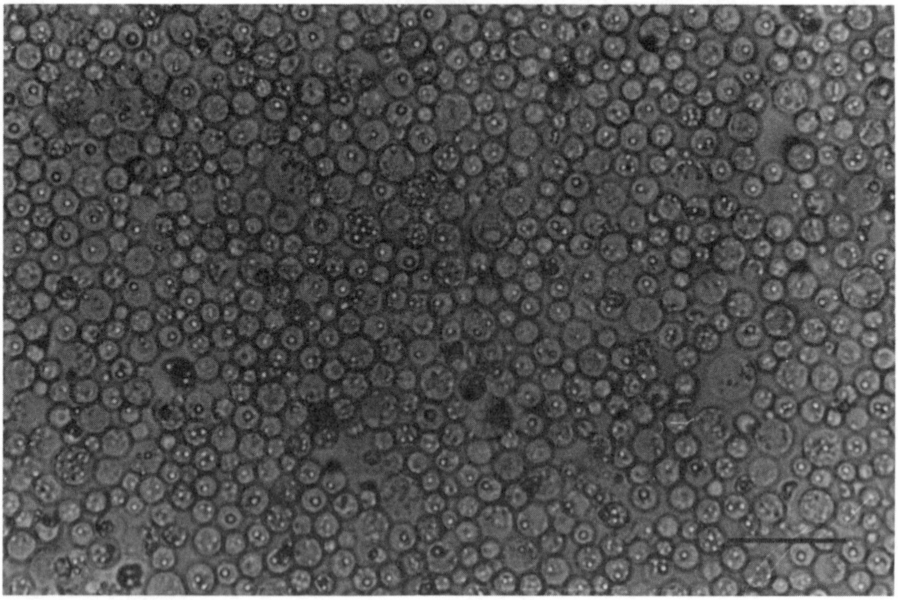

Fig. 4. Light micrograph of intact protoplasts of *R. solani* after purification with the two-phase system (×780). Bar represents 20 μm.

under UV light. Plasmids were also detected by a modification of the procedure of Birnboim and Doly (1979) or Kado and Liu (1981).

1.1.6. Detection of Plasmid DNA

Plasmid DNA was detected by a modification of the Hirt (1967). Sodium dodecyl sulfate lysates of protoplasts obtained from various isolates of *R. solani* were examined for the existence of plasmids by agarose gel electrophoresis. In investigating three weakly pathogenic isolates of *R. solani* (1668 RI-1, 1271 RI-64, and 1272 RI-1), a plasmid-like DNA band was found in addition to the top band, corresponding to nuclear and mitochondrial DNAs (Fig. 5*a*, lanes 2–4). No plasmid-like DNA was detected among three strongly pathogenic isolates of *R. solani* (1668, 1271, and 1272; Fig. 5*a*, lanes 5–7). All bands disappeared when treated with DNAase (Fig. 5*b*, lane 4), but were resistant to the RNAase treatment (Fig. 5*b*, lane 3). From these results, the DNA band found in addition to nuclear and mitochondrial DNA bands is interpreted as plasmid DNA and designated pRS64.

1.1.7. Structure and Size of DNA Plasmid

In electron microscopy, the pRS64 plasmid fractions in isolates 1668 RI-1, 1271 RI-64, and 1272 RI-1 purified from gels were found to consist of a homogeneous

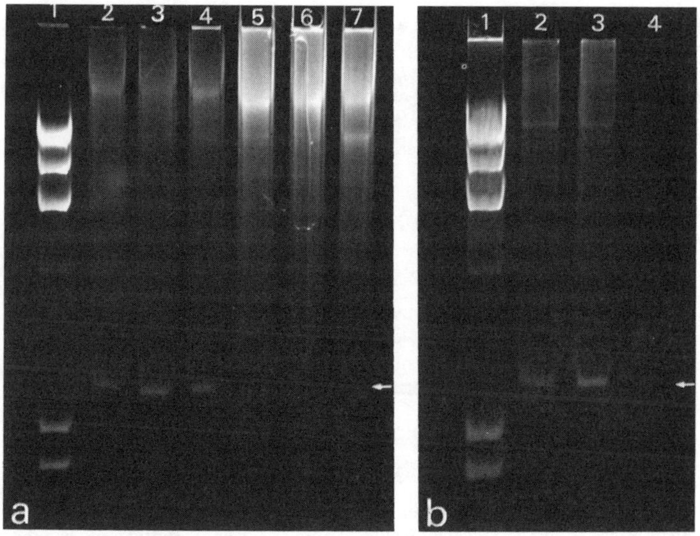

Fig. 5. (*a*) Electrophoresis of DNA extract from six isolates of *R. solani* in 1% agarose gels. Lane 1, *Hind* III fragments of phage λ DNA; lane 2, DNA from isolate 1668 RI-1; lane 3, DNA from isolate 1271 RI-64; lane 4, DNA from isolate 1272 RI-1; lane 5, DNA from isolate 1668; lane 6, DNA from isolate 1271; lane 7, DNA from isolate 1272. The arrow indicates plasmid DNA. (*b*) Agarose gel electrophoresis of DNA isolated from isolate 1271 RI-64 of *R. solani*. Lane 1, *Hind* III fragments of phage λ DNA; lane 2, DNA without treatment; lane 3, DNA digested with 10 μg RNAase (bovine pancreatic ribonuclease A)/ml for 1 hr at 37°C; lane 4, DNA digested with 10 μg DNAase (RNAase-free deoxyribonuclease)/ml for 1 hr at 37°C. The arrow indicates plasmid DNA.

population of linear DNA (Fig. 6). No circular molecules as observed for the marker pBR322 were detected. The size of the linear molecules of plasmids was determined by measuring their length relative to the length of open-circular pBR322, which was used as an internal standard in the same spreading. The ratio of the plasmid DNA length in the three weakly pathogenic isolates to pBR322 DNA length was 0.6 ± 0.02 (mean of 106 molecules ± SD). The length of the DNA was $2,617 \pm 87$ bp, assuming a length of 4,361 bp for pBR322. The DNA was shown to have a molecular weight of $1.68 \pm 0.06 \times 10^6$ by this method.

The molecular weight of plasmid DNA was also estimated on the basis of mobility by electrophoresis in comparison with linear DNA fragments obtained from *Hind* III digestion of DNA. When the logarithm of the molecular weight was plotted against distance migrated, the plasmid DNA in the three weakly pathogenic isolates was shown to have a molecular weight of $1.70 \pm 0.02 \times 10^6$ (mean of 30 determinations). The value was in close agreement with the size estimated by electron microscopy and evidence of the linear structures of the pRS64 plasmids. Data on the nucleotide sequences of the linear DNA plasmid pRS64 from *R. solani* will be reported in a subsequent paper.

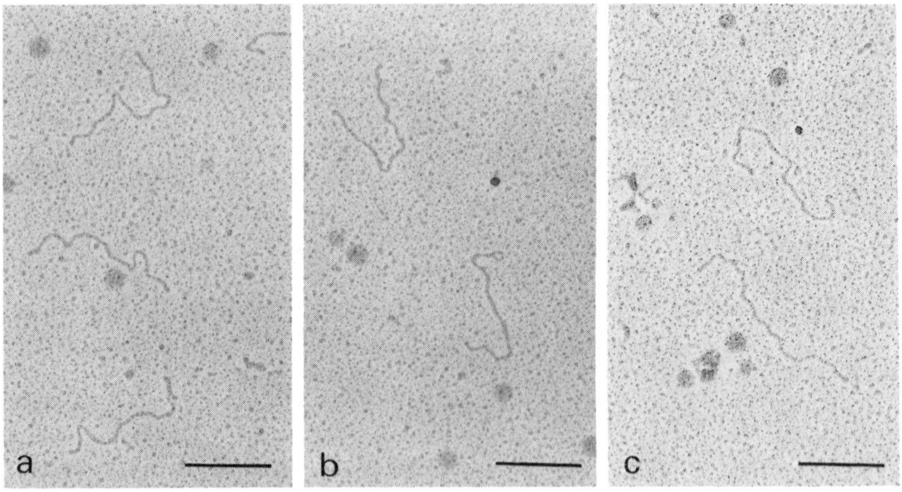

Fig. 6. Electron micrographs of DNA molecules from three weakly pathogenic isolates of *R. solani* by the formamide and cytochrome c technique. (*a*) DNA molecule from isolate 1668 RI-1; (*b*) DNA molecule from isolate 1271 RI-64; (*c*) DNA molecule from isolate 1272 RI-1. The bar represents 1 kb.

1.2. Plasmids of *Ceratocystis ulmi*

The effects of viruses on the pathogenicity of *C. ulmi*, which causes Dutch elm disease, have been studied by many investigators. The less aggressive strains may be virus-infected and the more aggressive strains are probably free of viruses (Atanasoff, 1973). The D-factor in the aggressive strain of *C. ulmi* was transmitted through mycelia by hyphal fusion, and severely reduced the growth and reproductive fitness of the fungus (Braiser, 1983). More recently, Pusey and Wilson (1981) suggested that dsRNA can influence pathogenicity in *C. ulmi.* It appears, however, that the various dsRNA segments differ in their contribution to diminished pathogenicity and that other unknown factors influence pathogenicity as well.

Five isolates of *C. ulmi*, aggressive or nonaggressive in pathogenicity, were examined for the presence of DNA plasmids (Takai et al., 1984). At least six different plasmids were observed after gel electrophoresis. Approximate sizes of these plasmids are 22, 18, 3, 2, and 0.54 kb. Plasmids were not identified in one isolate of the nonaggressive strains. A common plasmid species of about 22 kb was found in four other isolates, irrespective of their aggressiveness and/or cerato-ulmin (CU) productivity. Two isolates of the aggressive strain and one nonsporulating isolate, all of which were distinct in CU production, showed the presence of plasmids that were smaller than 22 kb. Neither of the two nonaggressive isolates that were negligible in CU production had any plasmids.

1.3. Plasmids of *Gaeumannomyces graminis*

Gaeumannomyces graminis is the cause of take-all disease of cereals. Several hypotheses have been advanced to explain take-all decline, including the expression of a latent mycovirus (Férault et al., 1979) "vegetative death" associated with a cytoplasmic element (plasmid) (Jinks, 1959), and a shift in ratios of genetically different nuclei if the fungus is heterokaryotic (Asher, 1981).

The discovery of virus-like particles in weakly pathogenic isolates of *G. graminis* led to the suggestion that take-all decline was the result of virus infection (Lapierre et al., 1970). Further tests with many field isolates of *G. graminis*, however, indicated a more complex situation. At least two viruses (27 and 35 nm in diameter) were present, but no direct evidence was obtained that virus infection impaired pathogenicity (Rawlinson et al., 1973).

Naiki and Cook (1983) suggested that the fungus either is commonly heterokaryotic, with a low frequency of nuclei lacking genes for virulence and/or carrying genes that can suppress expression of virulence, or has a cytoplasmic determinant (plasmid) that affects the expression of virulence.

Recently, two linear DNA elements were detected in *G. graminis* var. *tritici* (Honeyman and Currier, 1983), but more work is required to show the involvement of these elements in take-all decline.

2. DNA PLASMID IN NONPHYTOPATHOGENIC FUNGI

2.1. Plasmids of *Kluyveromyces lactis*

Two linear DNA plasmids, designated pGKL1 and pGKL2, were isolated from the yeast *K. lactis* (Gunge et al., 1981) with molecular weights of 5.4×10^6 and 8.4×10^6, respectively. All *K. lactis* strains lacking the pGKL1 were nonkillers. The terminal nucleotide sequences of pGKL2 and the complete nucleotide sequence of pGKL1 were determined (Hishinuma et al., 1984). The inverted terminal repetitions of 202 bp and 182 bp were found in pGKL1 and pGKL2, respectively.

2.2. Plasmids of *Morchella conica*

Altogether 18 different strains of the morel *M. conica* contained DNA plasmids of comparable size (6 and 8 kb, respectively) (Meinhardt and Esser, 1984). By restriction analyses and electron microscopy it was confirmed that the plasmid was linear, having a molecular weight of 6 kb. The plasmid carried inverted repeats of 0.75 kb at both ends.

2.3. Plasmids of *Ascobolus immersus*

Some DNA plasmids were isolated from the fungus *A. immersus* (Francou, 1981). The molecule had a density of 1.708 g/cc and a molecular weight of 4.1 $\times$

10^6 daltons. Digestion simultaneously by two restriction enzymes demonstrated the linearity of the molecule. Electron microscopy of alkaline-denatured and self-renatured molecules revealed that the ends are inverted repeats of 1,200 base pairs in length.

3. BIOLOGICAL CONTROL OF PATHOGENICITY BY USING PLASMIDS

3.1. Biological Control of *Rhizoctonia Solani*

The use of weakly pathogenic isolates of plant pathogenic fungi in disease control is attractive, but few examples have been reported.

The incidence of postemergence damping-off of Japanese radish, caused by *R. solani*, is significantly lower when weakly pathogenic isolates and strongly pathogenic isolates occur together in soil (Hyakumachi et al., 1984). The incidence of damping-off of Japanese radish seedlings was extensive in soil infested with the strongly pathogenic isolates, whereas little or no damping-off occurred in soil infested with weakly pathogenic isolates alone or in noninfested soil. Using mixtures of strongly pathogenic and weakly pathogenic isolates as inocula, the incidence of damping-off of Japanese radish seedlings decreased when the inoculum of strongly pathogenic isolate was small (Table 2, Fig. 7). Thus the incidence of damping-off was controlled by coinfestation with weakly pathogenic isolates.

Castanho and Butler (1978) successfully used nonpathogenic strain 189a of *R. solani* AG-1 to prevent damping-off of sugar beet seedlings by virulent strain 189HT5. They noted that broadcasting the hypovirulent strain had no effect on damping-off, but placing it in the seed furrows was effective.

TABLE 2. Degree of Disease Incidence of Japanese Radish Seedlings when Weakly Pathogenic and Strongly Pathogenic Isolates Occur Together in the Soil

Inoculum Density of Weakly Pathogenic Isolate (%) (v/v)	*Inoculum Density of Strongly Pathogenic Isolate (%) (v/v)*				
	0.0	*0.1*	*0.3*	*0.5*	*1.0*
0.0	0.0a	86.5d	100.0b	100.0b	100.0b
0.1	13.0b	76.0c	90.0ab	97.0b	96.5ab
0.3	19.0b	55.5ab	85.0a	84.0a	88.0a
0.5	18.0b	49.5a	88.0a	89.5ab	94.5ab
1.0	34.0c	67.0bc	90.5ab	89.0ab	92.0ab

[a]Disease rating: 0 = no disease; 100 = most severe disease.

[b]Values followed by a letter do not differ significantly ($P = 0.05$) *according to Duncan's multiple range test.*

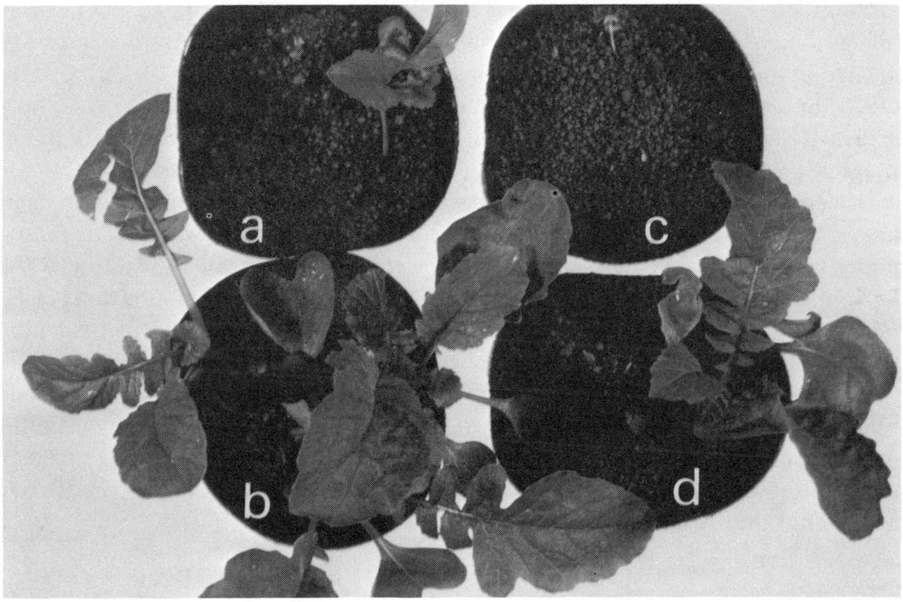

Fig. 7. Degree of disease incidence of Japanese radish seedlings when weakly pathogenic and strongly pathogenic isolates infested together in soil. (*a*) Inoculum density of strongly pathogenic isolate, 0.1% (v/v), and weakly pathogenic isolate 0.0% (v/v). (*b*) Inoculum density of strongly pathogenic isolate, 0.1% (v/v), and weakly pathogenic isolate, 0.5% (v/v). (*c*) Inoculum density of strongly pathogenic isolate, 0.3% (v/v), and weakly pathogenic isolate, 0.0% (v/v). (*d*) Inoculum density of strongly pathogenic isolate, 0.3% (v/v), and weakly pathogenic isolate, 0.5% (v/v).

This method might offer an ideal system for biological control of damping-off of radish and sugar beet seedlings, but there are practical problems. The hypovirulent strain examined survived for less than 1 month in soil; moreover, the disease agent did not transfer to other AG groups of *R. solani* via hyphal anastomosis. However, procedures to obtain protoplasts from AG-1 isolates, as well as to make intraspecific protoplast fusions, have been developed (Hashiba and Yamada, 1982, 1984). This may facilitate the transfer of nuclei and cytoplasm between isolates, and be useful in genetical and biological control studies.

3.2. Biological Control of Chestnut Blight

Since it is highly unlikely that the European chestnut had spontaneously become more resistant to *E. parasitica*, an explanation for this phenomenon had to be sought in the fungus. Grente (1965) made a thorough study of strains recovered from trees showing this healing reaction in Italy and France. Compared to isolates from expanding cankers, isolates from healing cankers tended to be less virulent, produced cankers that were more superficial, sporulated poorly if at all, and were commonly nonpigmented (white) in culture. Normal strains are orange in culture. The unusual strains were termed "hypovirulent" by Grente (1965) and are characteristically unstable in culture.

The H strains were introduced into cankers by removing from four to six plugs of bark, each 10 mm in diameter, around the circumference of the canker with a cork borer, and filling the holes with mycelium of one or more H strains in agar (Anagnostakis, 1982). Then the holes are covered with waterproof brown paper tape to prevent desiccation. As reported by Jaynes and Elliston (1980), 86% of the cankers were controlled in the first year. Hypovirulent strains of *E. Parasitica* (a causal agent of chestnut blight) can prevent more virulent strains from attacking the host when they coexist in the same canker.

This might seem to offer an ideal system for biological control of chestnut blight disease, but there are practical problems. There would seem to be no mechanism for long-term protection against future infection with a virulent race; moreover, the ascospores from hypovirulent isolates do not contain dsRNA and do not give rise to aggressively pathogenic mycelia.

4. CONCLUSIONS

The linear mitochondrial DNAs, named S1 and S2, present in some cytoplasmic male-sterile strains of maize, were detected by Pring et al. (1977). Subsequently, it became evident that these genetic elements are also present in other plants (*S. bicolor*, Pring et al., 1982; *B. campestris* and *B. napus*, Palmer et al., 1983) and in fungi (*A. immersus*, Francou, 1981; *K. lactis*, Gunge et al., 1981; *R. solani*, Hashiba et al., 1983; Hashiba and Yamada, 1984; *G. graminis*, Honeyman and Currier, 1983; *C. ulmi*, Takai et al., 1984; and *M. conica*, Meinhardt and Esser, 1984).

Linear plasmids may be very common in eukaryotes. Naturally this raises a question about the function of these genetic elements. The linear plasmids of *K. lactis* are the cause of the killer phenomenon in this yeast (Gunge et al., 1981). Certainly there are not sufficient examples to show a general function of the linear plasmids. However, many of them seem to share one common feature, that is, the existence of inverted repeats at both ends. This structure is of particular interest because it is also present in prokaryotic and eukaryotic transposons (Calos and Miller 1980; Roeder et al., 1980; Potter et al., 1980). One may speculate that the linear plasmids act as transposons.

A transposable genetic element can move from one place to the other within the genome of an organism. It is excised from one site and inserted at another site either on the same chromosome or on a different one. The movement of a transposable element can generate mutations or chromosomal rearrangements and thus affect the expression of other genes. It may be postulated that the ability of transposons to modify the expression of a virulence gene may cause a fungus to change virulence. It will be of interest to learn how pathogenicity is controlled by transposable genetic determinants, and how linear plasmids may be useful for the biological control of pathogenic fungi.

REFERENCES

Anagnostakis, S. A. (1982). Biological control of chestnut blight. *Science*, **215:**466–471.

Asher, M. J. C. (1981). In *Biology and Control of Take-All* (M. J. C. Asher and P. J. Shipton eds.). Academic, London, p. 538.

Atanasoff, D. (1973). The effect of superparasites on the pathogenicity of their hosts. *Phytopathol. Z.*, **78:**182–186.

Birnboim, H. C. and J. Doly (1979). A rapid alkaline extraction procedure for screening recombinant plasmid DNA. *Nucl. Acids Res.*, **7:**1513–1523.

Brasier, C. M. (1983). A cytoplasmically transmitted disease of *Ceratocystis ulmi. Nature*, **305:**220–223.

Calos, M. P. and J. H. Miller (1980). Transposable elements. *Cell*, **20:**579–595.

Castanho, B. and E. E. Butler (1978). Rhizoctonia decline: Studies on hypovirulence and potential use in biological control. *Phytopathology*, **68:**1511–1514.

Castanho, B., E. E. Butler, and R. J. Shepherd (1978). The association of double-strand RNA with Rhizoctonia decline. *Phytopathology*, **68:**1515–1519.

Férault, A. C., B. Tivoli, J. M. LeMaire, and D. Spire (1979). Etude de l'evolution comparee du niveau d'agressivité et du contenu en particules de type viral d'une souche de *Gaeumannomyces graminis* (Sacc.) Arx et Olivier (*Ophiobolus graminis* Sacc.). *Ann. Phytopathol.*, **11:**185–191.

Finkler, A., M. Auster, I., Barash, Y. Koltin, and B. Sneh (1983). Characterization of double stranded RNA (dsRNA) and virus-like particles in virulent and hypovirulent strains of *Rhizoctonia solani*. Fourth International Congress of Plant Pathology (Abstract), p. 114.

Francou, F. (1981). Isolation and characterization of a linear DNA molecule in the fungus *Ascobolus immersus. Mol. Gen. Genet.*, **184:**440–444.

Grente, J. (1965). Les formes hypovirulentes d'*Endothia parasitica* et les espoirs de lutte contre le chance du châtaignier. *C. R. Acad. Agric.*, **51:**1033–1037.

Gunge, N., A. Tamaru, F. Ozawa, and K. Sakaguchi (1981). Isolation and characterization of linear deoxyribonucleic acid plasmids from *Kluyveromyces lactis* and the plasmid associated killer character. *J. Bacteriol.*, **145:**382–390.

Hashiba, T. (1982a). Sclerotial morphogenesis in the rice sheath blight fungus (*Rhizoctonia solani*). *Bull. Hokuriku. Natl. Agric. Exp. Stn.*, **24:**29–83.

Hashiba, T. (1982b). Formation and fusion of protoplasts from *Rhizoctonia solani. Jpn. J. Med. Mycol.*, **23:**143–150.

Hashiba, T. (1982c). Protoplasts isolated from the rice sheath blight fungus (*Rhizoctonia solani*). *Plant Protect.*, **36:**324–328.

Hashiba, T. and T. Ishikawa (1978). Effect of adenosine 3′,5′-cyclic monophosphate on induction of sclerotia in *Rhizoctonia solani. Phytopathology*, **68:**1723–1727.

Hashiba, T. and M. Yamada (1982). Formation and purification of protoplasts from *Rhizoctonia solani. Phytopathology*, **72:**849–853.

Hashiba, T. and M. Yamada (1984). Intraspecific protoplast fusion between auxotrophic mutants of *Rhizoctonia solani. Phytopathology*, **74:**398–401.

Hashiba, T., M. Hyakumachi, Y. Homma, and M. Yamada (1983). Isolation and characterization of a plasmid DNA in the fungus *Rhizoctonia solani*. Fourth International Congress of Plant Pathology (Abstract), p. 202.

Hashiba, T., I. Uno, T. Ishikawa, Y. Homma, and M. Hyakumachi (1984a). Characterization of an isolate of the fungus *Rhizoctonia solani* carrying plasmid DNA. Third International Congress on Cell Biology (Abstract), p. 561.

Hashiba, T., Y. Homma, M. Hyakumachi, and A. Matsuda (1984b). Isolation and characterization of a DNA plasmid in the fungus *Rhizoctonia solani. J. Gen. Microbiol.*, **130:**2067–2070.

Hirt, B. (1967). Selective extraction of polyoma DNA from infected mouse cell cultures. *J. Mol. Biol.*, **26:**365–369.

Hishinuma, F., K. Nakamura, K. Hirai, R. Nishizawa, N. Gunge, and T. Maeda (1984). Cloning and nucleotide sequence of the linear DNA killer plasmids from yeast. *Nucl. Acids Res.*, **12:**7581–7597.

Homma, Y., Y. Yamashita, T. Kubo, and M. Ishii (1981). Suppressive factors to Japanese radish damping-off in the soil reinoculated with *Rhizoctonia solani. Ann. Phytopath. Soc. Jpn.*, **47:**388 (Abstract).

Homma, Y., Y. Ohbi, and M. Ishii (1986). Suppressive factors to Japanese radish damping-off in soil reinoculated with living mycelium of *Rhizoctonia solani. Ann. Phytopath. Soc. Jpn.*, (submitted for publication).

Honeyman, A. L. and T. C. Currier (1983). The isolation and characterization of 2 linear DNA elements from *Gaeumannomyces graminis* var. *tritici*, the causative agent of take-all disease of wheat. *Annu. Meet. Am. Soc. Microbiol.*, **83:**175 (Abstract).

Hyakumachi, M., Y. Homma, and T. Ui (1984). Properties of Diseased isolates of *Rhizoctonia solani. Ann. Phytopath. Soc. Jpn.*, **50:**281–285.

Jaynes, R. A. and J. E. Elliston (1980). Pathogenicity and canker control by mixtures of hypovirulent strains of *Endothia parasitica* in American chestnut. *Phytopathology*, **70:**453–456.

Jinks, J. L. (1959). Lethal suppressive cytoplasms in aged clones of *Aspergillus glaucus. J. Gen. Microbiol.*, **21:**397–409.

Kado, C. I. and S. T. Liu (1981). Rapid procedure for detection and isolation of large and small plasmids. *J. Bacteriol.*, **145:**1365–1373.

Lapierre, H., J.-M. Lemaire, B. Jouan, and G. Molin (1970). Mise en évidence des particules virals associées á une perte de pathogénicité chez le piétin-échaudage des céréales, *Ophiobolus graminis* Sacc. *CR. Acad. Sci. (Paris), Sér. D*, **271:**1833–1836.

Meinhardt, F. and K. Esser (1984). Linear extrachromosomal DNA in the morel *Morchella conica. Curr. Genet.*, **8:**15–18.

Naiki, T. and R. J. Cook (1983). Factors in loss of pathogenicity in *Gaeumannomyces graminis* var *tritici. Phytopathology*, **73:**1652–1656.

Palmer, J. D., C. R. Shields, D. B. Cohen, and T. J. Orton (1983). An unusual mitochondrial DNA plasmid in the genus *Brassica. Nature*, **301:**725–728.

Potter, S., M. Truétt, M. Phillips, and A. Maher (1980). Eucaryotic transposable genetic elements with inverted terminal repeats. *Cell*, **20:**639–647.

Pring, D. R., C. S. Levings, W. W. L. Hu, and D. H. Timothy (1977). Unique DNA associated with mitochondria in the "s"-type cytoplasm of male-sterile maize. *Proc. NAS USA*, **74:**2904–2908.

Pring, D. R., M. F. Conde, K. F. Schertz, and C. S. Levings (1982). Plasmid-like DNAs associated with mitochondria of cytoplasmic male-sterile sorghum. *Mol. Gen. Genet.*, **186:**180–184.

Pusey, P. L. and C. L. Wilson (1982). Detection of double-stranded RNA in *Ceratocystis ulmi. Phytopathology*, **72:**423–428.

Rawlinson, C. J., D. Hornby, V. Pearson, and J. M. Carpenter (1973). Virus-like particles in the take-all fungus, *Gaeumannomyces graminis. Ann. Appl. Biol.*, **74:**197–209.

Roeder, G. S., P. J. Farabaugh, D. T. Chaleff, and G. R. Fink (1980). The origins of gene unstability in yeast. *Science*, **209:**1375–1384.

Takai, S., T. Iizuka, and W. C. Richards (1984). Discovery of plasmids in *Ceratocystis ulmi. Phytopathology*, **74:**833 (Abstract).

Uno, I., T. Ishikawa, and T. Hashiba (1985a). Cyclic AMP-dependent phosphorylation of proteins in mycelia of *Rhizoctonia solani* forming sclerotia. *Ann. Phytopath. Soc. Jpn.*, **51:**183–189.

Uno, I., T. Ishikawa, and T. Hashiba (1985b). Detection of cyclic AMP levels and cyclic AMP-receptor protein in mycelia of *Rhizoctonia solani* forming sclerotia. *Ann. Phytopath. Soc. Jpn.*, **51:**190–198.

Zanzinger, D. H., B. P. Bandy, and S. M. Tavantzis (1984). High frequency of finding double-stranded RNA in naturally occurring isolates of *Rhizoctonia solani. J. Gen. Virol.*, **65:**1601–1605.

17

GENETIC ENGINEERING FOR PLANT DISEASE RESISTANCE

LOTHAR WILLMITZER

Max Planck Institut für Züchtungsforschung, Cologne, West Germany

During the last decades conventional plant breeding has been of utmost importance for the production of better and more useful plants. It will undoubtedly continue to play this role in the future; although, it is also no matter of dispute that it has certain limitations. The main restrictions are twofold:

1. Only closely related species can be crossed sexually because of incompatability problems leading to a rather limited gene pool available for new combinations.
2. Sexual crossing means a massive randomization of two complete genomes therefore making extensive backcrossing a necessity. The need for backcrossing considerably slows down the production of new and better varieties, the time needed for breeding a new variety being somewhere in between 8 and 15 yr.

A special problem of importance envisaged in plant breeding is disease resistance. The main problem here is that, owing to the extensive use of monocultures with a high degree of genetic uniformity, pathogen-resistance relying on either one major or a limited number of genes is overcome by new pathogen varieties which evolve rapidly because of the high selection pressure.

According to Flor's gene-for-gene hypothesis (Flor, 1947) it should be expected that the vast majority of disease resistances relying on oligogenic traits will be overcome by new pathogen varieties. Again the restriction of classical plant breeding to a limited gene pool represents a severe problem.

Therefore techniques allowing the shortening of the time needed to produce a new variety and/or to enlarge the gene pool available for plant breeding have received high interest during the last years.

Recent advances in plant molecular biology have led to the development of methods allowing the introduction of a single gene or a set of genes irrespective of its origin, into the chromosome of higher plants with minimal disturbance of the existing stable structure of the genome. These methods, therefore, have the potential of overcoming both severe limitations of conventional plant breeding mentioned above, that is, the limited gene pool available as well as the need for backcrossing due to the extensive mixing of two complete genomes.

Central to the modification of plant genomes by genetic engineering is the availability of efficient methods to introduce foreign genetic information into the plant cell. Therefore, in the first part of this chapter I briefly review the different established methods of introducing DNA into plant cells. In the second part I outline the different possibilities of using genetic engineering methods for creating disease-resistant plants. Whereas the first part is based on well-established data, the second part is a much more speculative discussion of the potential use of gene technology in plant breeding.

1. SYSTEMS ALLOWING TRANSFER OF ISOLATED GENES INTO PLANTS

1.1. RNA-Viruses

To the best of my knowledge no reports have yet been published about the use of cDNA copies of RNA plant viruses as potential gene vectors. Rather, at least in the case of brome mosaic virus, cDNA copies have been found to be noninfectious. However, as RNA synthesized *in vitro* from cDNA copies of BMV has been conclusively demonstrated to be infectious, it remains to be seen whether a virus in its RNA form itself might become a useful vector (Ahlquist et al., 1984).

1.2. DNA-Viruses

Only very few DNA plant viruses are known. The best studied DNA virus is Cauliflower mosaic virus (CaMV) which infects some cruciferous plants. It consists of a small (±8000 bp) double-stranded circular DNA. Recent work has demonstrated that this virus can principally be used to transfer and express foreign DNA in plants (Brisson et al., 1984).

However, at present, CaMV vectors are not generally used because sequences, to be stably propagated via CaMV vectors in plant cells, must be inserted into the CaMV genome in a very precise manner, owing to the "relay-race" translation system. Furthermore, size constraints exist on the amount of DNA to be inserted (Gronenborn, 1984).

Gemini viruses, a class of single-stranded DNA viruses infecting both dicot- as well as monocotyledonous (here especially cereal) plants, have recently received a high degree of attention because of their potential use as gene vectors (Goodman, 1981). Several gemini viruses have been cloned and sequenced (Hamilton et al., 1984; Mallineaux et al., 1984) and at least for the class of white-fly transmitted viruses, cloned sequences have been proven to be infectious using mechanical inoculation (Hamilton et al., 1983).

1.3. Direct Gene Transfer

Direct gene transfer without the help of virus- or Ti-based vector systems has been conclusively demonstrated recently (Paszkowski et al., 1984; Hain et al., 1985). This method has a potentially large application because, at least in principle, there is no limitation of the host range. Thus this method for the first time also allows the stable transformation of gramineae species (Lörz et al., 1985; Potrykus et al., 1985).

The main requirement for the successful use of this method, however, is the availability of an efficient protoplast–plant regeneration system which at the present time has only been developed for a few species (Chaleff, 1981).

1.4. *Agrobacterium tumefaciens* Ti-Plasmid Based Vectors

The most developed, plant-oriented gene vector system is the Ti-plasmid of *Agrobacterium tumefaciens*, which is described in somewhat greater detail in the following section.

Most dicotyledonous plants are susceptible to tumor formation as a result of infection of wounded sites by the gram-negative bacterium *Agrobacterium tumefaciens* (for a general review see Caplan et al., 1983). The bacterium is necessary for tumor induction but not for tumor maintenance and growth. The ability of the bacteria to induce tumor formation is strictly limited to those Agrobacteria species harboring a large extra-chromosomal DNA-element, the so-called Ti-(Tumor-Inducing) plasmid, which has a size of 120–180 kb. The molecular basis underlying the neoplastic transformation is the transfer and stable integration of a well-defined part of the Ti-plasmid, the T-DNA (T = Transfer or Tumor) into the plant nuclear DNA, leading to the formation of a gall. The size of this transferred DNA (T-DNA) is dependent upon the tumor-inducing bacteria. It varies between 12 kb (for octopine-producing tumors) and ±20 kb (for nopaline-type tumors). The transformed plant cells display three typical characteristics, that is, (1) they grow in the absence of phytohormones in axenic tissue culture, (2) they produce certain low-molecular weight compounds, the so-called opines (depending upon the tumor-inciting *Agrobacterium* strains either members of the octopine, nopaline, or agropine family are produced), and (3) they cannot be redifferentiated into normal plants by changing the level of phytohormones present in the medium.

On the Ti-plasmid the T-DNA is bordered on the left- and right-hand side by a nearly perfect direct repeat of 25 bp. Genetic as well as biochemical data clearly demonstrate that only this 25-bp direct repeat determines what part of the Ti-plasmid is transferred to the plant cell (Wang et al., 1984). Thus foreign DNA inserted in between these direct repeats will be cotransferred to the plant cell and there be stably integrated into the plant chromosomal DNA.

The Ti-plasmid, therefore, represents a natural gene vector to plant cells. The only other function on the Ti-plasmid necessary for the gene transfer to happen is the virulence region. This region has been shown by transposon mutagenesis to contain at least eight transcription units coding for diffusible factors being involved in the transfer and/or integration of the T-DNA of *Agrobacterium tumefaciens* to plant nuclear DNA (Klee et al., 1983). However, whereas the 25-bp repeat represents cis-acting elements concerning the T-DNA, the vir-region encoded products act in trans. This has led to the development of binary vector systems that are discussed in another section.

Extensive biochemical as well as genetic analysis of the T-DNA has shown that the T-DNA encoded functions in the plant cell can be divided into two groups:

1. One group of genes encodes enzymes involved in the synthesis of the opines.
2. Another group of genes encodes enzymes involved in the biosynthetic pathway of phytohormones.

Whereas the products of the first set of genes do not interfere with the normal differentiation process of the plant cell, the products of the second group do interfere as the endogenous hormone level is heavily disturbed. Their activity leads to the formation of plant cells with abnormal growth and differentiation properties. As it is essential for the practical use of this system for gene transfer to obtain normal plants, mutants of the Ti-plasmid have been constructed where all the T-DNA genes of the second group have been deleted, while keeping genes for opine biosynthesis intact. Normal plants with functional opine synthase have been obtained. These plants were fertile and sexually transmitted the introduced genes as a single, dominant Mendelian locus (Zambryski et al., 1983).

Analysis of the T-DNA encoded genes by nucleic acid sequencing revealed the presence of sequences in the 5′-upstream as well as the 3′-downstream region typical of eukaryotic promotor sequences. These have therefore been used to construct chimeric genes consisting of promoter and polyadenylation sequences of the nopaline synthase gene and protein-coding sequences from bacterial genes, such as the neomycin phosphotransferase from Tn5, the methotrexate-resistant dihydrofolate reductase of the plasmid R67 and the chloramphenicol acetyl transferase of pBR325. These chimeric genes were subsequently transferred to tobacco cells using Ti-plasmid based vectors and were shown to actively express functional enzymes (Herrera-Estrella et al., 1983a). As plant cells are sensitive to drugs such as methotrexate and kanamycin, these chimeric genes do represent potent dominant selectable markers for following transformation events in plant cells (Herrera-Estrella et al., 1983b).

The results described have led to the development of systematically designed Ti-plasmid based vectors, which have the following properties:

1. They contain the pair of the 25-bp direct repeats.
2. The T-DNA genes present in the wild-type Ti-plasmid in between these borders and coding for phytohormone disturbing enzymes (onc-genes) have been deleted and replaced by a pBR322 sequence.
3. In between the 25-bp repeats is a screenable and/or selectable marker gene allowing easy recognition of the transformed plant cells.
4. The Ti-plasmid contains an intact vir-region.

The presence of the pBR-sequence in between the 25-bp direct repeats accounts for another feature of the Ti-plasmids that makes their use as experimental gene vectors rather cumbersome, that is, the large size of the plasmids. This makes it nearly impossible to find adequate unique restriction sites within the T-region for the cloning of foreign DNA sequences. Therefore, "intermediate" vectors have been developed on the basis of pBR322 plasmids which allow the cloning of the gene to be transferred in *E. coli*. This plasmid is subsequently mobilized to an *Agrobacterium* strain containing an acceptor Ti-plasmid with the properties described above. A single *in vivo* cross-over between the intermediate vector and the pBR sequence on the Ti-plasmid will result in the insertion of the foreign gene in between the 25-bp direct repeats bordering the T-region (Van Haute et al., 1983). This *Agrobacterium* strain is then ready for use to transfer the gene of interest to plant cells using either the method of cocultivation (Marton et al., 1979) or the more recently developed leaf disk infection system (Horsch et al., 1985).

A number of foreign genes have meanwhile been transferred to plant cells using this system and normal and fertile plants have been regenerated in the case of petunia, tobacco, tomato, and potato. A slightly different system takes into account the observation that the only cis-standing elements necessary for a gene transfer are the 25-bp direct repeats (see discussion given above). This has led to the development of the binary vector systems, where the gene to be transferred is first cloned in *E. coli* in between the direct repeats present on a plasmid containing a broad host-range replicon. This plasmid is subsequently mobilized to an *Agrobacterium* strain containing a Ti-plasmid where all the T-region, including the 25-bp repeats, is deleted; however, the vir-region is left intact. Thus T-region and vir-region are present on two independent replicons; nevertheless, the DNA contained in between the 25-bp repeats is efficiently transferred to the plant cell due to the trans-acting functions supplied by the vir-region. The binary vector system as compared to the cointegration system discussed above, seems to lead to higher copy numbers of the transferred DNA and should be advantageous in "shot-gun" experiments where whole (plant) genomes should be transferred to an acceptor plant (Bevan, 1984).

In conclusion the *Agrobacterium*-Ti-plasmid system represents a well-established plant-oriented gene vector, which after cloning of a particular gene, allows its transfer to plant cells and subsequent regeneration of a whole plant containing the foreign genes within several weeks.

The main limitation at the present time concerning the use of Ti-plasmid mediated gene transfer is the limited host range of *Agrobacterium* species. Thus whereas nearly all dicotyledonous plants are susceptible to *Agrobacterium* infection, monocotyledonous plants as a rule (despite certain exceptions reported recently, Hernalsteens et al., 1984;; Hooykaas et al., 1984) and especially gramineae do not seem to be susceptible to *Agrobacterium* infection. However, this conclusion is mainly based on the observation that no tumors are formed after *Agrobacterium* treatment. It should be critically reexamined using the now available selectable markers.

2. STRATEGIES FOR PLANT DISEASE CONTROL USING GENETIC ENGINEERING AND POSSIBLE CANDIDATES FOR GENES CONFERRING RESISTANCE

2.1. Use of Resistance Genes Present in Plant Species

Resistance genes present in either wild or cultivated plant species have been used extensively in classical plant breeding. For a large number of resistancies the genetics are well understood and again in a reasonable number of plant species, certain resistance genes have been mapped on the respective chromosomes. However, to be used for genetic engineering the respective gene has to be cloned, which—using classical molecular genetic techniques—requires some information about the biochemical product of the gene. To the best of my knowledge, this is not known for any gene conferring pathogen resistance. Thus classical molecular cloning techniques do not look very promising at the present time concerning the physical isolation of a resistance gene. Therefore methods have to be used that allow the isolation of genes by knowledge of their phenotype only. Recently isolation and characterization of plant transposable elements have been described (see Saedler and Nevers, 1985, for a recent review). They offer the possibility of using such techniques to mutagenize resistance genes, thereby changing the phenotype from resistant to susceptible case resistance is dominant. As in parallel to insertion, the resistance gene also gets ear-marked, and can thus be cloned molecularly, using the transposable element as probe. This type of approach should allow the isolation of resistance genes in cases where resistance is conferred by a single dominant gene. It should be specifically advantageous in cases where resistance and/or susceptibility can already be screened on the single-cell callus level, such as for host-specific toxins produced by phytopathogenic fungi (Scheffer and Livingstone, 1984).

An alternative approach would involve the transfer of the complete (fragmented) genome of a plant possessing a dominant resistance gene to a susceptible acceptor plant and subsequent screening/selection for the resistance trait. Although this approach is possible, it is hampered because the large size of most plant genomes does not make the complete transfer of a genome feasible.

2.2. Possibilities of Using Unconventional Approaches Conferring Resistance

In this section I discuss several possibilities of using unconventional approaches to achieve pathogen resistance. It should be emphasized that the examples discussed are definitively of a rather speculative nature; however, they should help in outlining possible future developments. As stated, one of the factors limiting classical plant breeding is the rather narrow gene pool to be used. Genetic engineering principally overcomes this limitation. Therefore, the strategies discussed below try to take advantage of this fact as much as possible.

2.2.1. *Viral Infections*

There seem to be at least three possibilities for interference with virus multiplication.

Interference with the uncoating step. To multiply and to be expressed in the plant cell, the infecting virus has to strip off its coat protein. One of the explanations for the long-standing observation of cross-protection between serologically related viruses has been the hypothesis that the surplus of coat protein made by the first infecting virus prevents the uncoating of the second infecting virus, thereby preventing multiplication (Sherwood and Fulton, 1982). If this hypothesis is true, it should be possible, by introducing a chimeric gene composed of the cDNA copy for the coding sequence of the coat protein and regulatory sequences derived from plant genes, to achieve high levels of coat protein in the plant cell, thereby leading to a "cross-protection"-like situation.

Inhibition of translation of subgenomic/genomic RNAs. Plant RNA viruses use different expression strategies involving the translation of either genomic and/or subgenomic RNAs on the plant cell's ribosomes (Davies and Hull, 1982). Double-stranded RNA is prevented from being translated into proteins, a fact that is used extensively for control of gene expression in bacteria. Recently it has been shown in animal systems that the injection of RNA complementary to the mRNA (termed "antisense RNA") has led to a dramatic decrease in the translation efficiency of the corresponding mRNA (Weintraub et al., 1985).

A similar strategy might be feasible for suppressing the translation of certain parts of the RNA virus genome, preferably those that are crucial for multiplication (e.g., coding for protein involved in replication). This would again involve the construction of a chimeric gene composed of plant regulatory sequences and the cDNA copy of a part of the virus genome, which then has to be introduced into the plant chromosome. Though this strategy seems to be feasible and would be an elegant approach applicable to nearly all viral diseases, the main problem is the high level of antisense RNA necessary to achieve substantial inhibition of translation.

Inhibition of the processing of the polyprotein. The formation of a polyprotein derived from a genomic or subgenomic RNA is one expression strategy used by RNA plant viruses. The polyprotein is subsequently cleaved by specific virus-encoded proteins to give rise to the mature proteins (Davies and Hull, 1982). The cleavage step is therefore critical for the further multiplication and spread of the virus. If it were possible to inhibit the specific virus-encoded protease, this should lead to the development of resistance. Proteinase inhibitors are often of a proteinaceous nature themselves. Screening for proteinase inhibitors could easily be done in cell-free systems. Once a proteinaceous inhibitor is found, the introduction of its coding sequence, again under the control of plant regulatory sequences, into the plant genome might result in resistance against virus infection.

The general feasibility of this strategy is supported by the recent finding that the immunity of a certain cowpea line against cpMV-infection is very likely related to a plant-encoded proteinaceous factor interfering with the processing of the polyprotein encoded by the viral RNA (Sanderson et al., 1985).

2.2.2. Fungal Infections

Resistance against host-selective toxins producing fungi. Toxins produced by phytopathogenic organisms play an important role in pathogenesis. They are divided into two classes—host-selective (specific) and unspecific toxins.

Whereas the second group of unspecific toxins is not essential for virulence of the pathogen, the first group of host-selective toxins are absolutely essential for the pathogenicity of the phytopathogen, a tox- mutant being avirulent.

Fourteen different fungi, mainly from the Helminthosporum and Alternaria species, are known to produce host-selective toxins (Scheffer and Livingstone, 1984). A possible scheme to introduce unconventional resistance into plants susceptible to these toxins would be to screen for able enzymatic activities (e.g., in microorganisms) to detoxify the toxin by either modification and/or degradation. The subsequent introduction of this gene under plant regulatory sequence control should confer resistance to the plant as the toxin is absolutely essential for virulence of the pathogen.

Inhibition of formation resp. degradation of fungal cell wall. Fungal cell walls are mainly composed of chitin microfibrils, polysaccharides, and their chemical complexes with proteins; in addition, cell-wall structure is supported by β-glucans and crystalline polysaccharides.

As there is no evidence of the presence of chitin in plant cells, compounds inhibiting chitin synthase may work as effective antifungal agents. These inhibitors again ideally should be either of protein nature or produced from precursors already present in plant cells with one or a few modifications.

An alternative approach would take advantage of the extensive competition among soilborne fungi. Thus certain *Trichoderma harzianum* strains are very effective antagonists against different phytopathogenic fungi such as *Scleroticum rolfsii* (Chet et al., 1979). Very probably the antagonism is due to active secretion of β-1,3-glucanase as well as chitinase by the *Trichoderma* strain, both enzymes attacking and degrading the cell walls of the respective phytopathogenic fungi. The gene-encoding chitinase, therefore, represents another candidate to be transferred to plant cells which, via degradation of the fungal cell wall, might result in protection against fungal attack. In this connection, however, it should be mentioned that endogenous chitinase activity is found in certain plants (Boller et al., 1983). Obviously this endogenous activity though stimulated upon pathogen attack, is not sufficient to confer resistance against the fungal attack. A possible explanation might be that it is normally compartmentalized in the vacuole. Thus a more refined strategy should be used to direct added chitinase activity either into the cell wall or the extracellular space using leader peptides. A similar strategy has been used to direct certain nuclear-encoded proteins into chloroplasts (Schreier et al., 1985).

2.2.3. Bacterial Infections

Concerning bacterial infections, approaches similar to the ones described for fungal infections might be followed.

Antibacterial compounds. Numerous microorganisms are known that elaborate compounds inhibiting the growth of other microorganisms and also phytopathogenic bacteria. These compounds and the genes encoding their synthesis therefore represent possible candidates to be transferred to plants. However, a serious problem is evident because in cases where the bactericidal compound is known, genetic evidence points out that its synthesis might be encoded by a number of genes, therefore making it a less suitable candidate for gene transfer. A different situation is found in antibacterial compounds synthesized by certain insects (Lee et al., 1983). These compounds are proteinaceous in nature, which would result in a decrease in the number of genes to be transferred.

Degradation of bacterial cell walls. Bacterial cell walls of a number of gram-positive as well as gram-negative bacteria are known to be susceptible to attack by the lytic enzyme lysozyme.

Proteins displaying lysozyme activity are produced by a number of different organisms. In at least two classes (T4 bacteriophage and egg-white lysozyme), the corresponding genes have been cloned and sequenced. Thus they represent primary candidates for transfer to plants.

2.2.4. Insect Attack

The prime organism serving as a pool for genes conferring resistance to plants against attack by insects is the *Bacillus thuringiensis*, a gram-positive, spore-forming bacterium. A common feature of *B. thuringiensis* strains is their capacity to synthesize parasporal proteinaceous crystals during sporulation. This crystal is composed of a monomeric glycoprotein (termed δ-endotoxin) which is toxic to larvae of many lepidopterous insects.

The gene encoding the δ-endotoxin has been cloned from different varieties of *B. thuringiensis* (Klier et al., 1982; Sekar and Carlton, 1985). Because of its high activity and specificity, as well as its proteinaceous nature, it represents a nearly ideal candidate to achieve resistance against pathogen attack by using gene technology. Actually a chimeric gene composed of T-DNA regulatory elements and the coding sequence for the δ-endotoxin has already been introduced into tobacco plants and shown to display insecticidal activity (J. Leemans, Plant Genetic System, Belgium, personal communication).

3. CONCLUDING REMARKS

As outlined in the first part of this chapter, plant-oriented gene technology has reached a state that allows the introduction, expression, and stable transmission of genes into plants, independent of their origin. The main constraint of this technology is the limited host range, excluding especially the gramineae from being used in such experiments.

In the second part I tried to briefly outline different possibilities that might result in resistance of the plant against pathogen attack. It should be emphasized here again, that this discussion is mainly of a speculative nature. However, there is no doubt that defense against pathogen by either conventional plant breeding or by using chemicals has reached a critical point due to the rapid evolvement of new pathogen races and environmental pollution, respectively. The technology of plant genetic engineering offers completely new approaches to the problem. By purposely describing speculative strategies, I hope to promote discussion on using these new technologies to achieve the old goal of obtaining pathogen-resistant plants.

REFERENCES

Ahlquist, P., R. French, M. Janda, and S. L. Loesch-Fries (1984). Multicomponent RNA plant virus infection derived from cloned viral DNA. *Proc. Natl. Acad. Sci.*, **81:**7066–7070.

Bevan, M. (1984). Binary Agrobacterium vectors for plant transformation. *Nucl. Acids Res.*, **12:**8711–8721.

Boller, T., A. Gehri, F. Mauch, and U. Vögeli (1983). Chitinase in bean leaves: Induction by ethylene, purification, properties and possible function. *Planta*, **157:**22–31.

Brisson, N., J. Paszkowski, J. R. Penswick, B. Gronenborn, I. Potrykus, and T. Hohn (1984). Expression of a bacterial gene in plants by using a viral vector. *Nature*, **310:**511–514.

Caplan, A., L. Herrera-Estrella, D. Inze, E. Van Haute, M. Van Montagu, J. Schell, and P. Zambryski (1983). Introduction of genetic material into plant cells. *Science*, **222:**815–821.

Chaleff, R. S. (1981). Genetics of higher plants—Application of cell culture. Cambridge University Press, Cambridge.

Chet, J., Y. Hadar, Y. Elad, J. Katan, and Y. Henis (1979). In *Soil-borne Plant Pathogens* (B. Schippers and W. Gams, eds.). Academic, New York, pp. 585–591.

Davies, J. W. and R. Hull (1982). Genome expression of plant positive-strand RNA viruses. *J. Gen. Virol.*, **61:**1–14.

Flor, H. H. (1947). Inheritance of reaction to rust in flax. *J. Agric. Res.*, **74:**241-262.

Goodman, R. M. (1981). In *Handbook of Plant Virus Infections and Comparative Diagnosis* (E. Kirstak, ed.). Elsevier, Amsterdam, pp. 879-910.

Gronenborn, B. (1984). Cauliflower mosaic virus: A plant gene vector. In *The Impact of Gene Transfer Techniques in Eucaryotic Cell Biology* (J. Schell and P. Starlinger, eds.). 35th Colloquium-Mosbach 1984, Springer-Verlag, Berlin, pp. 108-120.

Hain, R., P. Stabel, A. P. Czernilofsky, H. H. Steinbib, L. Herrera-Estrella, and J. Schell (1985). Uptake, integration, expression and genetic transmission of a selectable chimeric gene, by plant protoplasts. *Mol. Gen. Genet.*, **199:**161-168.

Hamilton, W., D. Bisaro, R. Coatts, and K. Buck (1983). Demonstration of the bipartite nature of the genome of a single-stranded DNA plant virus by infection with cloned DNA. *Nucleic Acids Res.*, **11:**7387-7391.

Hamilton, W., V. Stein, R. Coatts, and K. Buck (1984). Complete nucleotide sequence of the infectious cloned DNA components of tomato golden mosaic virus: Potential coding regions and regulatory sequences. *EMBO J.*, **3:**2197-2205.

Hernalsteens, J. P., L. Thia-Toong, J. Schell, and M. Van Montagu (1984). An Agrobacterium-transformed cell culture from the monocot *Asparagus officinalis. EMBO J.*, **3:**3039-3041.

Herrera-Estrella, L., A. Depicker, M. Van Montagu, and J. Schell (1983a). Expression of chimeric genes transferred into plant cells using a Ti-plasmid derived vector. *Nature*, **303:**209-213.

Herrera-Estrella, L., De Block, E. Messens, J. P. Hernalsteens, M. Van Montagu, and J. Schell (1983b). Chimeric genes as dominant selectable markers in plant cells. *EMBO J.*, **2:**987-995.

Hooykaas-Van Slogteren, G. M. S., P. Hooykaas, and R. Schilperoort (1984). Expression of Ti-plasmid genes in monocotyledonous plants infected with *Agrobacterium tumefaciens. Nature*, **311:**763-764.

Horsch, R. B., J. E. Fry, N. L. Hoffmann, D. Eichholtz, S. G. Rogers, and R. T. Fraley (1985). A simple and general method for transferring genes into plants. *Science*, **227:**1229-1231.

Klee, H. J., F. F. White, V. N. Iyer, M. P. Gordon, and E. W. Nester (1983). Mutational analysis of the virulence region of an *Agrobacterium tumefaciens* Ti plasmid. *J. Bacteriol.*, **153:**878-883.

Klier, A., Fargette, F., J. Ribier, and G. Rapoport (1982). Cloning and expression of the crystal protein genes from *Bacillus thuringiensis* strain berliner 1715. *EMBO J.*, **1:**791-799.

Lee, J.-Y., T. Edlund, T. Ny, I. Faye, and H.-G. Boman (1983). Insect immunity. Isolation of cDNA clones corresponding to attacins and immune protein P4 from *Hyalophora cecropia. EMBO J.*, **2:**577-581.

Lörz, H., B. Baker, and J. Schell (1985). Gene transfer to cereal cells mediated by protoplast transformation. *Mol. Gen. Genet.*, **199:**183-188.

Mallineaux, P. M., J. Donson, B. Morris-Krsinich, M. Boulton, and J. Davies (1984). The nucleotide sequence of maize streak virus DNA. *EMBO J.*, **3:**3063-3068.

Marton, L., G. J. Wullems, L. Molendijk, and R. A. Schilperoort (1979). In vitro transformation of cultured cells from *Nicotiana tabacum* by *Agrobacterium tumefaciens. Nature*, **277:**129-131.

Paszkowski, J., R. D. Shillito, M. Saul, V. Mandak, T. Hohn, B. Hohn, and J. Potrykus (1984). Direct gene transfer to plants. *EMBO J.*, **3:**2717-2722.

Potrykus, I., M. W. Saul, J. Petruska, J. Pazkowski, and R. D. Shillito (1985). Direct gene transfer to cells of a graminaceous monocot. *Mol. Gen. Genet.*, **199:**183-188.

Saedler, H. and P. Nevers (1985). Transposition in plants: A molecular model. *EMBO J.*, **4:**585-590.

Sanderson, J. L., G. Bruening, and M. L. Russell (1985). Possible molecular basis of immunity of cowpeas to cowpea mosaic virus. In *Cellular and Molecular Biology of Plant Stress.* UCLA Symposia, pp. 401-412.

Scheffer, R. D. and R. S. Livingston (1984). Host-selective toxins and their role in plant diseases. *Science*, **222:**17-21.

Schreier, P. H., E. Seftor, J. Schell, and H. Bohnert (1985). The use of nuclear-encoded sequences to direct the light-regulated synthesis and transport of foreign protein into chloroplasts. *EMBO J.*, **4**:25–32.

Sekar, V. and B. Carlton (1985). Molecular cloning of the delta-endotoxin gene of *Bacillus Thuringiensis* var. israelensis. *Gene*, **33**:151–158.

Sherwood, J. L. and R. W. Fulton (1982). The specific involvement of coat protein in tobacco mosaic virus cross protection. *Virology*, **119**:150–158.

Van Haute, E., H. Joos, M. Maes, H. G. Warren, M. Van Montagu, and J. Schell (1983). Intergenic transfer and exchange recombination of restriction fragments cloned in pBR322: A novel strategy for the reversed genetics of the Ti-plasmids of *Agrobacterium tumefaciens. EMBO J.*, **2**:411–417.

Wang, K., L. Herrera-Estrella, M. Van Montagu, and P. Zambryski (1984). Right 25 bp terminus sequence of the nopaline T-DNA is essential for and determines direction of DNA transfer from Agrobacterium to the plant genome. *Cell*, **38**:455–462.

Weintraub, H., J. G. Izant, and R. M. Harland (1985). Anti-sense RNA as a molecular tool for genetic analysis. *Trends Genet.*, **1**:22–25.

Zambryski, P., H. Joos, C. Genetello, M. Van Montagu, and J. Schell (1983). Ti-plasmid vector for the introduction of DNA into plant cells without altering their normal regeneration capacity. *EMBO J.*, **2**:2143–2150.

INDEX